FRICTION and WEAR of CERAMICS

MATERIALS ENGINEERING

1. Modern Ceramic Engineering: Properties, Processing, and Use in Design. Second Edition, Revised and Expanded, *David W. Richerson*
2. Introduction to Engineering Materials: Behavior, Properties, and Selection, *G. T. Murray*
3. Rapidly Solidified Alloys: Processes · Structures · Applications, *edited by Howard H. Liebermann*
4. Fiber and Whisker Reinforced Ceramics for Structural Applications, *David Belitskus*
5. Thermal Analysis of Materials, *Robert F. Speyer*
6. Friction and Wear of Ceramics, *edited by Said Jahanmir*

Additional Volumes in Preparation

Mechanical Properties of Metallic Composites, *edited by Shojiro Ochiai*

FRICTION and WEAR of CERAMICS

edited by

SAID JAHANMIR

National Institute of Standards and Technology
Gaithersburg, Maryland

Marcel Dekker, Inc. New York • Basel • Hong Kong

Library of Congress Cataloging-in-Publication Data

Friction and wear of ceramics / edited by Said Jahanmir.
p. cm. — (Materials engineering)
Includes bibliographical references and index.
ISBN 0-8247-9115-0
1. Ceramic materials—Mechanical properties. 2. Mechanical wear.
3. Friction. 4. Tribology. I. Jahanmir, Said. II. Series:
Materials engineering (Marcel Dekker, Inc.)
TA455.C43F75 1993
620.1'404292—dc20 93-11618
CIP

The publisher offers discounts on this book when ordered in bulk quantities. For more information, write to Special Sales/Professional Marketing at the address below.

This book is printed on acid-free paper

Marcel Dekker, Inc.
270 Madison Avenue, New York, New York 10016

Current printing (last digit):
10 9 8 7 6 5 4 3 2 1

PRINTED IN THE UNITED STATES OF AMERICA

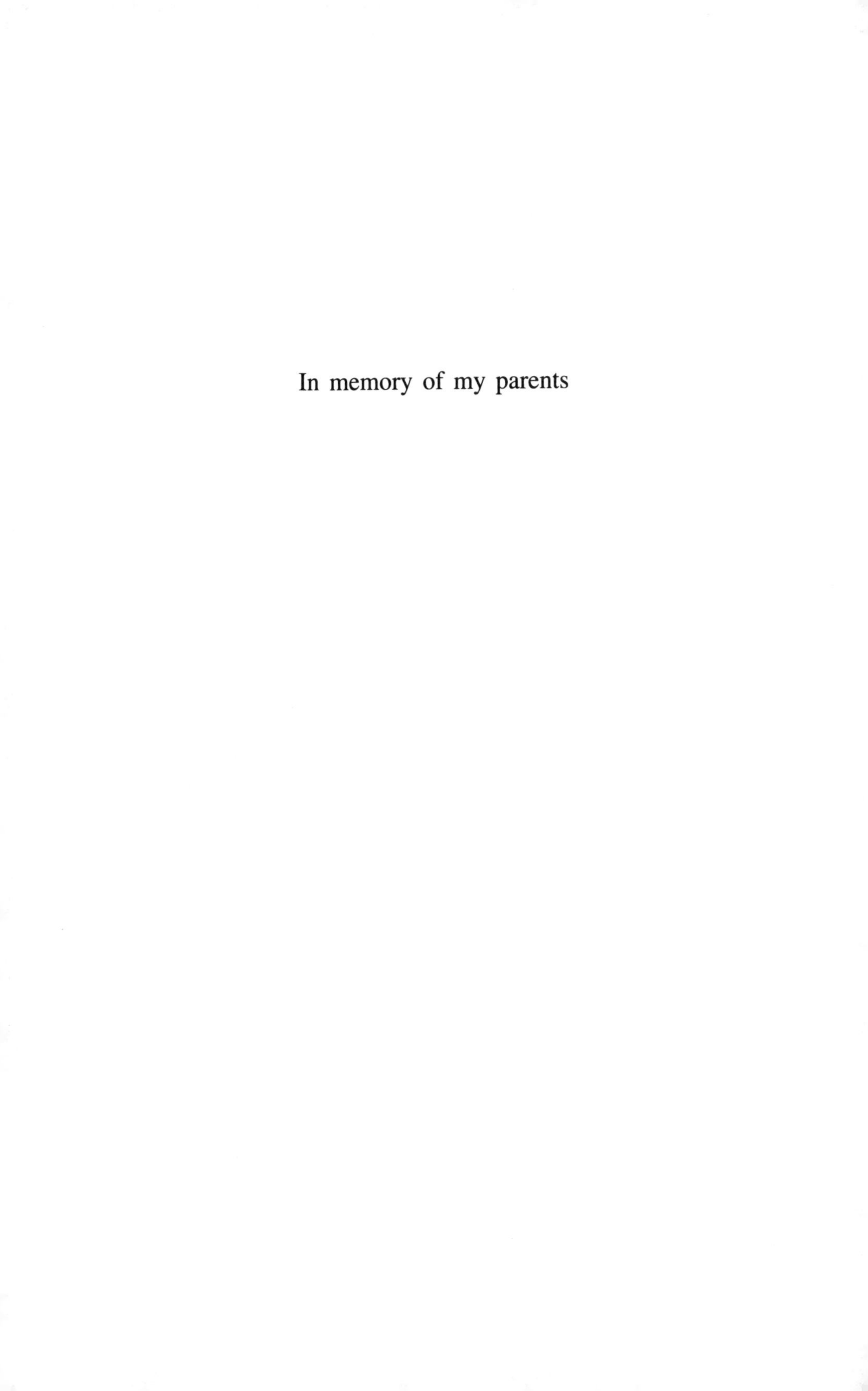

In memory of my parents

Preface

Advanced structural ceramics are presently used in diverse tribological applications such as tips for ballpoint pens, precision instrument bearings, and cutting tool inserts. Applications under development include prosthetic articulating joints, scratch-resistant watchbands, and cylinder liners and other components in advanced heat engines. Design and selection of ceramics for these applications require reliable data and mechanistic information on the tribological behavior of these materials. During the past ten years, research results have been reported in various scholarly journals and conference proceedings. However, this information is not easily accessible to practicing engineers. The purpose of this book is to summarize our current knowledge of the subject in a single volume. This book can also be used in an advanced tribology course dealing with the materials aspects of friction and wear.

The book is divided into four parts: "Introduction," "Fundamental Mechanisms of Friction and Wear," "Solid Lubrication and Composites," and "Tribological Applications." Each part is divided into several chapters written by experts currently involved in advanced research in the subject matter. The first part covers an overview of structure, processing techniques, and properties of advanced structural ceramics, followed by a brief description of present and potential future tribological applications. The second part discusses wear mechanisms, surface damage, wear models, the effect of microstructure on wear, and tribochemical effects. The third part discusses friction and lubrication of monolithic and composite ceramics, and the last part summarizes the current research activities in the utilization of advanced ceramics in various applications such as cutting tools, bearings, seals, automotive cam roller-followers, and read/write magnetic recording heads.

Many colleagues and friends have been instrumental in developing and completing this project. First, I would like to thank my mentor, Professor Nam P. Suh, for introducing me to the subject of tribology, and for staying in touch as a colleague and friend. I have also had the privilege of working with many great tribologists who unknowingly had a tremendous influence on my career. I started my tribology research by reading pioneering publications by Professor David Tabor, who is now a good friend and a colleague. Later, I had the pleasure of working with Professor Traugott Fischer and Mr. Marshall Peterson. Their friendship and guidance through the years has been tremendous.

This book was originally suggested by Ms. Deborah J. Hope, Associate Acquistions Editor at Marcel Dekker Inc. From there, Andrew Berin and Rod Learmonth, Production Editors, and Mr. Russell Dekker, Executive Editor, provided the necessary guidance and encouragement for the completion of this project. I am grateful to the authors who took time away from their jobs and family to complete their chapters. This book would have not been possible without the sacrifices made by my wife, Feri, and my children, Sam and Farid, who spend many weekends and evenings alone while I worked on this project.

Said Jahanmir

Contents

Preface *v*

Contributors *ix*

PART I: INTRODUCTION

1 Advanced Ceramics in Tribological Applications 3
Said Jahanmir

PART II: FUNDAMENTAL MECHANISMS OF FRICTION AND WEAR

2 Wear Mechanisms of Aluminum Oxide Ceramics 15
Said Jahanmir and X. Dong

3 Relation Between Surface Chemistry and Tribology of Ceramics 51
Traugott E. Fischer and William M. Mullins

4 Friction and Wear of Silicon Nitride Exposed to Moisture at High Temperatures 61
Steven Danyluk, M. McNallan, and D. S. Park

5 Surface Damage and Mechanics of Fretting Wear in Ceramics 79
P. J. Kennedy, A. A. Conte, Jr., E. P. Whitenton, L. K. Ives, and Marshall B. Peterson

6 Abrasive Wear of Ceramics 99
Jorn Larsen-Basse

PART III: SOLID LUBRICATION AND COMPOSITES

7 A Review of the Lubrication of Ceramics with Thin Solid Films 119
Ali Erdemir

8 Self-Lubricating Ceramic Matrix Composites 163
Arup Gangopadhyay, Said Jahanmir, and Marshall B. Peterson

9 Tribological Behavior of Whisker-Reinforced Ceramic Composite Materials 199
Charles S. Yust

10 Tribological Characteristics of Silicon Carbide Whisker-Reinforced Alumina at Elevated Temperatures 225
Christopher DellaCorte

11 Microstructure and Wear Resistance of Silicon Nitride Composites 261
Steven F. Wayne and S. T. Buljan

PART IV: TRIBOLOGICAL APPLICATIONS

12 Ceramic Cutting Tools 289
Ranga Komanduri

13 Ceramic Materials for Rolling Element Bearing Applications 313
R. Nathan Katz

14 Tribological Performance of Ceramic Cam Roller Followers 329
Arup Gangopadhyay, H. S. Cheng, Joseph F. Braza, Stewart Harman, and John M. Corwin

15 Ceramics in Mechanical Face Seal Applications 357
Ramesh Divakar

16 Ceramics for Magnetic Recording Applications 383
S. Chandrasekar, T. N. Farris, and B. Bhushan

Index *425*

Contributors

Joseph F. Braza Advanced Technology Center, Torrington Company, Torrington, Connecticut

B. Bhushan Ohio State University, Columbus, Ohio

S. T. Buljan Superabrasives, Norton–St. Gobain, Worcester, Massachusetts

S. Chandrasekar Purdue University, West Lafayette, Indiana

H. S. Cheng Northwestern University, Evanston, Illinois

A. A. Conte, Jr. Naval Air Development Center, Warminster, Pennsylvania

J. M. Corwin Advanced Materials Technology, Unique Solutions, Inc., Royal Oak, Michigan

Steven Danyluk Department of Mechanical Engineering, Georgia Institute of Technology, Atlanta, Georgia

Christopher DellaCorte Materials Division, Lewis Research Center, NASA, Cleveland, Ohio

Ramesh Divakar Niagara Falls Technology Center, The Carborundum Co., Niagara Falls, New York

X. Dong Department of Mechanical Engineering, University of Maryland, College Park, Maryland

Ali Erdemir Materials and Components Technology Division, Argonne National Laboratory, Argonne, Illinois

T. N. Farris Purdue University, West Lafayette, Indiana

Traugott E. Fischer Department of Materials Science and Engineering, Stevens Institute of Technology, Hoboken, New Jersey

Arup Gangopadhyay Chemistry Department, Scientific Research Laboratory, Ford Motor Company, Dearborn, Michigan

Stewart Harman Chrysler Corporation, Detroit, Michigan

L. K. Ives National Institute of Standards and Technology, Gaithersburg, Maryland

Said Jahanmir Materials Science and Engineering Laboratory, National Institute of Standards and Technology, Gaithersburg, Maryland

R. Nathan Katz Materials Directorate, Army Research Laboratory, Watertown, Massachusetts

P. J. Kennedy Naval Air Development Center, Warminster, Pennsylvania

Ranga Komanduri Department of Mechanical Engineering, Oklahoma State University, Stillwater, Oklahoma

Jorn Larsen-Basse Surface Engineering and Tribology Program, National Science Foundation, Washington, D.C.

M. McNallan Department of Civil Engineering, Mechanics, and Metallurgy, University of Illinois at Chicago, Chicago, Illinois

William M. Mullins Purdue University, West Lafayette, Indiana

D. S. Park Ceramics Laboratory, Korea Institute of Machinery and Metals, Chang-Won, Korea

Marshall B. Peterson National Institute of Standards and Technology, Gaithersburg, Maryland

Steven F. Wayne Advanced Technology Center, Valenite Inc., Madison Heights, Michigan

E. P. Whitenton National Institute of Standards and Technology, Gaithersburg, Maryland

Charles S. Yust Metals and Ceramics Division, Oak Ridge National Laboratory, Oak Ridge, Tennessee

I

INTRODUCTION

1

Advanced Ceramics in Tribological Applications

Said Jahanmir

National Institute of Standards and Technology
Gaithersburg, Maryland

ABSTRACT

Advanced ceramics have been the focus of research and development in the past 20 years as materials for tribological applications. This chapter briefly reviews the structure, processing techniques, and properties of five important classes of advanced structural ceramics: alumina, zirconia, silicon nitride, sialon, and silicon carbide. This review is followed by a discussion of some typical applications in which these materials are being used or may be used in the near future. These applications are classified into five categories based on the inherent properties of the ceramics that are exploited in the application: resistance to abrasion and erosion, resistance to corrosive wear, wear resistance at elevated temperatures, low density, and thermal, electrical, and magnetic properties.

I. INTRODUCTION

Ceramics are defined [1] as solid materials that have as their essential component, and are composed in large part of, inorganic nonmetallic materials. This definition includes not only materials such as pottery, porcelains, refractories, cements, abrasives, and glass, but also nonmetallic magnetic materials, ferroelectrics, and a variety of other products that were not in existence until a few years ago. Ceramics play a major role in industrial applications because of their unique and varied properties, abundance in the earth's crust, and low cost. The renewed interest in ceramics is rooted in recent advances made in materials science, which have led to the introduction of new ceramic materials with unique properties [2–5]. These new materials are classified into electronic and optical ceramics, and advanced structural ceramics.

Selection of materials for tribological components or applications is based not only on the tribological behavior (i.e., controlled friction and wear) but also on other application requirements such as strength, fatigue resistance, corrosion resistance, dimensional stability, thermal properties, reliability, ease of fabrication, and cost. In this chapter we first review the structure, processing techniques, and properties of five important classes of advanced structural ceramics: alumina, zirconia, silicon nitride, sialon, and silicon carbide. Then, we examine some typical tribological applications in which ceramics are either being used or may be used in the future.

II. STRUCTURE AND PROPERTIES OF ADVANCED CERAMICS

Structural ceramics are those ceramic materials that possess sufficient mechanical properties for use as load-bearing components. In this section, five classes of advanced structural ceramics are discussed. Much of the information presented in this section is based on three publications [6–8]. The readers should consult these publications for further information.

A. Alumina Ceramics

Ceramics based on alumina have been used in commercial applications for many years because of their availability and low cost. Alumina ceramics are often classified into high alumina, having more than 80% aluminum oxide, and porcelains, having less than 80% aluminum oxide. High aluminas are used in many mechanical devices and electronics.

Pure aluminum oxide, Al_2O_3, has one thermodynamically stable phase at room temperature with a hexagonal crystal structure, which is designated as the α phase. Often the mineralogical term corundum is used for α-alumina. Commercial high purity ($> 95\%$) alumina ceramics usually contain MgO as sintering aid, and SiO_2 impurities. In less expensive lower grades, silicates are usually used as the sintering aid. Strength and other properties generally improve as the percentage of alumina is increased (Table 1). However, cost increases because of the difficulty in processing high aluminas. Commercially available aluminas are usually processed by sintering a mixture of alumina powder with the sintering aids. Generally, a satisfactory densification is obtained without the application of pressure.

B. Zirconia Ceramics

Zirconia ceramics are an important class of ceramic materials that are characterized by a high strength and toughness at room temperature. Zirconia, or zirconium

Table 1 Representative Mechanical and Physical Properties of Advanced Structural Ceramics[a]

Material	Density (g/cm^3)	Flexural strength (MPa)	Fracture toughness ($MPa \cdot m^{1/2}$)	Elastic modulus (GPa)	Poisson's ratio	Hardness (GPa)	Thermal expansion coefficient ($\times 10^{-6}$ °C^{-1})	Thermal conductivity (W m^{-1} °C^{-1})
Silicon nitride								
sintered	3.2	600	4.5	276	0.24	14	3.4	28
hot-pressed	3.2	800	5.0	317	0.28	20	3.2	30
reaction-bonded	2.5	210	3.6	165	0.22	10	2.8	6
sintered reaction-bonded	3.3	825	—	297	0.28	19	3.5	30
HIPed	3.2	1000	6.0	310	0.28	20	3.5	32
Sintered sialon	3.2	650	5.0	297	0.28	18	3.2	22
Silicon carbide								
hot-pressed	3.2	550	3.9	449	0.19	25	4.5	70
sintered	3.1	400	3.0	427	0.19	27	4.8	80
reaction-bonded	3.0	350	3.5	385	0.19	17	4.4	90
CVD	3.2	500	2.6	450	—	30	5.5	150
Alumina (%)								
85	3.4	296	3.5	221	0.22	10	5.3	15
90	3.6	338	3.5	276	0.22	11	6.1	17
96	3.7	352	4.5	303	0.21	12	6.3	23
99.8	3.9	552	4.5	386	0.22	15	7.1	30
Zirconia								
cubic	5.9	245	2.4	150	0.25	11	8.0	1.7
TZP	6.1	1020	11.0	210	0.24	13	10.6	0.4
PSZ	6.0	750	8.1	205	0.23	12	8.3	2.1

[a]The data in this table were compiled from several sources including material data sheets supplied by the producers. Certain wide variations in the data are due to the differences in processing and composition of the materials, as well as measurement methods. These data represent average values of the data obtained from all the sources.

oxide, is a refractory oxide obtained from abundant zircon sand. Pure zirconia exists in three different crystal structures: monoclinic, tetragonal, and cubic. The monoclinic phase is stable at room temperature and up to about 1170 °C, where it transforms to the tetragonal phase. The tetragonal phase is stable up to 2370 °C; it transforms to the cubic phase at this temperature.

An important characteristic of zirconia ceramics is that the microstructure of the material can be controlled by addition of various cubic oxides such as MgO, CaO, Y_2O_3, and CeO_2. The amount of the additive and the thermomechanical processing history can be chosen such that the tetragonal and cubic phases become stable at room temperature. Zirconia ceramics used in technical applications are classified into three types: cubic, partially stabilized, and tetragonal zirconia.

The cubic zirconia is obtained by fully stabilizing the high temperature cubic phase by addition of about 10% cubic oxides. Although cubic zirconia is used in certain applications, its relatively low fracture toughness and strength (Table 1) prevents its use in tribological applications.

Partially stabilized zirconia (PSZ) has a two-phase structure, consisting of cubic grains with tetragonal and/or monoclinic precipitates, depending on the thermal history during processing. By careful process and microstructural control, PSZ can be made with a larger fracture toughness, and is therefore, of more importance in structural applications. The metastable tetragonal precipitates undergo a stress-induced phase transformation near an advancing crack tip. The compressive stress associated with the increase in volume in the transformation of tetragonal to the monoclinic phase reduces the stress at the crack tip and results in a high strength and toughness. It is also believed that the transformation absorbs energy, which would otherwise be used for crack propagation, thereby reducing the total energy available for advancing the crack and resulting in a toughening effect. Typical commercial PSZ materials contain about 8% MgO or CaO and have a composition of about 58% cubic, 37% tetragonal, and 5% monoclinic.

Tetragonal zirconia polycrystal (TZP) is made by addition of about 2–3% Y_2O_3 or CeO_2 to stabilize the tetragonal phase. This material is nearly 100% tetragonal at room temperature and exhibits the highest toughness and strength among zirconia ceramics and other monolithic structural ceramic materials (Table 1). The toughening mechanism in this material is similar to that for PSZ, that is, tetragonal to monoclinic transformation. The TZP materials are not used as widely in tribological applications because of severe degradation and decrease in strength at intermediate temperatures (200–400 °C) in humid atmospheres.

C. Silicon Nitride Ceramics

Silicon nitride is one of the strongest structural ceramics; it has emerged as a promising material for tribological applications. It has an excellent oxidation resistance as a result of the presence of a protective oxide layer. Its thermal shock resistance is very good because of its low thermal expansion coefficient, low elastic modulus, and high strength. Because of strong covalent bonds, silicon nitride does not melt, but decomposes at temperatures about 1900 °C. As the temperature is increased above this value, an increasingly high overpressure of nitrogen is necessary to prevent decomposition during processing. In oxidizing environments, silicon nitride is stable only at very low partial pressures of oxygen; but in air it rapidly forms a silicon oxide layer on the surface. This oxide layer is protective against further oxidation, but if it is removed—for example, by wear—oxidation occurs rapidly.

Pure silicon nitride, Si_3N_4, exists in two crystallographic forms: α and β, both of which have a hexagonal crystal structure. The α phase has a unit cell approximately twice as large as the β phase. It is generally accepted that the β phase is thermodynamically more stable than the α phase. Because of this, silicon nitride materials formed by conventional sintering processes are primarily in the β phase, although the starting powder is usually in the α phase.

Silicon nitride materials are classified according to the processing techniques used to prepare the solid form. The categories include sintered, hot-pressed, reaction-bonded, and hot isostatically pressed. It is important to recognize that the composition, microstructure, and properties of silicon nitrides vary strongly depending on the processing route used in the fabrication of the product.

Silicon nitride powder compacts can be sintered to full density, without the application of any pressure, using combinations of rare earth oxides and/or alumina as sintering aids. However, the mechanical properties of commercially available sintered silicon nitrides are inferior to those processed by hot-pressing (Table 1). Hot-pressed silicon nitride is usually made with MgO or Y_2O_3 sintering aids. Application of pressure during sintering is instrumental in achieving nearly full density, resulting in very good properties. Hot isostatic pressing (HIP) improves the properties of silicon nitride materials by means of the application of a uniform pressure, which is generally higher than those used during hot-pressing. For example, fracture toughness values as high as 11 MPa $m^{1/2}$ have been obtained in some experimental HIPed silicon nitride materials. However, the shapes that can be formed by hot-pressing and hot isostatic pressing are limited and the processing cost is relatively high.

Reaction-bonded silicon nitride is usually made by pressing pure silicon powder at room temperature, then reacting the preform with nitrogen at temperatures up to 1400 °C. The resulting material, although much less expensive than hot-pressed or sintered materials, has a porosity greater than 10%, which results in poor mechanical properties (Table 1). By adding a high temperature sintering step, it is possible to improve the properties of reaction-bonded silicon nitride materials. This material, however, requires addition of oxide sintering aids to the starting silicon powder. The two principal advantages of this type of material are low cost of starting powder and small shrinkage during sintering.

D. Sialon Ceramics

Sialons, solid solutions of Si, Al, O, and N with the β-silicon nitride crystal structure, are usually made by addition of AIN, MgO, BeO, Y_2O_3, or other metal oxides to silicon nitride. Addition of metal oxides causes a distortion of the lattice; therefore, these solid solutions are sometimes referred to as the β′ silicon nitride. Most mechanical and physical properties of sialons are intermediate to those of silicon nitride and alumina (Table 1). The primary advantage of sialons over alumina and silicon nitride is that the cost of processing is lower than silicon nitride, since densification can be achieved by pressureless sintering at lower temperatures and the mechanical properties are better than those of alumina.

E. Silicon Carbide Ceramics

Silicon carbide ceramics are widely used in applications requiring wear resistance, high hardness, retention of mechanical properties at elevated temperatures, and resistance to corrosion and oxidation. The oxidation resistance of silicon carbide is due to a protective SiO_2 surface layer, similar to silicon nitride. Its thermal shock resistance is considered to be good, although it is somewhat lower than that of silicon nitride.

Silicon carbide exists in hexagonal and cubic crystallographic forms, termed α and β, respectively. Silicon carbide ceramics can be grouped into four types depending on the processing methods used in the production of the solid forms: reaction-bonded or reaction-sintered, hot-pressed, sintered, and chemical vapor deposition (CVD).

In the reaction-bonding process a mixture of silicon carbide powder, graphite, and a plasticizer is pressed in a mold to prepare a preform or a “green” compact. The plasticizer is then burned off, producing a somewhat porous product, and silicon metal, as a liquid or vapor, is infiltrated into the pores. The reaction between silicon and carbon forms SiC. However, this reaction is not complete, leaving some residual Si and C. Usually excess Si is used to fill the pores. The finished product is almost fully dense, and it contains a mixture of Si, C, and reaction-formed SiC between the original SiC particles. The mechanical properties of reaction-bonded silicon carbide depend on the amounts of free Si and C. The densification process does not produce any shrinkage, so dimensional tolerances are more easily achieved than with other processes such as sintering. The primary advantage of this type of silicon carbide is the relatively low cost, since near-net-shape components can be made with little machining required after densification.

Hot-pressing is used to produce high strength silicon carbide to nearly full density. In this process, boron and carbon, and sometimes alumina are used as sintering aids for processing of both α- *and* β-silicon carbide components. Although this type of silicon carbide exhibits very good mechanical properties (Table 1), it is not widely used. The major drawback of hot-pressed silicon carbide is the high cost of finished components, which is due the difficulty in machining after densification.

Silicon carbide components are also produced by sintering without the application of pressure, using carbon and boron sintering aids. The major advantage of this process is that most of the machining can be easily done on the green compact. The densified component is then finish machined by diamond grinding and polishing.

Chemical vapor deposition is used to produce relatively pure and dense silicon carbide. This material is highly anisotropic, as a result of the columnar structure developed during the deposition process. In addition to anisotropy, the high cost and residual stresses are major drawbacks against widespread use of this material. Nevertheless, CVD silicon carbide is an excellent material as a coating where resistance to wear, erosion, and oxidation is required.

III. TRIBOLOGICAL APPLICATIONS

The properties of advanced ceramics that make them suitable for tribological applications are high hardness and compressive strength, retention of mechanical properties at elevated temperatures, resistance to chemical reactions, low density, high stiffness, and good electrical, thermal, and magnetic properties. Their disadvantages include low fracture resistance, cost of processing raw materials such as powders, and fabrication cost.

Advanced ceramics are presently used in diverse applications [9–13] such as tips for ballpoint pens, precision instrument bearings, and cutting tool inserts. Applications under development include prosthetic articulating joints, watchbands, and cylinder liners in advanced low heat rejection engines. Each specific application capitalizes on a unique property or a set of specific properties. Tribological applications of advanced ceramics can be divided into five categories based on the properties of ceramics: resistance to abrasion and erosion, resistance to corrosive wear, wear resistance at elevated temperatures, low density, and electrical, thermal, and magnetic properties.

A. Resistance to Abrasion and Erosion

The resistance of materials to abrasive wear is generally related to the hardness of the material. Therefore, a simple solution against abrasive wear in tribological applications is to increase the hardness of the component. This is normally done by hard surface coatings or by selection of hard materials. Although there are some controversies regarding the wear mechanisms in solid particle erosion, experimental data and field experience show that erosion at grazing angles is also related to hardness. Because of their high hardness, ceramics are suitable for applications requiring resistance to abrasive and erosive wear. PSZ, for instance, is used for thread guides and process knives in textile fiber processing; and silicon carbide is used in rocket nozzles, spray-drying nozzles, and sandblasting nozzles. Chromium carbide and tungsten carbide coatings are applied to the turbine blades in jet engines and steam turbine generators. Silicon carbide and sialon are used as seals, bearings, and bushings for slurry and particulate handling equipment.

B. Resistance to Corrosive Wear

Although ceramics are not totally inert, they are generally more resistant to chemical reactions and degradation than metals. This is of particular importance in the process industry. Advanced ceramics such as PSZ, silicon carbide, and silicon nitride are used as pump sleeves, seals, bushings, and valve components in the chemical process industry to combat corrosive wear in harsh environments. In some applications the components must operate at temperatures as high as 850 °C. Perhaps the most common usage of ceramics is in flow control operations, where soda lime silica glass, borosilicate glass, PSZ, alumina, or silicon carbide balls are used in check valves. PSZ is used as the tip of ballpoint pens to resist abrasive wear by the paper and corrosion by the ink. Other applications include diesel injector needle valves and seals for coal particle slurry pumps, which require resistance to abrasive/erosive wear and corrosion resistance.

C. Wear Resistance at Elevated Temperatures

Advanced ceramics are currently used in metal forming and high speed metal cutting operations [14]. In high speed cutting applications, the temperature at the cutting point can reach 1000 °C. Therefore, high hardness, fracture strength, and wear resistance at these temperatures are required. Examples of cutting tool inserts for high speed machining of metals include sialon, TiB_2, PSZ, silicon nitride, and composites such as silicon carbide whisker-reinforced alumina, Al_2O_3/B_4C, and Al_2O_3/TiC. Sialon, silicon nitride, and PSZ are also used in metal forming operations such as extrusion, drawing, bending, and tube expanding. These operations require high strength and wear resistance at high temperatures.

An important potential application for advanced structural ceramics is in automotive engines [13,15]. The tribological components presently under development are cylinder liners, piston rings, valves, valve seats, valve guides, tappet inserts, wrist pins, cam followers, and rocker arms. Ceramics are also being considered for cylinder head plates, pistons, and turbocharger rotors, although these are not considered to be tribological components. Various advanced ceramic materials such as silicon nitride, PSZ, silicon carbide, and composites are being evaluated for these applications.

D. Low Density

Ceramic rolling element bearings are under development for use in abrasive, corrosive, and high temperature environments [16–19]. One particular application is the main bearing for gas turbines [16]. An important advantage of ceramic materials, besides high strength and resistance to abrasion and corrosion, is their lower density. The ceramic balls and rollers reduce the centrifugal force and skidding in ultra-high-speed operations. This has allowed rotational speeds of up to 1600 revolutions per second (100,000 rpm) under high thrust or radial loads at temperatures reaching 1000 °C.

All-ceramic bearings and hybrid bearings (ceramic rollers with steel races) based on silicon nitride are in commercial use [17]. The high strength of silicon nitride, together with the low coefficient of thermal expansion, which minimizes distortions and thermal stresses, allow a longer life than conventional bearings, even under conditions of starved lubrication [19].

An example of a potential application for silicon nitride bearings is in machine tool spindles, where a 30% increase in rotational speed is possible without a substantial design change. In this application, the high modulus of ceramics provides added rigidity for high precision operation. This could translate into large improvements in manufacturing productivity. Another example is found in prosthetic articulating joints [19], where alumina is used because of its low weight, excellent wear resistance, and biocompatibility. The low density of ceramics is also exploited in automotive turbochargers to reduce turbolag. Other examples in which the low density of ceramics is advantageous include exhaust cones in rocket engines and tribological components (bearings, seals, and bushings) for space applications.

E. Thermal, Electrical, and Magnetic Properties

The unique thermal, electrical, and magnetic properties of advanced ceramics are of great importance in certain applications. For example, the primary utilization of ceramic cylinder liners in low heat rejection engines [15] is based on the excellent thermal resistivity of ceramics. Ceramic bearings are indispensable in some instruments, where the magnetic and electrical properties of other materials would interfere with the operation of the instrument [19]. Recent advances in high T_c superconducting ceramics have renewed interest in the development of bearings based on magnetic levitation. Such concepts would allow the evolution of new and unconventional tribological components in the future.

REFERENCES

1. W. D. Kingery, H. K. Bowen, and D. R. Uhlmann, *Introduction of Ceramics*, 2nd ed., Wiley: New York, 1976.
2. D. W. Richerson, *Modern Ceramic Engineering*, Dekker, New York, 1982.
3. *Ceramic Source '89*, American Ceramic Society: Westerville, OH, 1989.
4. D. E. Clark (Ed.), *Ceramic Engineering and Science Proceedings*, American Ceramic Society: Westerville, OH, 1988.
5. Advanced ceramics, *Ceram. Bull. 67*(2), 341–402 (1988).
6. J. B. Wachtman and D. E. Niesz, Commercial structural ceramics, in *Handbook of Structural Ceramics*, M. Schwartz (Ed.), McGraw-Hill, New York, 1992, pp. 3.1–3.44.
7. G. L. Leatherman, and R. N. Katz, Structural ceramics: Processing and properties, in *Superalloys, Supercomposites and Superceramics*, Academic Press: Orlando, FL, 1989, pp. 671–696.
8. R. Stevens, *Zirconia and Zirconia Ceramics*, Magnesium Elektron Publication No. 113, Magnesium Elektron Ltd.: Manchester, U.K., 1986.
9. S. Jahanmir (Ed.), *Tribology of Ceramics*, Vol. 1, *Fundamentals*, Special Publication SP-23, Society of Tribologists and Lubrication Engineers: Park Ridge, IL, 1987.
10. S. Jahanmir (Ed.), *Tribology of Ceramics*, Vol. 2, *Applications*, Special Publication SP-24, Society of Tribologists and Lubrication Engineers: Park Ridge, IL, 1987.
11. *Ceramic Tribological Materials*, (January 1970–January 1988), Citations from the U.S. Patent Bibliographic Database, PB88-85238, U.S. Department of Commerce, National Technical Information Service: Springfield, VA, 1988.
12. *Ceramic Bearings*, Citations from the International Aerospace Abstracts Database, PB88-854708, U.S. Department of Commerce, National Technical Information Service: Springfield, VA, 1987.
13. P. Wray, Advanced structural ceramics, *TechMonitoring*, SRI International, July 1991.
14. Machining issue, *Ceram. Bull. 67*(6), 991–1052 (1988).
15. R. P. Larsen and A. D. Vyas, The outlook for ceramics in heat engines, 1900–2010, SAE Paper No. 880514, Society of Automotive Engineers: Dearborn, MI, 1988.
16. E. V. Zaretsky, Ceramic bearings for use in gas turbine engines, ASME Paper No. 88-GT-138, American Society of Mechanical Engineers: New York, 1988.
17. R. N. Katz and J. G. Hannoosh, Ceramics for high performance rolling element bearings: A review and assessment, *Int. J. High Technol. Ceram. 1*, 69–79 (1985).
18. J. F. Dill, Rolling element bearing technology: Sizing up the Japanese, *Mech. Eng.* December 1987, pp. 37–40.
19. S. Jahanmir, *Ceramic Bearing Technology*, NIST Special Publication No. 824, U.S. Department of Commerce, National Institute of Standards and Technology, Washington, DC: Government Printing Office, 1991.
20. J. A. Davidson and G. Schwartz, Wear, creep and frictional heat of femoral implant articulating surfaces, *J. Biomed. Mater. Res.* 21(A3), 261–285 (1987).

II

FUNDAMENTAL MECHANISMS OF FRICTION AND WEAR

2

Wear Mechanisms of Aluminum Oxide Ceramics

Said Jahanmir

National Institute of Standards and Technology
Gaithersburg, Maryland

X. Dong

University of Maryland
College Park, Maryland

ABSTRACT

Recent results on the wear behavior of aluminum oxide are reviewed, and the effects of contact load and temperature on the wear mechanisms are described. It is shown that the results of wear tests can be consolidated into a single diagram, which can be used to delineate the effects of load and temperature, and to identify the boundaries between the regions dominated by different mechanisms. The tribological characteristics of alumina can be divided into four distinct regions, based on the fundamental mechanisms involved in the wear process. These regions are tribochemical reaction, plastic flow and plowing, microfracture, and formation of a glassy surface film. The transition from mild to severe wear observed at the intermediate temperature range is analyzed using a fracture mechanics model. The model indicates that this transition is controlled by propagation of microcracks from preexisting near-surface flaws in the material. The utility of the "wear transition diagram" is discussed, and it is shown that this type of diagram is useful for design and material selection for tribological applications.

I. INTRODUCTION

Advanced engineering ceramics offer unique capabilities as tribomaterials. These materials are being used currently in diverse applications requiring wear resistance and chemical stability at elevated temperatures [1]. They are also being considered for advanced applications in future high efficiency engines and other mechanical systems. The design and selection of ceramics for these applications require reliable data and mechanistic information on the effects of temperature, load, and environment on the tribological behavior of these materials.

Data obtained by Peterson and Murray show that the tribological behavior of ceramics is greatly influenced by the contact load, temperature, and environment [2]. These authors concluded that at low stress levels, ceramics can slide effectively without surface damage. The presence of adsorbed moisture and the formation of oxide films were also reported as having strong effects on the frictional behavior. The effect of temperature on the friction and wear of various ceramics has been studied by several investigators [3–14]. The reported results clearly indicate that the tribological behavior of ceramics is complex and depends on the material's composition and structure, as well as on the test conditions. For example, the friction coefficient of alumina sliding against alumina was found to increase from 0.6 at room temperature to 0.9 at 420 °C, then to decrease to 0.7 at 640 °C [6]. Similar behavior was recently reported by Woydt and Habig [13] for a high purity alumina and by Yust and DeVore [14] for zirconia-toughened alumina. These investigators also reported that further increase in temperature to 800 °C was accompanied by a decrease in the friction coefficient to 0.3–0.4. Although the reduction in friction was followed by a decrease in wear rate, the mechanisms for the observed behavior were not explored.

Wear rates of several ceramics have been reported to increase suddenly by several orders of magnitude as a result of a slight increase in one of the test variables. These wear transitions occur as a result of increasing the normal contact load [15,16], sliding distance [10,13,16,17], or sliding speed [10,13,16]. In certain cases wear transitions, which also accompany an increase in the coefficient of friction, have been explained by the formation or removal of tribochemical reaction films [12]. However, the exact mechanism for the removal of these films, or the mechanism of transition in the absence of tribochemical films, has not been clarified.

This chapter reviews recent findings on the fundamental wear mechanisms of a high purity alumina. Our investigation is focused on a systematic study of the wear process as a function of temperature and contact load to define the conditions for each specific wear mechanism. To achieve this objective, friction and wear tests were conducted under various test conditions and the wear tracks were examined using scanning electron microscopy (SEM). Particular attention was focused on the mechanisms of transition from mild to severe wear. A contact mechanics analysis was used to determine whether crack initiation or crack propagation controls the observed fracture process leading to severe wear. We show that the results of the tests and the observations on the mechanisms can be consolidated in a simple wear transition diagram.

II. EXPERIMENTAL PROCEDURE

Wear tests were conducted in a linear reciprocating ball-on-flat high temperature tribometer. The balls were 99.5% pure alumina and the flats were 99.8% pure alumina. The impurities primarily consisted of silicon oxide and other oxide sintering aids. The alumina balls (12.7 mm in diameter) were used in the as-received condition with a highly polished surface of 0.01 μm rms. The flats (19 mm in diameter, thickness of 3 mm) were polished to a final roughness of 0.1 μm rms. The composition and properties of the specimens are listed in Table 1.

The experiments were conducted at a low speed of 1.4 mm/s to minimize thermal effects due to sliding. A stroke length of 10 mm was used in all the tests. The normal load was varied from 10 N to 100 N; and a temperature range of 23 °C (room temperature) to 1000 °C was used. The tests were run for different durations ranging from 10 minutes to 24 hours (approximately 1–120 m sliding distance) in laboratory air with a relative humidity ranging from 30 to 75%. Before the initiation of sliding, the specimens were allowed to stabilize at the test temperature. Complete details of the tribometer and the test procedure can be found in Reference 18.

After the wear experiments, the specimens were ultrasonically cleaned in hexane for 5 minutes to remove the loose wear debris. The wear volume on the flat specimens was calculated from surface profile traces (usually three to five)

Table 1 Composition and Properties of Alumina Used in the Experiments

	Form of alumina[a]	
	Balls (Coors AD 995)	Flats (Coors AD 998)
Impurities, wt %	0.16 SiO_2, 0.11 CaO, 0.11 MgO, 0.03 Fe_2O_3, 0.04 Na_2O	<0.1 SiO_2, 0.03 Na_2O, 0.02 TiO_2, 0.01 K_2O
Grain size μm	17	3
Porosity, %	~ 0	0.2
Vickers hardness, GPa	14.7[b]	15.0 (at 1000 g)[b]
Elastic modulus, GPa	372	345
Compressive strength, MPa	2620	2071
Flexural strength, MPa	379	331
Thermal conductivity, W m^{-1} K^{-1}	35.6	29.4
Fracture toughness, MPa · $m^{1/2}$	3.0[b]	3.2[b]
Thermal expansion coefficient, °C^{-1}	7.1×10^{-6}	6.7×10^{-6}

[a]Information on the manufacturer or supplier is included only to identify the materials. This does not imply endorsement of the products.
[b]These properties were determined as a part of our research; the rest were supplied by the manufacturer.

across the wear track perpendicular to the sliding direction. The wear volume on the alumina balls was calculated from measurements of the wear scar diameter, assuming that the worn volume is a flat segment of a sphere. These data were then used to calculate the wear coefficient, which is a dimensionless quantity defined as wear volume per unit sliding distance per unit load normalized by indentation hardness at the test temperature.

The term "severe wear" is used here to indicate a wear coefficient larger than 10^{-4}; and "mild wear" is used to reflect wear coefficients much smaller than this value. An attempt was made to measure the wear coefficients in the mild wear regime. An estimate of 10^{-6} was obtained for the wear coefficient using surface profilometry. However, this value is not accurate, since the wear depth was comparable to the original surface roughness.

III. WEAR TRANSITION DIAGRAM FOR α-ALUMINA

The results of the wear tests were consolidated in a single diagram [18], shown in Figure 1, where the abscissa and the ordinate show the temperature and the contact load used. The boundaries between mild and severe wear regimes are shown by solid lines. Since these boundaries separate the mild and severe wear regions, the figure can be designated as the wear transition diagram for alumina used in this investigation. The contact stress (i.e., maximum pressure), based on elastic calculations for a ball/flat contact, is also shown for particular loads. The figure shows that the tribological performance of α-alumina can be divided into four regions. It should be noted that this diagram is based on the experimental

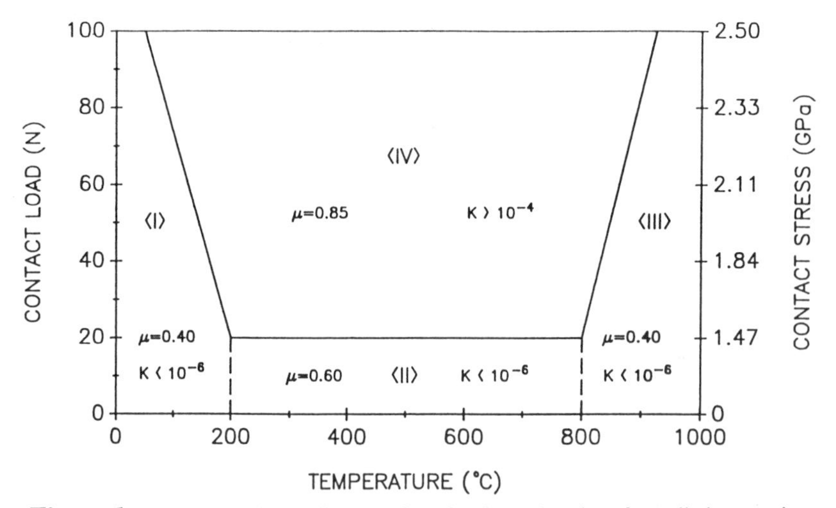

Figure 1 Wear transition diagram for alumina, showing four distinct regions and indicating the friction coefficient μ and the wear coefficient K for each region.

data for a particular material combination tested in the present investigation. The solid lines correspond to transitions in the mechanisms that result in a large change in the wear rate and friction coefficient. The dashed lines at 200 and 800 °C designate changes in the mechanism alone, without much effect on the friction coefficient or the wear rate.

A. Wear Mechanism Under Ambient Conditions

In region I, tribochemical reaction between the alumina surface and water vapor in the environment controls the tribological performance. The coefficient of friction ranges from 0.35 to 0.45, with a mean value of 0.40. Friction force measurements for two typical tests run at room temperature and at two different loads are shown in Figure 2a. In both cases, the friction coefficient trace is fairly smooth and reaches a steady state value rapidly. Figures 2 and 3 show that the coefficient of friction is independent of load and sliding distance. The test results also showed that the wear coefficient is independent of load and sliding distance; and it is less than 10^{-6} [19].

Examination of the wear tracks for the tests conducted at room temperature, with the exception of the tests conducted under large loads (> 30 N), showed no microcracks or any evidence for fracture. Typical scanning electron micrographs are shown in Figure 4: the surface pits are associated with grain pullout during polishing and porosity of the material. Many small cylindrical particles appear in the micrograph. These particles, which were observed only on the wear track, range from 1 to 5 μm in length and are approximately 0.2 μm in diameter.

The SEM micrograph in Figure 4b, at a higher magnification, shows a few microcracks along the grain boundaries, as well as plowing marks due to polishing. These microcracks were observed only on the wear track of the specimens tested at contact loads larger than 30 N. But, these microcracks appeared to be stable and did not propagate to produce severe wear, even after 24 hours (120 m) of sliding.

The cylindrical wear particles are similar in shape to the debris formed by tribochemical reaction during sliding of silicon nitride in humid environments [20,21]. In our experiment, the reaction of water vapor with the alumina surface may produce a thin aluminum hydroxide film, which is rolled to a cylindrical form by the sliding and reciprocating action at the contact.

Micro-FTIR (Fourier-transformed infrared) microscopy was used [18] to examine the possibility of aluminum hydroxide formation on the wear track. The results of this analysis conducted on three different positions are shown in Figure 5. The spectra were obtained outside the wear track, on the wear track, and on the debris accumulated at one end of the wear track. All three spectra show a strong peak at 1035 cm^{-1}, which corresponds to α-alumina. It should be recognized that this spectral feature is not directly due to a vibrational mode

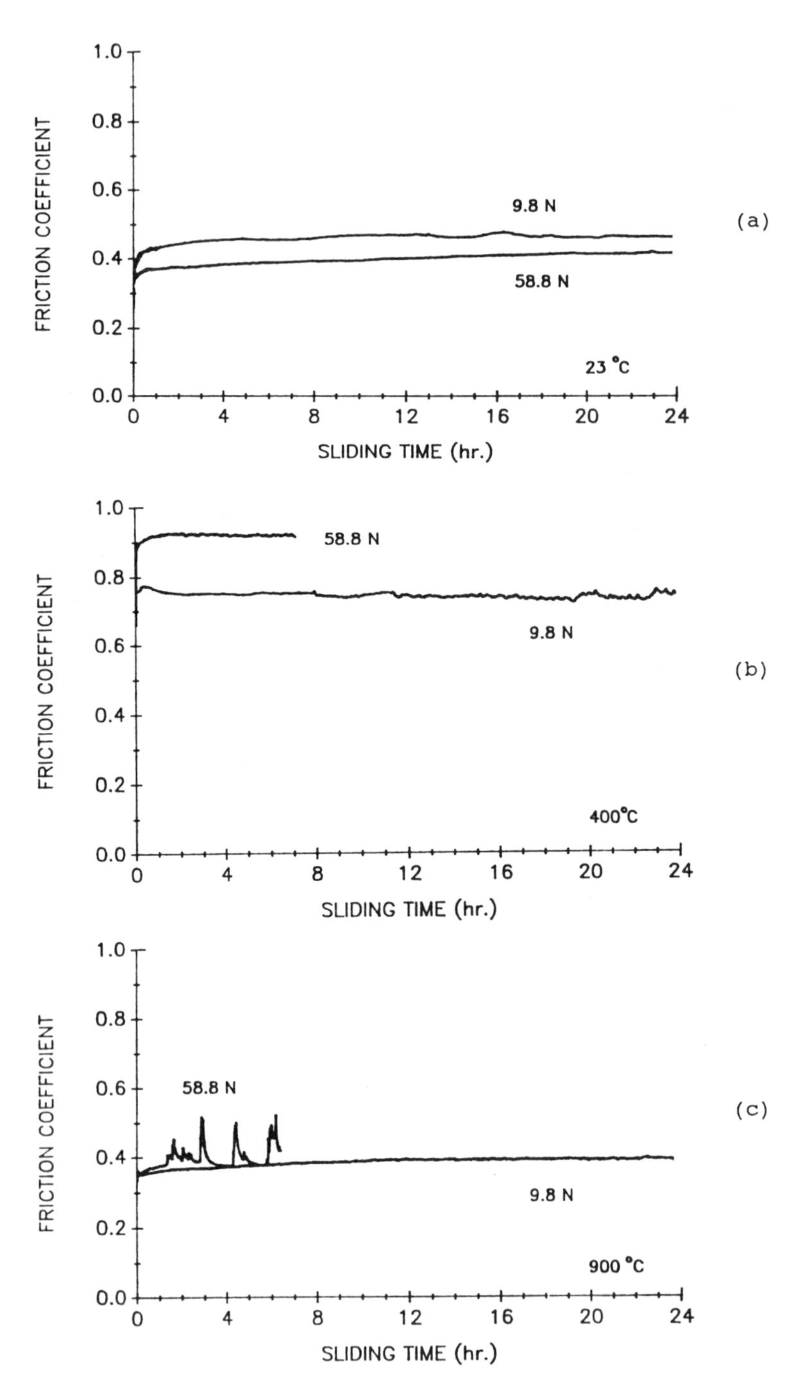

Figure 2 Friction coefficient traces obtained under different test conditions: (a) 23 °C, (b) 400 °C, and (c) 900 °C.

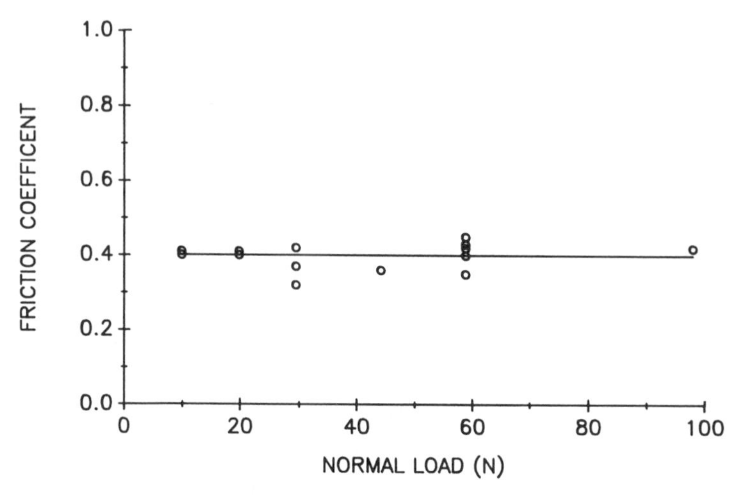

Figure 3 Friction coefficient as a function of normal load at room temperature.

energy absorption, but rather represents an IR reflectivity phenomenon of α-alumina related to its frequency-dependent dielectric constant. The three spectra indicate the presence of an –OH region around 3600 cm^{-1}. The absorption in the 3600 cm^{-1} region on trace c is stronger than either on a or b. Although a and b also have slight peaks in the 3600 cm^{-1} region suggesting –OH, these peaks may be due to the presence of water, as indicated by the peak at 1630 cm^{-1}. It appears that the water content in all three locations is similar, in which case the increase in absorption in the 3600 cm^{-1} region on trace c indicates –OH, other than water. Since this spectrum was obtained on accumulated wear debris, we might expect that a thin –OH surface film (i.e., aluminum hydroxide) was also present on the wear track.

Aluminum hydroxides have a layered structure [22] with relatively weak bonding between the layers and are much softer than the alumina substrate. Formation of such tribochemical films on the alumina surface [23] may explain the relatively low coefficient of friction of alumina at room temperature. The same film, which is expected to have lower elastic properties than the substrate, can reduce the magnitude of the contact stresses [24], resulting in stable microcracks and low wear rates.

B. Wear Mechanism at Low Loads

The controlling wear mechanism in region II is plastic flow and plowing. The coefficient of friction ranges from 0.45 to 0.80, with a mean value of 0.60. The friction traces for one test run at 400 °C under 9.8 N loads is shown in Figure 2b. The friction trace under this test condition is more variable than that of region

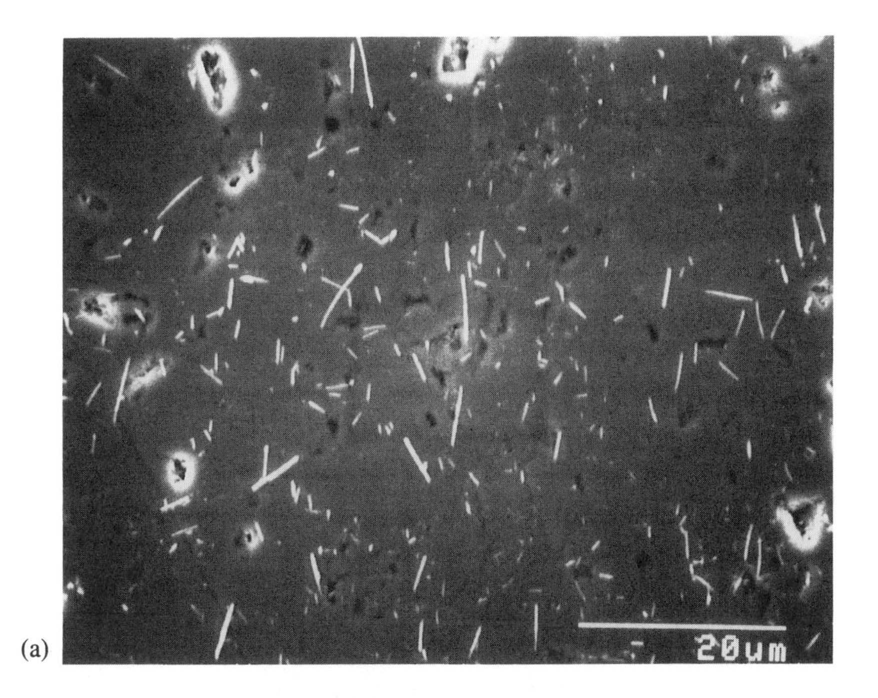

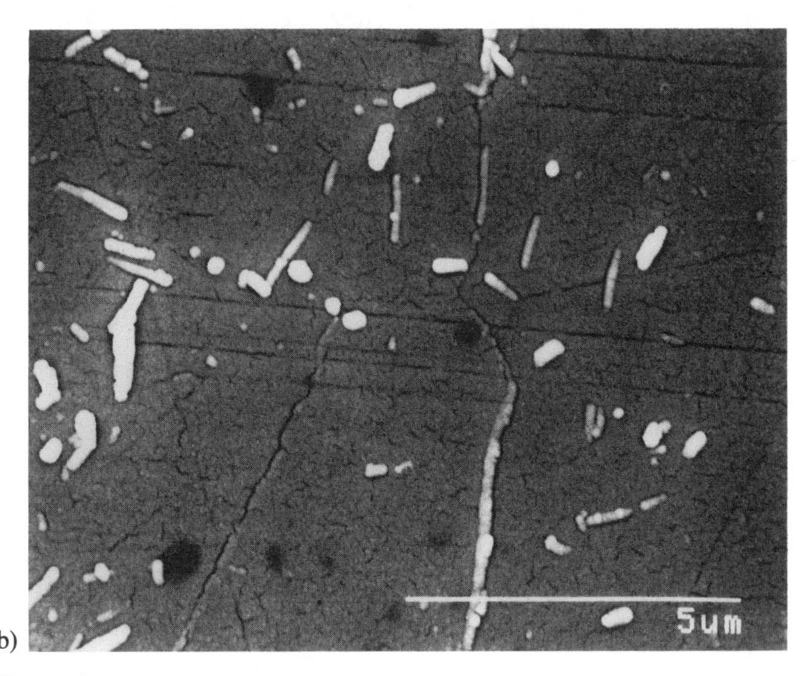

Figure 4 Cylindrical wear debris on the wear track of α-alumina tested at room temperature under loads of (a) 19.8 N and (b) 58.8 N.

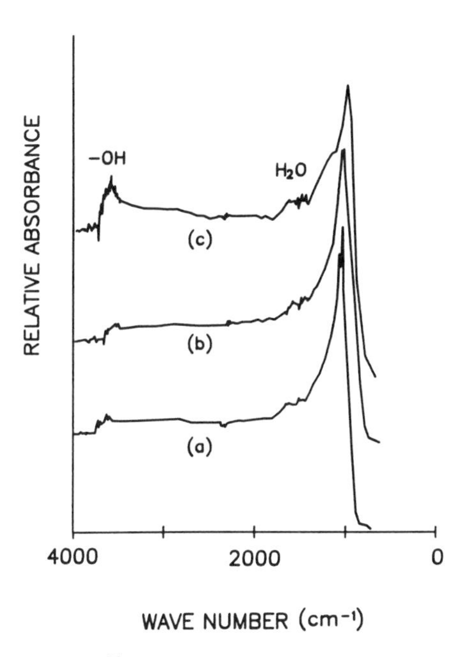

Figure 5 FTIR spectra on the surface of alumina tested at room temperature and 58.8 N load: a, outside wear track; b, on the wear track; c, on wear debris agglomerated at the end of the wear track.

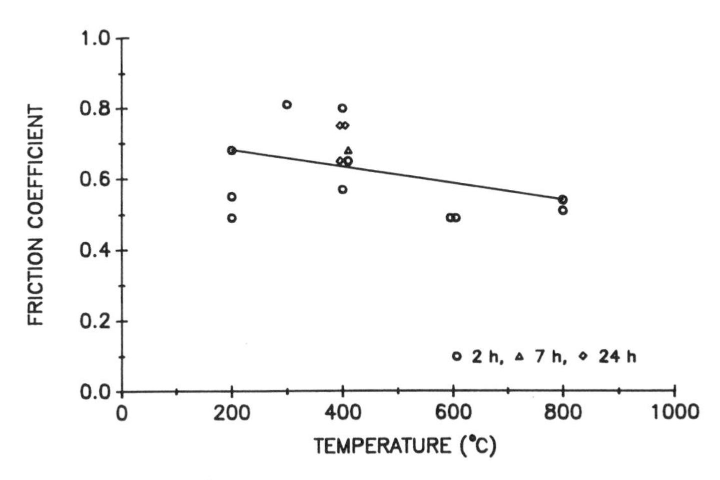

Figure 6 Friction coefficient as a function of temperature at a load of 9.8 N and test durations of 2, 7, and 24 hours.

I; but, similar to region I, steady state condition is reached rapidly. Considering the large scatter in the data in Figure 6, it can be stated that the friction coefficient is independent of temperature in region II. The wear coefficient under these conditions is less than 10^{-6}, similar to region I.

Figure 7a shows the typical appearance of the flat specimens tested under a low load (< 20 N), in the temperature range of 200–800 °C. As the micrograph shows, there is very little damage on the wear track. The surface appears smooth and no wear debris is observed. There are indications of plastic flow, grain boundary relief, and microcracks along the grain boundaries. SEM examination of specimens tested in region II also revealed a few shallow plowing marks oriented in the sliding direction (Fig. 7b). These observations suggests that under these conditions wear occurs by plastic flow and plowing.

C. Wear Mechanism at Elevated Temperatures

The tribological behavior in region III is primarily controlled by formation of a silicon-rich layer on the surface by diffusion and viscous flow of the glassy grain boundary phase. Similar to region I, the friction coefficient ranges from 0.35 to 0.45, with a mean value of 0.40. The friction traces for two typical tests conducted at 900 °C at two different loads are shown in Figure 2c. At 9.8 N, the coefficient of friction remains steady for the duration of the tests, whereas brief but sudden increases in friction are observed at 58.8 N. Although the magnitude of the wear coefficient for both tests was found to be in the mild wear regime, the test at the larger load produced a larger wear coefficient. Typical micrographs of the wear tracks are shown in Figures 8 and 9 for the tests run at 900 °C under a load of 9.8 N for 2 and 24 hours. The wear track is smooth, and only pits from the porosity or polishing are observed in Figure 8a. A close examination of the wear track at a higher magnification reveals that the wear track is covered with a thin film. This can be observed near the edge of a pit in Figure 8b and in Figure 9.

Analysis of the flat specimens tested at 900 °C by energy-dispersive spectroscopy (EDS) showed that the Si concentration was larger in the wear track as compared to the unworn regions outside the track. The effect of test temperature on the Si concentration on the wear track and outside the wear track is shown in Figure 10, where the data points correspond to different regions analyzed. The concentration of Si is low at temperatures below 800 °C but increases by more than an order of magnitude at higher temperatures. At 900 °C, the Si concentration is much higher in the track than the concentration outside the track. However, at 1000 °C, the Si concentration is approximately the same inside and outside the track. A possible source of Si is the glassy grain boundary phase used in the alumina. The furnace lining also contained silicates. To confirm that the Si found on the wear samples is associated with the alumina, one flat specimen

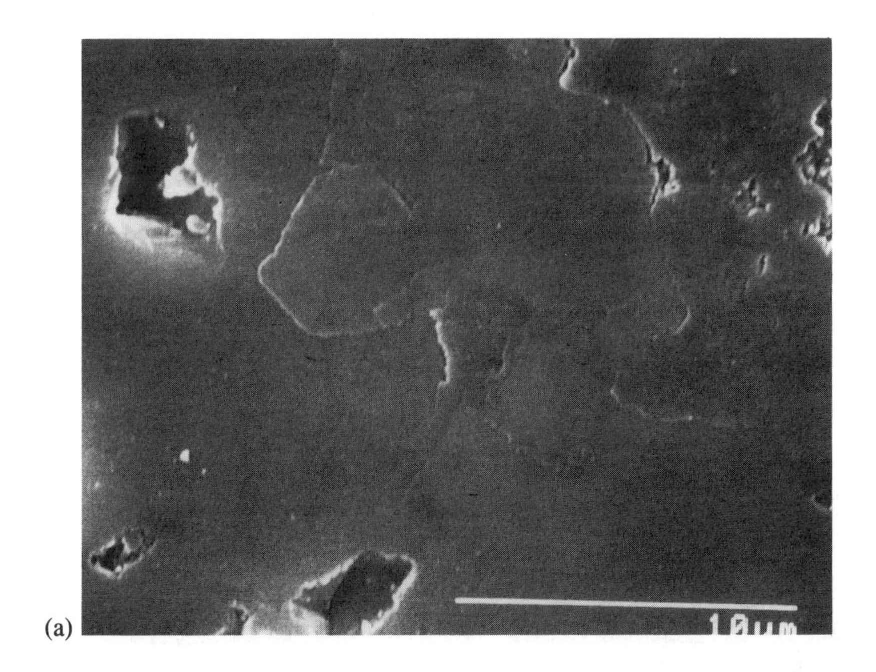

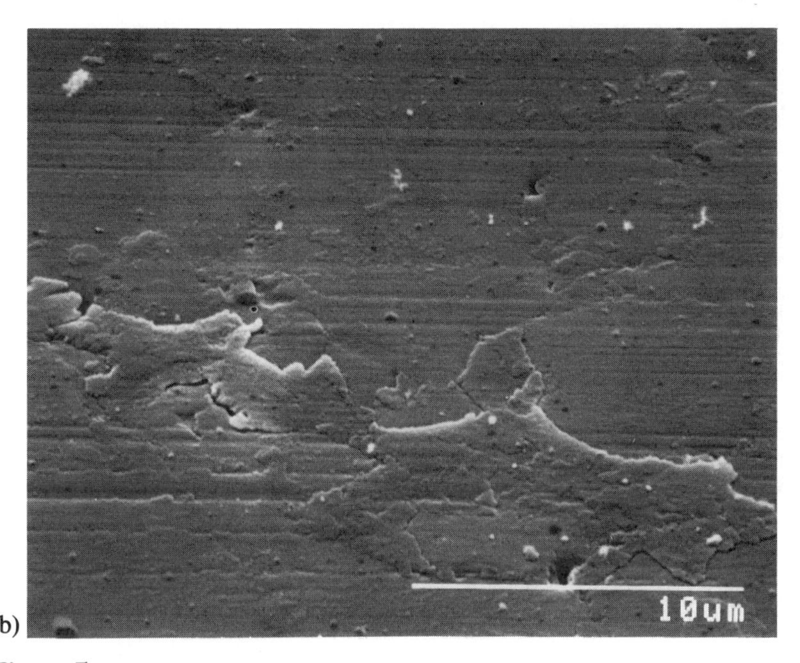

Figure 7 Scanning electron micrograph of the wear track of the flat specimen tested under a contact load of 9.8 N at 200°C: (a) Smooth surface and (b) shallow plowing marks and microcracks on grain boundaries in region II.

(a)

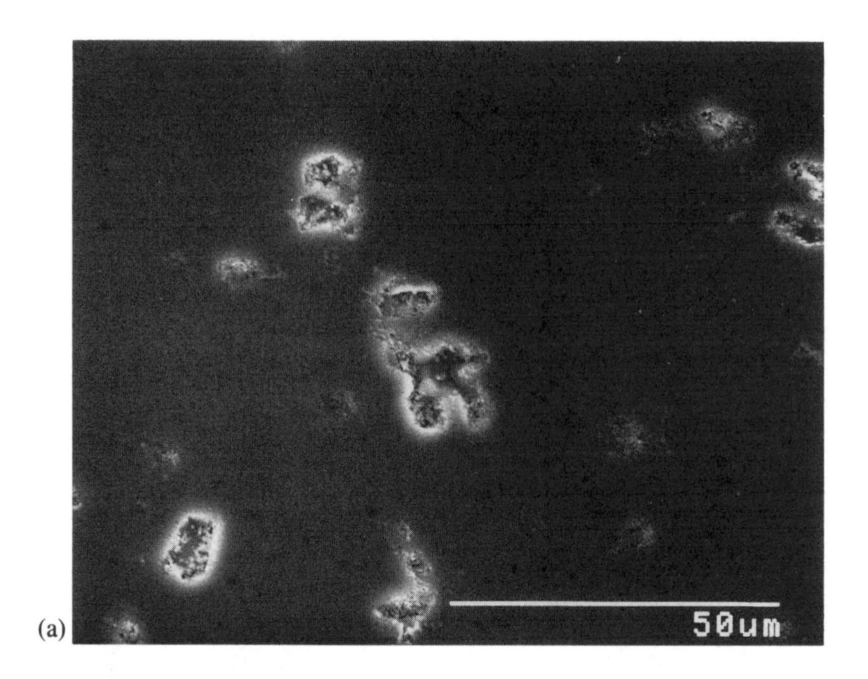

(b)

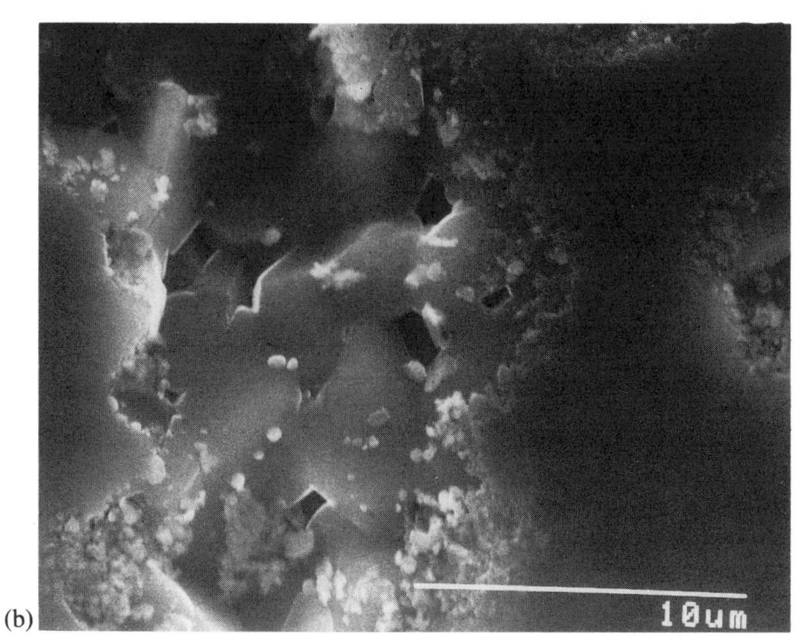

Figure 8 (a) Scanning electron micrograph of the wear track of the flat specimen tested at 900 °C, under a contact load of 58.8 N. (b) Higher magnification of an area in (a).

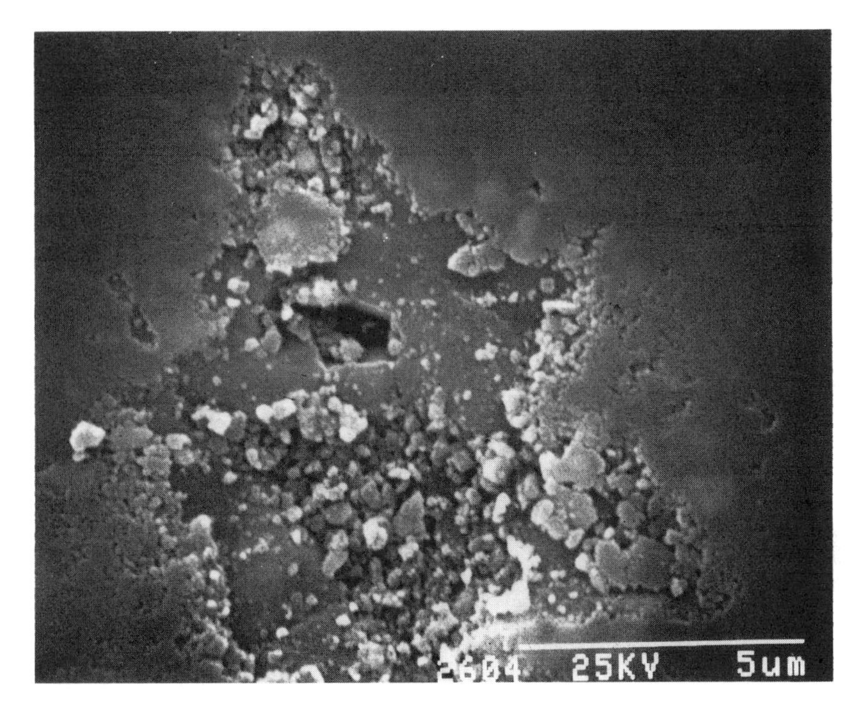

Figure 9 Scanning electron micrograph of the wear track of the flat specimen tested under a load of 9.8 N, at 900 °C, and for 24 hours.

was heated inside a stainless steel enclosure to 1200 °C in a vacuum of 10^{-5} torr for 2 hours. The surface was then examined by EDS. It was found that this heat treatment had produced a silicon-rich layer on the surface.

To determine the thickness of the Si-rich film, the wear track was analyzed with EDS by varying the electron accelerating voltage from 5 to 20 keV. The results of this analysis is shown in Figure 11 for samples tested at 900 °C. Considering the effect of density on the electron penetration, and assuming a specific gravity of 2.63 (for quartz), the thickness of the film was estimated to be 0.4 μm. Although this analysis is not very accurate, it does suggest that the surface film is very thin, which agrees with the appearance of the film in the scanning electron micrographs.

Both the alumina flats and the balls contain Si primarily in the form of glassy phase at the grain boundaries. This glassy material can undergo viscous flow and creep at 900 °C and above [25]. We, therefore, speculate that at 900 °C the glassy phase is forced out of the grain boundaries by the contact pressure, and spread on the wear track. Since the glassy material in the film can deform easily by viscous shear at this temperature, the friction coefficient is reduced. Presence

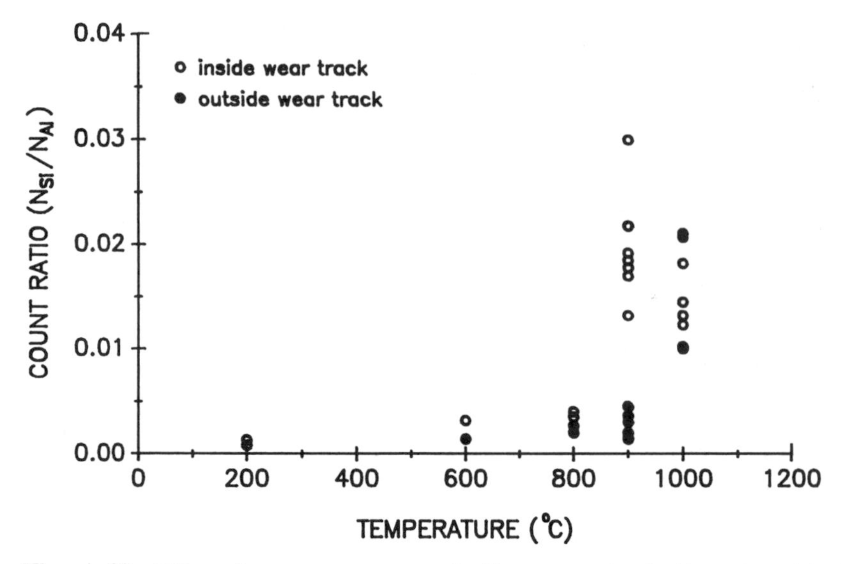

Figure 10 Effect of test temperature on the Si concentration inside and outside the wear track.

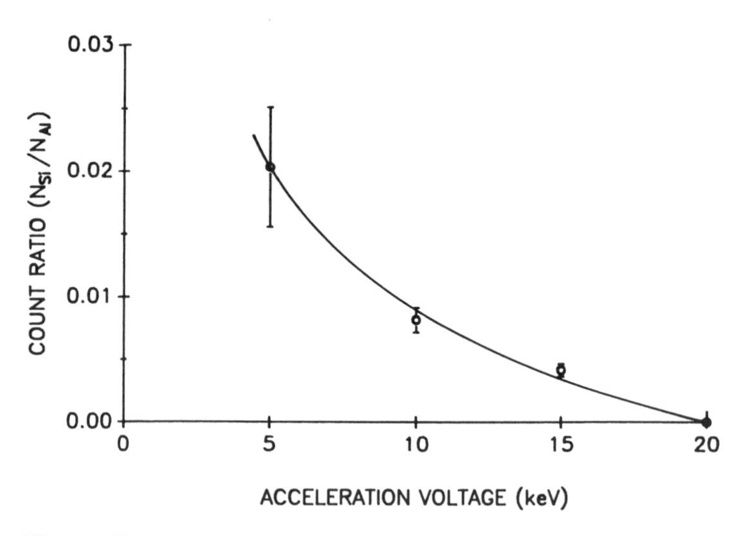

Figure 11 Ratio of Si to Al peaks versus the electron accelerating voltage used for EDS analysis.

of this film on the wear track can also alter the stress distribution at the contact. The net result is very low wear rates in the mild wear regime. Formation of a silicon-rich film can be assisted by diffusion at higher temperatures (i.e., $T \geq$ 1000 °C), since heating experiments in vacuum and analysis of off-track areas have shown that such films can form in the absence of mechanical forces.

The wear track of the specimen tested under a load of 58.8 N contained features that were similar to those observed under a load of 9.8 N. However, localized damage and debris were observed in certain locations on the wear track of this specimen (Fig. 12). The local damage and the rise in friction observed during the test are probably related. Under the larger load of 58.8 N, the silicon-rich lubricating layer may be unstable and is removed in localized areas, producing a larger coefficient of friction. The loss of the lubricating layer promotes localized fracture and wear debris formation. As the debris is pushed out of the contact, the friction coefficient returns to the steady value. The unstable behavior in this test was not unexpected, since the combination of load and temperature for this test falls close to the mild to severe wear transition boundary. The test was terminated after 7 hours.

D. Wear Mechanisms at High Loads

Both the coefficient of friction and the wear rate are large in region IV. In this region the wear process is primarily controlled by intergranular fracture, which results in wear coefficients larger than 10^{-4}. As shown in Figure 13, the coefficient of friction ranges from 0.75 to 0.90; but it drops off to 0.5 near the transition boundary at 800 °C. A typical friction trace for the test run at 400 °C at 58.8 N load is shown in Figure 2b.

The wear transition diagram in Figure 1 shows a transition from mild wear to severe wear as the contact load was increased above 20 N, in the temperature range between 200 and 800 °C. Examination of the flat specimens in the scanning electron microscope showed that in the severe wear regime the wear track is covered with compacted wear debris (Fig. 14a). This micrograph was taken on the wear track of the alumina flat specimen tested at 200 °C under a load of 44 N. Cleaning of the specimens in an ultrasonic bath with hexane removed some of the loose debris revealing the grain structure of alumina at certain locations on the wear track (Figs. 14b and c). This suggests that in this temperature range, the primary wear mechanism under high loads is intergranular fracture, which results in severe wear.

In this region of the transition diagram, it was found that the wear coefficient decreases with temperature (Fig. 15). The micrographs in Figure 16 were taken on the wear track of the flat specimens tested at 600 °C. Figure 16a is similar to Figure 14a, showing wear debris accumulation on the wear track. A comparison of Figures 16b and 14b, which were taken at a higher magnification

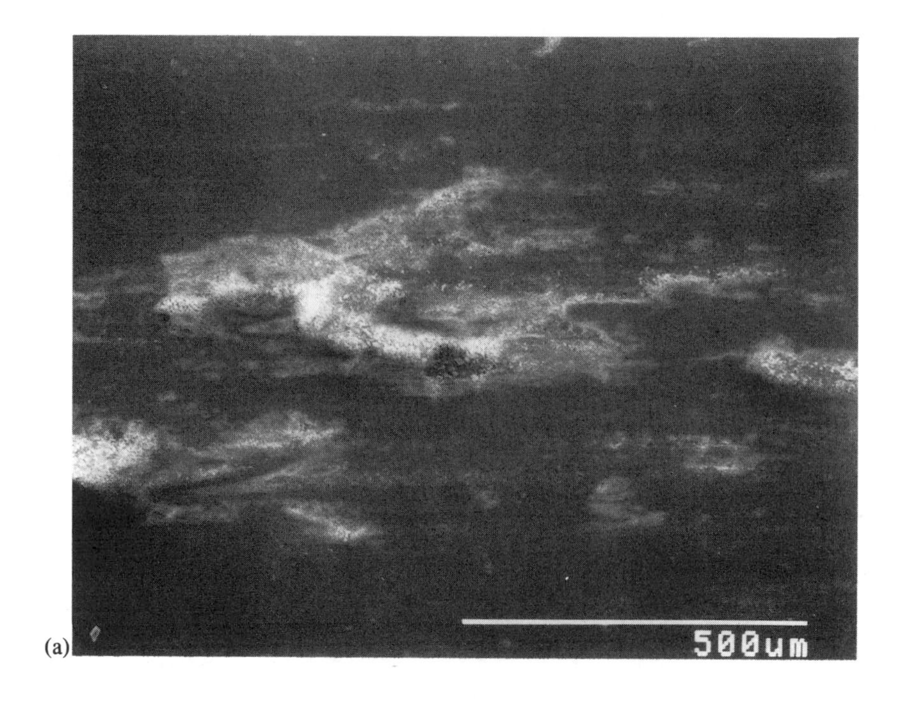

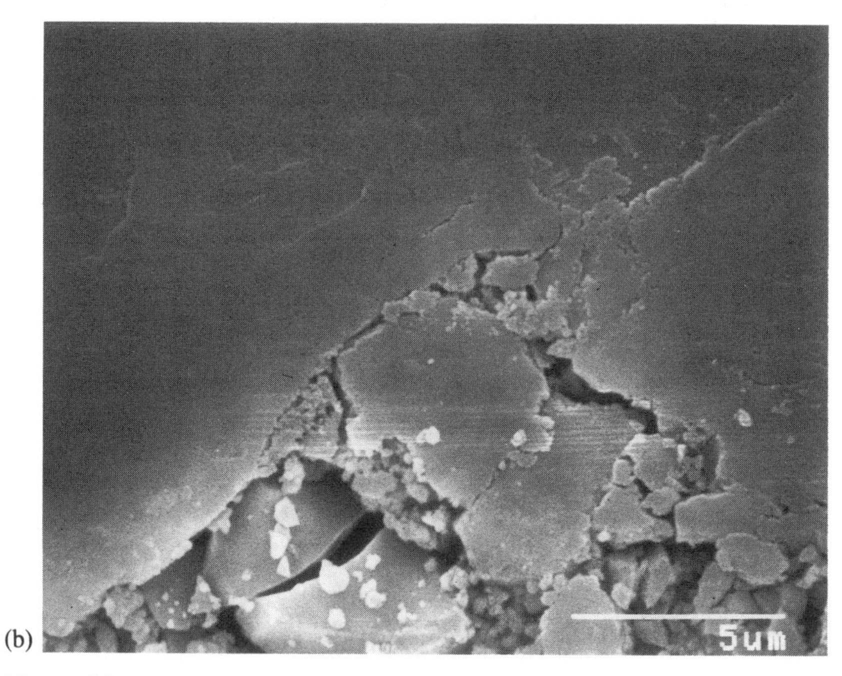

Figure 12 Typical scanning electron micrographs of the wear track of specimens tested at 900 °C, 58.8 N, and 7 hours at two different magnifications.

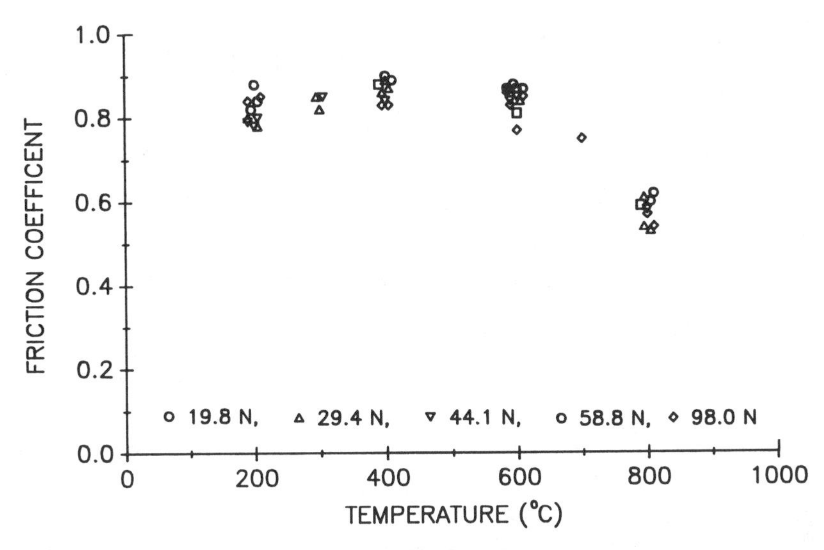

Figure 13 Friction coefficient as a function of temperature in region IV.

(a)
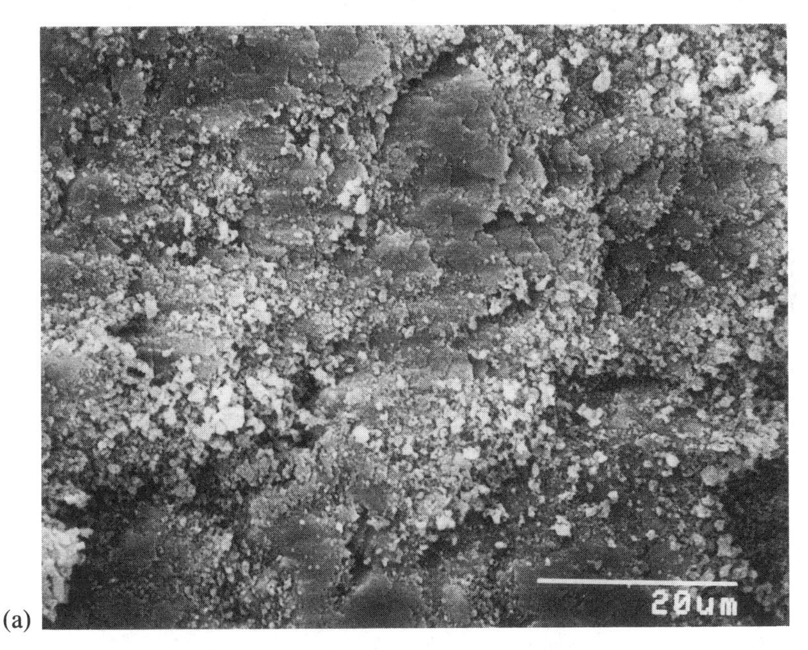

Figure 14 Scanning electron micrograph of the wear track of the flat specimens tested at 200 °C under contact loads of (a) 44.1 N, (b) 58.8 N, and (c) 29.4 N. The micrographs in (b) and (c) were taken after ultrasonic cleaning.

(b)

(c)

Figure 14 Continued

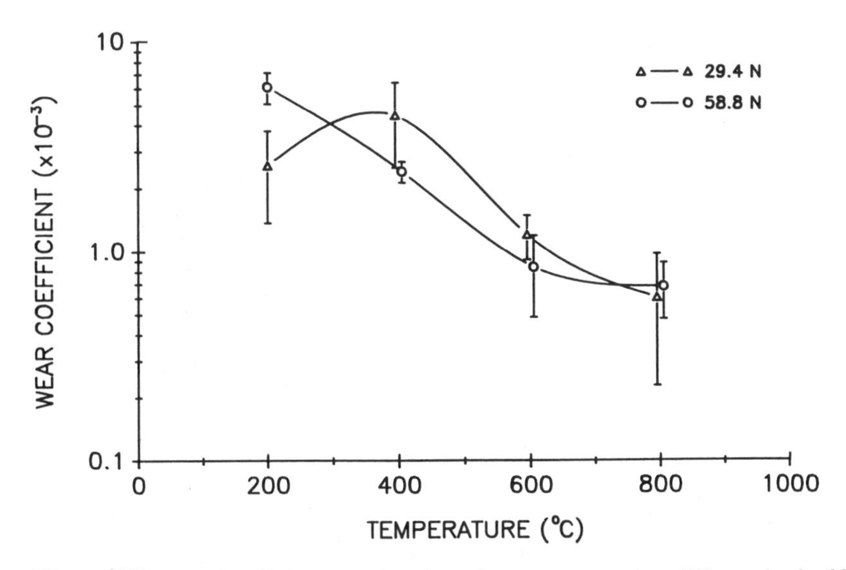

Figure 15 Wear coefficient as a function of temperature at two different loads: 29.4 and 58.8 N.

after ultrasonic cleaning, reveals that the two surfaces are similar in showing intergranular fracture. However, some evidence for transgranular fracture is also observed in Figure 16b. The micrographs in Figure 14b and 16c also show what appears to be plastic flow on the wear track. Since the wear track is covered with debris, it is the wear debris that is subjected to plastic flow, not the alumina substrate. A possible explanation for the reduction in wear at higher temperatures is an increased plasticity and flow of the debris, which would tend to absorb some of the frictional energy and reduce the stresses transmitted to the alumina substrate.

E. Mechanism of Transition from Mild to Severe Wear

To elucidate the mechanism of transition from mild to severe wear and to observe the initiation of fracture in the intermediate temperature range (200 °C $> T >$ 800 °C), a series of experiments was conducted at shorter test durations [19]. A typical friction coefficient trace is given in Figure 17. Under this test condition of 200 °C and load of 58.8 N, a large value is expected for the wear coefficient ($K > 10^{-4}$) and the coefficient of friction (0.85). The friction coefficient trace in Figure 17 indicates that initially the coefficient of friction is approximately 0.50. However, after 15 minutes of sliding it rises to 0.74. (Although not shown, continued sliding would increase the coefficient of friction to 0.85.) This behavior is typical of all test conditions resulting in a high wear coefficient, but the

(a)

Figure 16 Scanning electron micrograph of the wear track of the flat specimens tested at 600 °C, under contact loads of 44.1 N (a) and 58.8 N (b and c). The micrographs in (b) and (c) were taken after ultrasonic cleaning.

transition time is not well-behaved and does not correlate with the test variables. Repeat tests at 58.8 N and 200 °C were stopped at points approximately corresponding to A, B, C, and D in Figure 17.

The photomicrographs in Figure 18 were obtained from the wear tracks of the flat specimens for these tests. Figure 18a, which was obtained from the test terminated at point A, represents the condition of the wear track prior to the rise in friction coefficient; no wear damage is visible on the surface. Figure 18b, corresponding to point B, when the coefficient of friction begins to rise, indicates a few microcracks, which are generally oriented perpendicular to the sliding direction. Higher magnification of this area suggested that some of the microcracks are associated with the grain boundaries. A comparison between Figures 18c and d, corresponding to points C and D, and Figures 18a and b, clearly shows that the surface damage and the number of microcracks increase as the duration of sliding is increased. The increase in the level of surface damage seems to correlate with the coefficient of friction. This implies that two possible explanations for the rise in the coefficient of friction at the transition point are

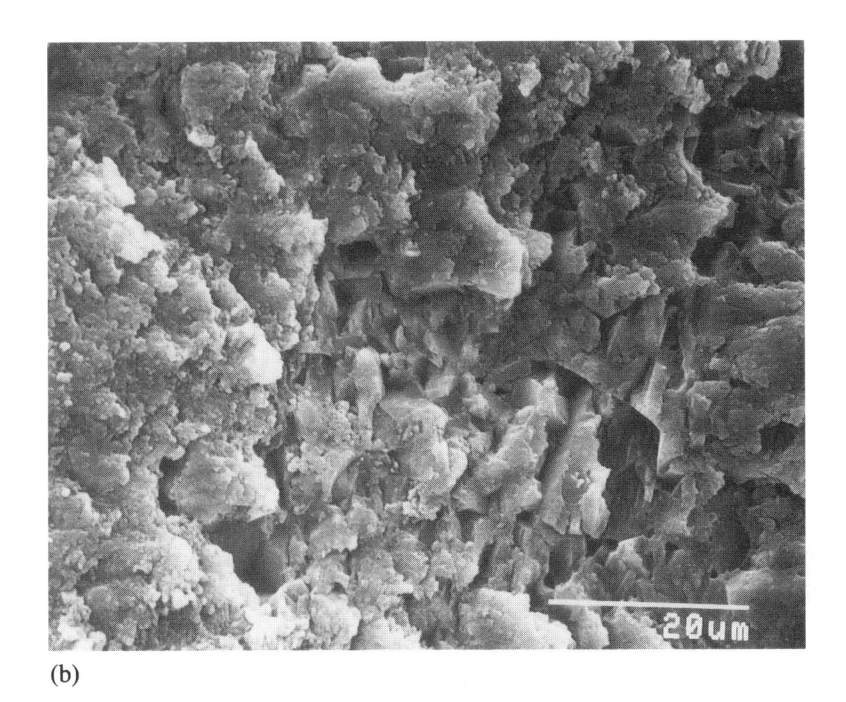

(b)

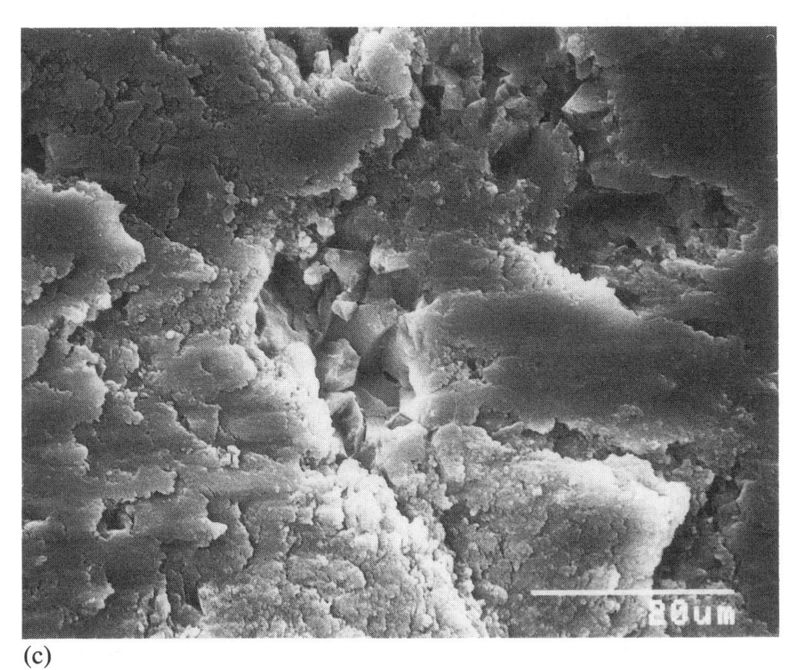

(c)

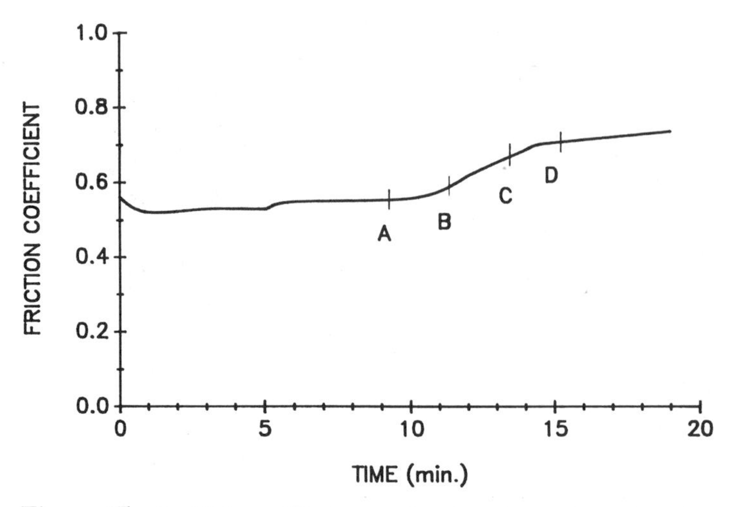

Figure 17 Coefficient of friction as a function of time obtained at 58.8 N normal load, 200 °C temperature, and 1.4 mm/s sliding speed. Points A, B, C, and D correspond to the termination points of repeat tests.

the larger surface roughness developed by wear, and injection of wear debris at the interface as a result of severe wear.

IV. THEORETICAL ANALYSIS OF FRACTURE AT A SLIDING CONTACT

The SEM observations suggest that the transition from mild to severe wear in alumina occurs by a fracture process in the intermediate temperature range. In a sliding contact, the surface is subjected to a set of normal and tangential stresses, which produce deformation and fracture of the material in a surface layer [26]. The process of fracture consists of plastic flow, formation of microcracks, and propagation of these crack to eventual failure. If the material does not contain any microcracks, the first step in the fracture process is the initiation of microcracks. However, if the material already contains microcracks or flaws, from which microcracks can be readily initiated, the fracture process is controlled by the propagation of these microcracks. The following theoretical analysis was performed to determine whether the wear transition is controlled by the initiation or the propagation of microcracks [19]. In this respect, the state of stress under a sliding contact is determined, and the conditions for plastic flow and microcrack initiation are analyzed. Linear elastic fracture mechanics is then used to analyze the process of fracture by the propagation of microcracks from preexisting flaws.

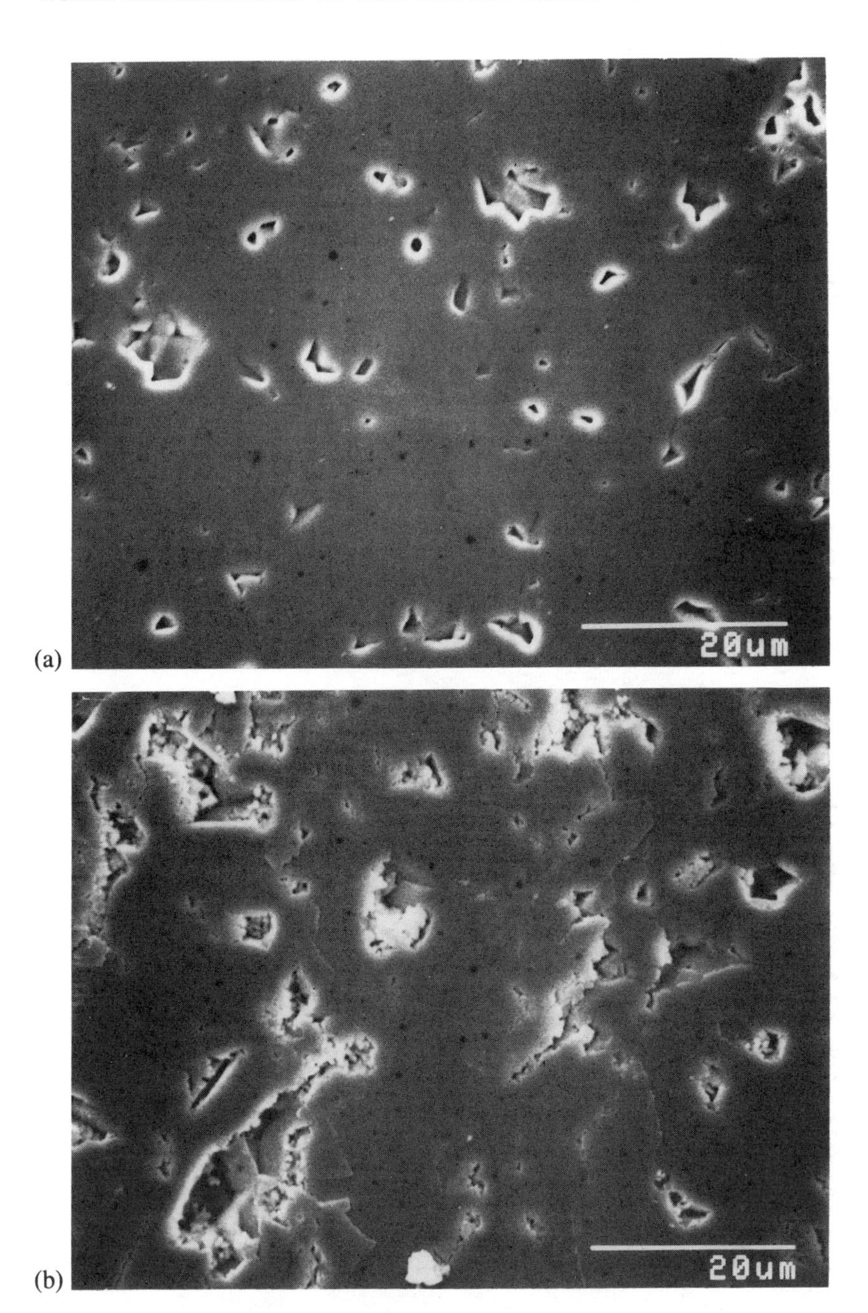

Figure 18 Scanning electron micrographs on the wear track of alumina flat specimens, terminated at different sliding durations designated in Figure 17. (Sliding direction is horizontal.)

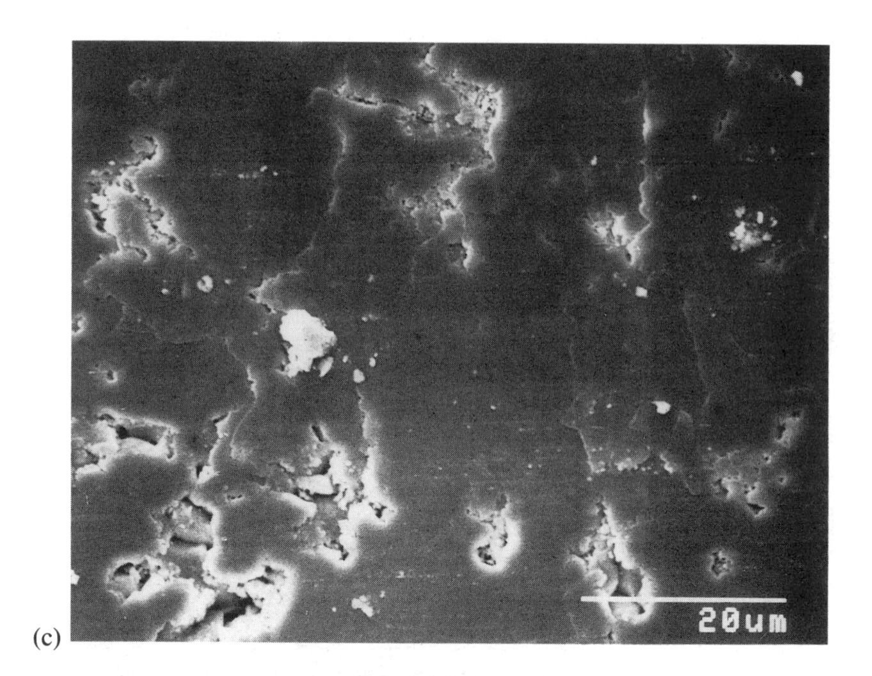

(c)

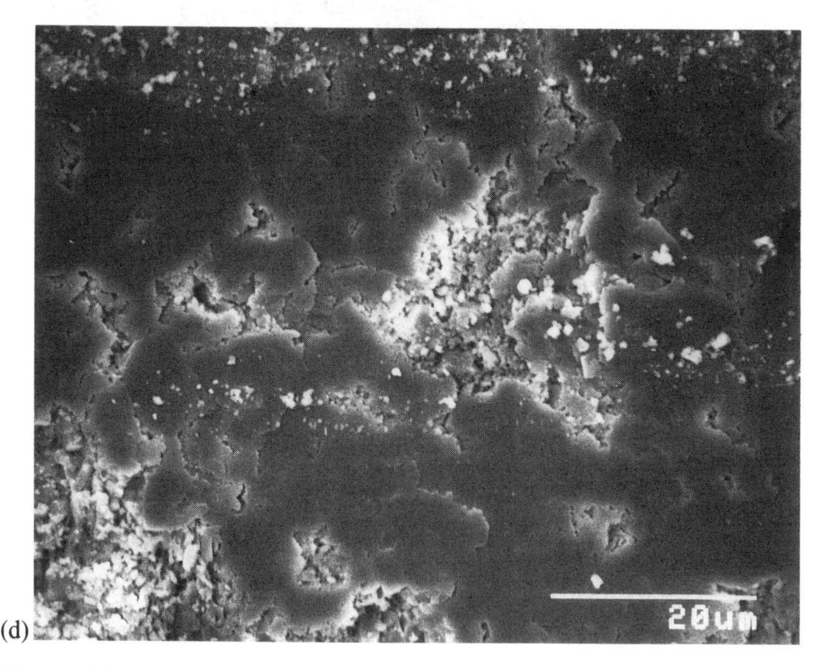

(d)

Figure 18 Continued

A. Analysis of Failure Initiated by Plastic Flow

Microcracks can be nucleated by several mechanisms in a plastically deforming field [26]; these can be grouped into three main categories: dislocation pile-ups, twin intersections, and strain incompatibility. The microcrack nucleation mechanism in a purely elastic field has not been clearly identified. It is possible that even in an elastically deforming region, localized microplastic deformation leads to microcrack nucleation by one of the plastic nucleation processes.

The process of microfracture at the sliding interface is controlled by the stresses imposed on the surface. However, in a real contact situation, asperities are always present; and the microfracture process must be examined by considering the state of stress at the asperity contacts. Recent microcontact models [27–31] have shown that the pressure at the real area of contact is related to the surface roughness. Rough surfaces produce microcontact pressures that are much larger than the calculated Hertzian pressure. However, the deviation of microcontact pressure from the Hertzian pressure calculation is small for relatively smooth surfaces [30], such as the ones used in our study.

Hamilton and Goodman [32] developed explicit equations for the stress field due to a circular contact region under a hemispherical normal pressure and a shear traction. These very long equations, which were refined by Hamilton [33], are not repeated here. However, it is instructive to examine certain important features of the stress state relevant to the wear process.

For a purely normal loading contact between a spherical body and a semi-infinite plane, a maximum tensile stress is predicted on the circumference of the contact circle. The effect of sliding with friction is to add a compressive stress to the front edge and to intensify the tensile stress at the trailing edge of the contact. The magnitude of the maximum tensile stress is directly related to the applied normal load and the coefficient of friction μ. It is also inversely related to the second power of the radius of contact circle. Experimental results on the formation of "ring" cracks in indentation and sliding [34,35] have confirmed the location of the maximum tensile stress calculated from these equations.

The stress field can be used to calculate the second stress invariant J_2 and examine the conditions for the initiation of plastic deformation. By invoking the von Mises yield criterion, one can evaluate the condition for plastic deformation, as well as the location for the initiation of plastic flow. Analysis of plastic flow for a sliding contact between a sphere and a flat plane indicates that the location of flow initiation depends on the magnitude of the coefficient of friction [33]. For $\mu = 0$, plastic flow initiates below the contact surface. The location for flow initiation moves toward the surface as μ increases. At $\mu \geq 0.3$, plastic flow initiates on the surface at the trailing edge of the contact circle (i.e., at the same location as the maximum tensile stress).

To analyze the conditions for initiation of plastic flow for alumina in the present study, the second stress invariant J_2 was calculated for a sliding contact between an alumina ball and a flat surface. The elastic constants for alumina, E and ν, were assumed to be independent of temperature. It was also assumed that the yeild stress can be approximated by a value equivalent to one-third of the indentation hardness determined experimentally at different temperatures [19]. This assumption is based on perfectly plastic slip-line field analysis, and there is ample evidence in its support for metallic alloys [36]. It is used here as a first approximation for ceramics, in the absence of any other reliable criteria.

The results of this analysis for the case of alumina ball sliding on a smooth alumina surface at different temperatures are presented in Figure 19 for different coefficients of friction. Figure 19a represents the critical contact load necessary for plastic flow, and Figure 19b is a plot of critical maximum contact pressure (i.e., maximum Hertzian pressure). The results show a strong influence of temperature and coefficient of friction on the initiation of plastic flow in alumina. As either the temperature or the coefficient of friction is increased, a lower value of applied load is needed to initiate plastic flow and a possible subsequent fracture by microcrack nucleation and propagation.

B. Analysis of Fracture from Preexisting Flaws

Preexisting flaws, such as microcracks at the grain boundary triple points in ceramics, may preempt the requirements for microcrack nucleation. Propagation or extension of preexisting flaws by the tensile component of stress has been analyzed by Chiang and Evans [37], using linear elastic fracture mechanics. This analysis applies to a homogeneous, isotropic, and linear elastic semi-infinite plane subjected to a combination of normal and tangential surface forces, similar to the stress state solution of Hamilton and Goodman [32]. The analysis procedure consists of prescribing the state of stress and assuming that a preexisting surface flaw with a semicircular crack plane is located at the trailing edge of the contact circle. The crack is assumed to be located on a plane normal to the surface. To predict the condition for crack extension, one first determines the stress intensity factor for such cracks. It is assumed that the crack propagates when K becomes equal to or exceeds a critical value of the stress intensity factor K_c. The stress intensity factor was calculated by integrating σ_x component of prior stress over the crack. (Prior stress is defined as the stress at the crack location, in the absence of the crack, from the Hamilton–Goodman solution.) Allowing $K = K_c$ for crack extension, an equation was obtained for the critical normal load for the occurrence of contact fracture. This equation is based on maximum flaw size for unstable crack growth and calculates a lower bound to the load necessary for fracture.

The model of Chiang and Evans [37] was used to calculate the critical load for the initiation of failure for an alumina ball sliding on a smooth alumina

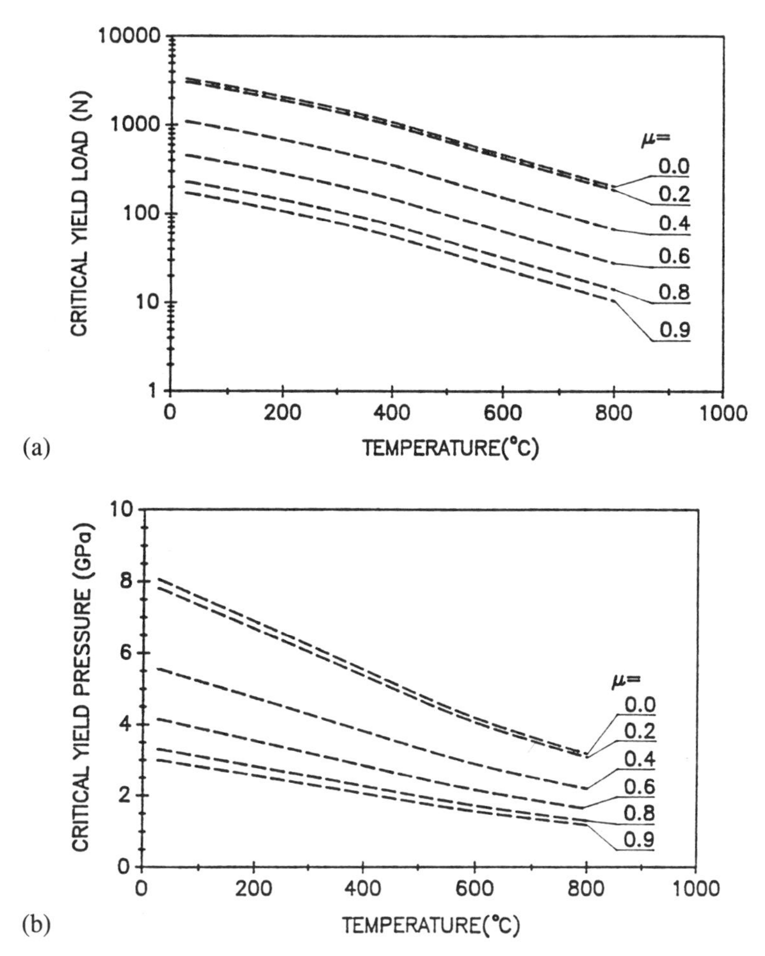

Figure 19 Calculated critical yield load (a) and critical yield pressure (b) of alumina as a function of temperature at different values of coefficient of friction.

surface. The values of critical stress intensity factors determined experimentally by indentation hardness measurements [19] were used in the calculations to predict contact fracture at different temperatures. The results of the analysis for the critical contact load and critical maximum contact pressure are presented in Figure 20. The critical failure load for the extension of preexisting flaws depends strongly on the coefficient of friction. The temperature dependence is very slight, because the small influence of temperature on the fracture toughness of alumina.

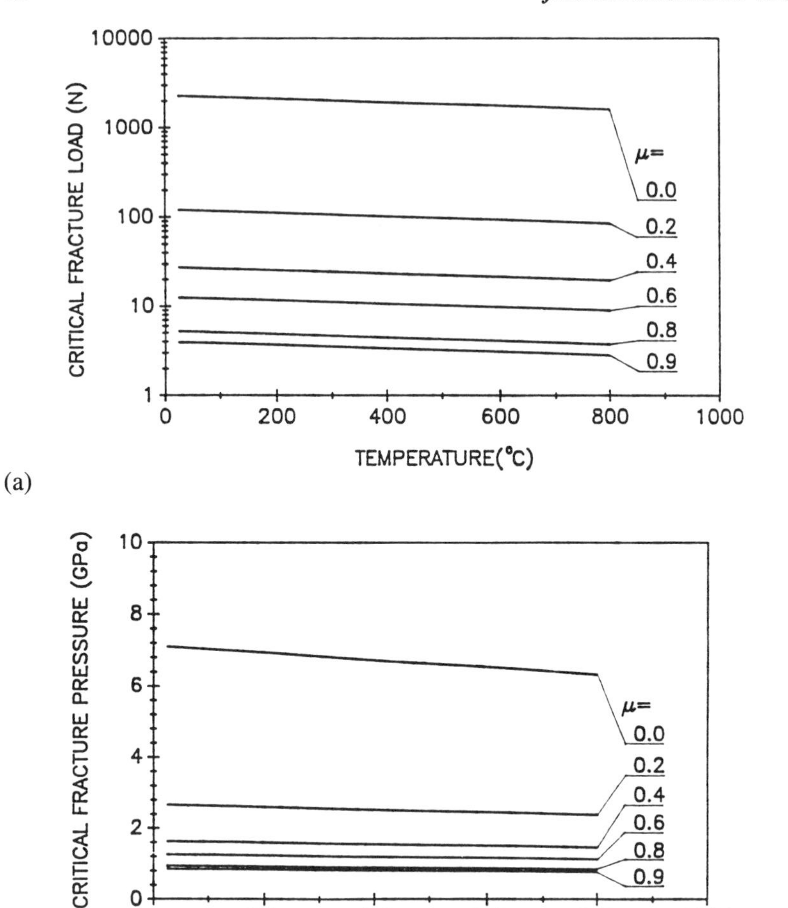

Figure 20 Calculated critical fracture load (a) and critical fracture pressure (b) of alumina as a function of temperature at different values of coefficient of friction.

C. Comparison Between Experimental Results and Theoretical Calculations

The results of the wear tests conducted in the intermediate temperature range for a duration of 2 hours are consolidated in a single diagram (Fig. 21); the abscissa and ordinate show the temperature and the contact load used in the tests. Therefore, the conditions (i.e., load and temperature) of each test are shown by a single data point. The open data points designate the conditions under which

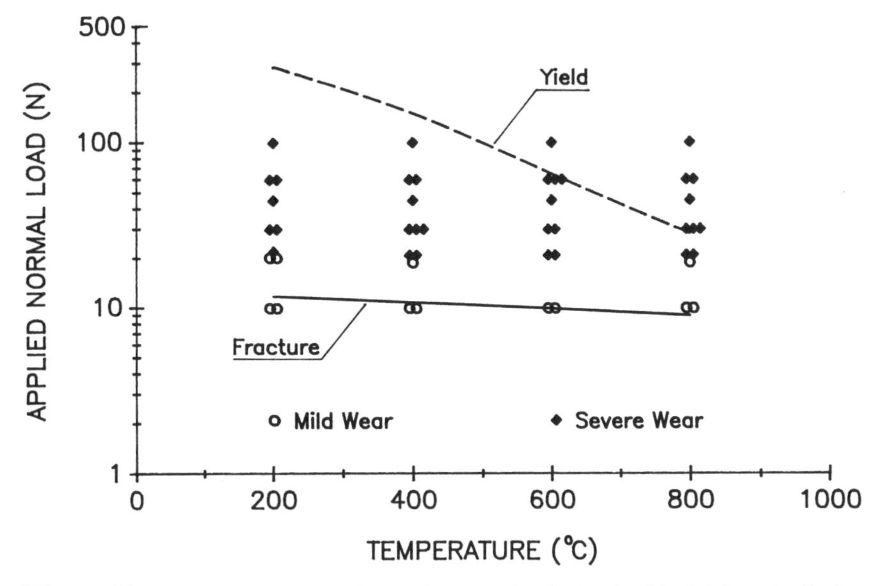

Figure 21 Comparison of experimental data and calculated critical failure loads for fracture and yield (for a coefficient of friction of 0.6) at indicated temperatures.

mild wear was observed, and the solid points correspond to the test conditions for severe wear. (To clearly identify the repeat tests, some of the data points were slightly shifted in the figure.) Figure 21 was constructed with the wear data on the flat specimens. It may also be used for the alumina balls, since the conditions of mild and severe wear were observed on both contacting specimens.

The results of theoretical calculations for the initiation of failure by plastic flow and by crack extension from preexisting flaws are plotted in Figure 21. A friction coefficient of 0.60 was used in these calculations, which corresponds to the average coefficient of friction prior to the transition. Comparison of the theoretical curves with the boundary between the solid data points and the open points indicates that microfracture from preexisting flaws, not plastic flow, is the dominating process for the observed transition from mild to severe wear in the intermediate temperature range.

A close examination of the experimental results in Figure 21 reveals that the actual transition loads are higher than the predicted values. This is expected, because the model of Chiang and Evans is a lower bound approximate solution to the contact fracture problem [37]. The repeat tests slightly above the theoretically predicted transition loads sometimes produce high wear and at other times low wear. This can be expected as a result of the instabilities near the transition point.

V. DISCUSSION

Reliable data and fundamental knowledge on the mechanisms of friction and wear can be instrumental in developing guidelines for design and material selection. Our study on alumina has shown that the tribological performance of this material can be consolidated in a wear transition diagram consisting of four separate regions. The tribological performance is controlled by a different set of mechanisms in each region. The critical region that must be avoided in practice is region IV, because of severe wear and a large coefficient of friction.

In region I, wear is controlled by tribochemical reaction between alumina and water vapor. The adsorption tendency of water is reduced as the temperature is increased and the formation of oxides are favored over the hydroxides. Therefore, the transition from region I to region IV occurs in the temperature range of 100–200 °C, because aluminum hydroxide either is not formed or is unstable at higher temperatures. This transition boundary may be shifted to lower temperatures at larger loads and higher speeds, due to increased contact temperatures.

The transition from region II to region IV is controlled by the fracture strength of the material. The contact mechanics model showed that propagation of cracks from preexisting flaws controls the onset of severe wear. This model also predicts that the transition from region II to region IV can be prevented by the reduction of the normal load, reduction of the friction coefficient, or an increase in the fracture toughness of the contacting materials. This transition boundary may be shifted to lower loads at higher speeds, by thermoelastic stresses developed at the contact as a result of the low thermal diffusivity of ceramics.

The transition from region III to region IV is controlled by diffusion and viscous flow of the glassy grain boundary phase. Since both these processes are thermally activated, an increase in the sliding speed or the contact load may shift the transition boundary to lower temperatures. However, the thermoelastic stresses generated at higher speeds and increased wear at larger loads may override the beneficial effects.

The wear transition diagram for alumina clearly shows that the tribological performance of this material can be controlled by a proper selection of operating conditions. This fundamental understanding also can be used to improve the performance of alumina by making changes in its composition and/or microstructure. First, modifications that increase the fracture toughness should result in beneficial effects. Second, changes in the composition of the grain boundary phase might produce a material that could spread easily on the surface under certain operating conditions. Finally, the tribological performance can be improved by lowering the coefficient of friction. This can be achieved by solid lubricant coatings [38], self-lubricating ceramic matrix composites [39,40], liquid lubrication [41], or reaction of the surfaces with the environment [21].

Several investigations have shown that surface films alter the state of stress both in the film and in the substrate [24,42,43]. In this respect, a surface film generated by tribochemical reactions or intentionally deposited on the surface may prevent the wear transition if its flow strength is lower than that of the substrate. This conclusion is quite important, since different investigators have confirmed the formation of tribochemical films on ceramic surfaces in sliding contact [12,20,21,23,41,44–46]. Therefore, by controlling the sliding environment one may be able to control the wear transition through tribochemical film formation.

The theoretical calculations and supporting experimental evidence clearly indicated that a simple linear elastic fracture mechanics analysis can be used to predict the onset of catastrophic wear in alumina, and perhaps in other ceramics. It is, however, important to recognize the limitations of the analysis. The contact model is based on the assumption of "smooth" contacts (i.e., the presence of asperities is not included); the analysis may not be applicable for surfaces rougher than the ones used in this investigation. Second, the contact stress calculations are based on the assumption of a homogeneous elastic material; it may not apply to composites. Furthermore, the possible existence of tribochemical films on the surface is not accounted for in the analysis. Of course, the model can be extended to include the effect of thin layers on the stress distribution below the contact. Finally, the analysis does not account for thermoelastic stresses that may develop at high sliding speeds, nor does it consider the possibility of residual stresses in the material. Again, it is possible to expand the model and incorporate these effects.

The wear transition diagram may be compared with the wear mechanism maps [47], wear maps [16], and transition diagrams [48]. The wear mechanism map, developed by Lim and Ashby [47] for unlubricated sliding of steel on steel, summarizes the wear behavior over a wide range of load and sliding velocities. These diagrams can be used to obtain wear coefficients and to identify the dominant wear mechanism.

The wear map concept proposed by Hsu et al. [16] is another approach to the representation of wear data. The wear coefficients are plotted against pressure and velocity, or pressure and time, in three-dimensional diagrams. An important aspect of these wear maps is that they are developed from test data obtained from a single material combination under controlled test conditions. The published wear maps for ceramics, such as alumina, provide useful design information. However, it is difficult to extract mechanistic information from these data unless two-dimensional contour maps are developed to delineate the wear mechanisms.

The transition diagrams for lubricated contacts developed by de Gee et al. [48] are useful for the analysis of the lubrication condition as a function of load and velocity. These diagrams provide an estimate of the load-carrying capacity of a particular lubricant. The transition diagrams are also used to determine the lubrication regime and the boundaries for each regime. When the contact condition is altered, the lubrication regime is changed and a large change in the wear rate or the friction coefficient is obtained. Thus, the term "transition" is used. An important aspect of these transition diagrams is that similar to the case of wear maps, the data are obtained using a controlled test procedure.

The wear transition diagram for alumina presented in this chapter is similar in concept to the transition diagram for a lubricated contact. It provides information on the load-carrying capacity of alumina as a function of temperature, and on the fundamental wear mechanisms. This approach is useful for design of tribological contacts, since it alerts the designer to conditions that would promote catastrophic failure.

VI. SUMMARY

The tribological characteristics of alumina sliding on a similar material under dry sliding conditions are divided into four distinct regimes. At low temperatures ($T < 200$ °C), tribochemical reaction between the alumina surface and water vapor in the environment controls the tribological performance. The coefficient of friction in this temperature range is approximately 0.40 and the wear coefficient is less than 10^{-6}, independent of contact load. At intermediate temperatures (200 °C $< T <$ 800 °C), the wear behavior depends on the contact load. At low loads, wear occurs by plastic flow and plowing; the coefficient of friction is approximately 0.60 and the wear coefficient is less than 10^{-6}. At loads higher than a threshold value, severe wear occurs by intergranular fracture. The coefficient of friction increases to 0.85 and the wear coefficient increases to 10^{-4}. A contact mechanics model based on linear elastic fracture mechanics indicated that propagation of cracks from preexisting flaws controls the onset of catastrophic wear in the intermediate temperature range. At temperatures above 800 °C, formation of a silicon-rich layer on the wear track by viscous flow and diffusion of the grain boundary phase reduces the coefficient of friction to 0.40, and the wear coefficient is reduced to less than 10^{-6}. The results of the wear tests and observations of the fundamental mechanisms controlling the tribological behavior of this material are consolidated in a wear transition diagram. This diagram can be used to identify the effect of load and temperature on the tribological characteristics of the material.

ACKNOWLEDGMENTS

Financial support received from the National Science Foundation under grant MSS-9005868 to the University of Maryland is gratefully acknowledged.

REFERENCES

1. S. Jahanmir, Tribological applications for advanced ceramics, *Mat. Res. Soc. Symp. Proc. 140*, 285–291 (1989).
2. M. B. Peterson and S. F. Murray, Frictional behavior of ceramic materials, *Met. Eng. Q. 7* (2), 22–29 (1967).
3. E. F. Finkin, S. J. Calabrese, and M. B. Peterson, Evaluation of materials for sliding at 600F–1800F in air, *Lubr. Eng. 29*(5), 197–204 (1973).
4. K. M. Taylor, L. B. Sibley, and J. C. Lawrence, Development of a ceramic rolling contact bearing for high temperature use, *Wear, 6*, 226–240 (1963).
5. M. F. Amateau and W. A. Glaeser, Survey of materials for high temperature bearing and sliding applications, *Wear, 7*, 385–418 (1964).
6. E. Rabinowicz, Friction and wear at elevated temperatures, WADC Technical Report No. 59-603 (1960).
7. C. S. Yust and F. J. Carignan, Observations on the sliding wear of ceramics, *ASLE Trans. 28*, 245–252 (1985).
8. J. Lankford, W. Wei, and R. Kossowsky, Friction and wear behavior of ion beam modified ceramics, *J. Mat. Sci. 22*, 2069–2078 (1987).
9. K. F. Dufrane, Wear performance of ceramics in ring/cylinder applications, *J. Am. Ceram. Soc. 72*, 691–695 (1989).
10. M. G. Gee and C. S. Matharu, The measurement of sliding friction and wear at high temperature, *Int. J. High Technol. Ceram. 4*, 319–331 (1988).
11. M. G. Gee, C. S. Matharu, E. A. Almond, and T. S. Eyre, The measurement of sliding friction and wear of ceramics at high temperature, in *Wear of Materials—1989*, K. C. Ludema (Ed.), American Society of Mechanical Engineers: New York, 1989, pp. 387–398.
12. H. Tomizawa and T. E. Fischer, Friction and wear of silicon nitride at 150 to 800 °C, *ASLE Trans. 29*, 481–488 (1986).
13. M. Woydt and K.-H. Habig, High temperature tribology of ceramics, *Tribol. Int. 22*, 75–88 (1989).
14. C. S. Yust and C. E. DeVore, Wear of zirconia-toughened alumina and whisker-reinforced zirconia-toughened alumina, *Tribol. Trans. 33*, 573–580 (1990).
15. D. E. Deckman, S. Jahanmir, S. M. Hsu, and R. S. Gates, Friction and wear measurements for new materials and lubricants, in *Engineered Materials for Advanced Friction and Wear Applications*, F. A. Smidt and P. J. Blau (Eds.), ASM International: Metals Park, OH, 1988, pp. 167–168.
16. S. M. Hsu, Y. S. Wang, and R. G. Munro, Quantitative wear maps as a visualization of wear mechanism transitions in ceramic materials, in *Wear of Materials*, K. C. Ludema (Ed.), American Society of Mechanical Engineers: New York, 1989, pp. 723–728.
17. S. Cho, B. J. Hockey, B. R. Lawn, and S. J. Bennison, Grain size and *R*-curve effects in the abrasive wear of alumina, *J. Am. Ceram. Soc. 72*, 1249–1252 (1989).
18. X. Dong, S. Jahanmir, and S. M. Hsu, Tribological characteristics of alumina at elevated temperatures, *J. Am. Ceram. Soc. 74*, 1036–1044 (1991).
19. S. Jahanmir and X. Dong, Mechanism of mild to severe wear transition in α-Alumina, *J. Trib. 114*, 403–411, (1992).

20. T. E. Fischer and H. Tomizawa, Interaction of tribochemistry and microfracture in the friction and wear of silicon nitride, *Wear*, *105*, 29–45 (1985).
21. S. Jahanmir and T. E. Fischer, Friction and wear of silicon nitride lubricated by humid air, water, hexadecane and hexadecane + 0.5 percent stearic acid, *Tribol. Trans.* *32*, 32–43 (1989).
22. R. S. Gates, S. M. Hsu, and E. E. Klaus, Tribochemical mechanism of alumina with water, *Tribol. Trans.* *32*, 357–363 (1989).
23. M. G. Gee, The formation of aluminum hydroxide in the sliding wear of alumina, *Wear*, *153*, 201–228 (1992).
24. R. B. King and T. C. O'Sullivan, Sliding contact stresses in a two-dimensional layered elastic half-space, *Int. J. Solids Struct.* *23*, 581–597 (1987).
25. S. Weiderhorn, B. J. Hockey, R. Krause, and K. Jakus, Creep and fracture of vitreous bonded aluminum oxide, *J. Mater. Sci.* *21*, 810–824 (1986).
26. S. Jahanmir, On mechanics and mechanisms of laminar wear particle formation, in *Advances in the Mechanics and Physics of Surfaces*, Vol. 3, R. M. Latanision and T. E. Fischer (Eds.), Harwood Academic: New York, 1986, pp. 261–331.
27. E. Ioannides and J. C. Kuijpers, Elastic stresses below asperities in lubricated contacts, ASME Reprint No. 85-Trib-3 (1985).
28. J. I. McCool, Comparison of models for the contact of rough surfaces, *Wear*, *107*, 37–60 (1986).
29. W. R. Chang, I. Etsion, and D. B. Bogy, An elastic–plastic model for the contact of rough surface, *J. Tribol.* *109*, 257–263 (1987).
30. T. Merriman and J. Kannel, Analyses of the role of surface roughness on contact stresses between elastic cylinders with and without soft surface coating, ASME Preprint No. 88-Trib-54 (1988).
31. R. S. Zhou, H. S. Cheng, and T. Mura, Micropitting in rolling and sliding contact under mixed lubrication, ASME Preprint No. 89-Trib-4 (1989).
32. G. M. Hamilton and L. E. Goodman, The stress field created by a circular sliding contact, *J. Appl. Mech.* *88*, 371–376 (1966).
33. G. M. Hamilton, Explicit equations for the stress beneath a sliding spherical contact, *Proc. Inst. Mech. Eng.*, *197C*, 53–59 (1983).
34. B. R. Lawn, Partial cone crack formation in a brittle material loaded with a sliding spherical indenter, *Proc. R. Soc. Ser. A*, *299*, 307–316 (1967).
35. B. R. Lawn and R. Wilshaw, Review—Indentation fracture, principles and applications, *J. Mater. Sci.* *10*, 1049–1081 (1975).
36. D. Tabor, Indentation hardness and its measurement, in *Microindentation Techniques in Materials Science and Engineering*, ASTM STP No. 889, P. J. Blau and B. R. Lawn (Eds.), American Society for Testing and Materials: Philadelphia, 1985, pp. 129–159.
37. S. Chiang and A. G. Evans, Influence of a tangential force on the fracture of two contacting elastic bodies, *J. Am. Ceram. Soc.* *66*, 4–10 (1983).
38. A. Gangopadhyay and S. Jahanmir, Control of friction and wear of α-alumina with a composite solid-lubricant coating, in *Tribology of Composite Materials*, P. K. Rohatgi, P. J. Blau, and C. S. Yust (Eds.), ASM International: Metals Park, OH, 1990, pp. 337–344.
39. A. Gangopadhyay and S. Jahanmir, Reduction of friction coefficient in sliding

ceramic surfaces by in-situ formation of solid lubricant coatings, in *Mechanics of Coatings*, D. Dowson, C. M. Taylor, and M. Godet (Eds.), Elsevier: Amsterdam, 1989, pp. 63–72.

40. A. Gangopadhyay and S. Jahanmir, Friction and wear characteristics of silicon nitride–graphite and alumina–graphite composites, *Tribol. Trans. 34*, 257–265 (1991).
41. R. S. Gates and S. M. Hsu, Effect of selected chemical compounds on the lubrication of silicon nitride, *Tribol. Trans. 34*, 417–425 (1991).
42. K. Komvopoulos, Finite element analysis of a layered elastic solid in normal contact with a rigid surface, *J. Tribol. 110*, 477–485 (1988).
43. K. Komvopoulos, N. Saka, and N. P. Suh, The role of hard layers in lubricated and dry sliding, *J. Tribol. 109*, 223–231 (1987).
44. B. E. Hegemann, S. Jahanmir, and S. M. Hsu, Microspectroscopy applications in tribology, in *Microbeam Analysis*, D. E. Newbury (Ed.), San Francisco Press–San Francisco, 1988, pp. 193–196.
45. D. C. Cranmer, Ceramic tribology—Needs and opportunities, *Tribol. Trans. 31*, 164–173 (1988).
46. T. Sugita, K. Veda, and Y. Kanemura, Material removal mechanism of silicon nitride during rubbing in water, *Wear*, *97*, 1–8 (1984).
47. S. C. Lim and M. F. Ashby, Wear mechanism maps, *Acta Metall. 15*, 1–24 (1987).
48. A. W. J. deGee, A. Begelinger, and G. Salmon, Failure mechanisms in sliding lubricated concentrated contacts, in *Proceedings of the 11th Leeds–Lyon Symposium on Tribology*, D. Dowson and C. Taylor (Eds.), Butterworths: London, 1985, pp. 105–116.

3

Relation Between Surface Chemistry and Tribology of Ceramics

Traugott E. Fischer

Stevens Institute of Technology
Hoboken, New Jersey

William M. Mullins

Purdue University
West Lafayette, Indiana

ABSTRACT

Tribochemical interactions of ceramics with gaseous environments and lubricants determine the wear of these materials. Such interactions can decrease wear or increase it, depending on the particular ceramic and environment. Water decreases wear of silicon nitride and silicon carbide by tribochemical reaction and increases that of zirconia and alumina by chemisorption embrittlement. Lubricants that reduce friction to low values are nevertheless capable of increasing the wear rate of certain ceramics. These phenomena are manifestations of the surface chemistry of ceramics, and the chemical reactions are related to the electronic structure of the materials.

I. INTRODUCTION

An important feature of ceramic tribology is the influence that chemical reactions, with the environment and with lubricants, exert on friction as well as on wear of these materials [1–6]. This behavior is related to the chemical properties of ceramics and represents an interaction of mechanical and chemical processes. This chemical interaction takes several forms, depending on the materials, the environment, and the mechanical parameters; it can consist in modifications of surface composition and topology that decrease wear [3,5], in a purely chemical form of wear (by dissolution in the liquid environment) [6,7], in the chemisorption and boundary lubrication effectiveness of inert hydrocarbon [4,5], and in chemically induced cracking, which increases wear rates [4]. The rates of these chemical reactions are very much influenced by simultaneous friction [8,9]. Because of this interaction with friction, these reactions are called "tribochemical." They have been observed in all materials, but they are particularly pronounced in the tribological behavior of ceramics; their understanding is a prerequisite for the successful applications of these materials in tribological service and can form a rational basis for the development of synthetic lubricants or lubricant additives suited for them.

II. CHEMICAL EFFECTS ON CERAMIC TRIBOLOGY

A. Wear Reduction by Tribochemical Oxidation

The ambient humidity has a pronounced effect on the wear of silicon nitride and other ceramics [1,3]; not only the amount of wear, but the wear mechanism itself is modified by humidity. Silicon nitride wears rapidly in dry argon, but if the environment contains various amounts of water vapor, its wear rate decreases by as much as two orders of magnitude. Under these conditions, the wear scar is much smoother than after sliding in dry gases and it is covered with an amorphous silicon oxide, which is probably strongly hydrated [3]. In the absence of friction, measurable oxidation rates are obtained only above 1000 K [10] and the oxidation rate of this material is increased up to one-thousandfold by the presence of humidity in the air [11]. The rate of oxidation of this material is accelerated by the simultaneous action of friction. Exactly how this occurs has not been determined yet. One can speculate that the reaction is accelerated because the hydroxide formed on the surface is continuously removed and a fresh surface is exposed by friction, but clear experimental evidence for any one mechanism is still lacking.

The friction coefficient is very nearly the same in a dry and in a humid environment; the total mechanical forces acting on the contacting surfaces are thus the same [3]. Why then is the wear decreased instead of increased when chemical attack is added to the friction force? The answer lies in the different topographies of the surfaces: tribochemical wear of contacting asperities creates a much smoother surface, and the load and friction forces are distributed over wider areas than in the absence of tribochemistry. As a consequence, the local stresses responsible for the mechanical wear of the material are reduced.

When silicon nitride slides in water, wear occurs by dissolution [7] and produces ultraflat surfaces that favor hydrodynamic lubrication [6]. Silicon nitride dissolves in water slowly but measurably above 400 K [12], but simultaneous friction induces relatively fast tribochemical dissolution at room temperature. Similar experiments with silicon carbide also produce dissolution of the material, but hydrodynamic lubrication is not achieved easily because the material is not as strong as silicon nitride and suffers local fracture as well as dissolution, so that ultraflat surfaces are not obtained.

B. The Formation of Lubricious Oxides

At elevated temperatures, sliding in humid air reduces friction as well as wear in a limited range of load and sliding velocities [13]. The surfaces are covered by a smooth layer of silicon oxide. As the severity of sliding (i.e., the product of load and sliding velocity) is increased beyond a certain value, friction and wear are high and the wear surfaces are rough. The passage from low to high friction is the result of a competition between the kinetics of formation of the lubricious oxide layer and its wear.

Recently lubricious oxides have been formed on silicon nitride by preoxidation [14]. To achieve low friction (friction coefficients μ as low as 0.05 have been obtained), it is necessary to prepare a very smooth and flat surface to avoid friction and wear by plowing. This is achieved by friction in water as described above [6]. Subsequent oxidation in air for a few hours produces a surface that presents a low friction coefficient.

Gates et al. [15] have recently shown that sliding in water causes a decrease of the friction coefficient of alumina against itself from 0.6 to 0.25. (By contrast, the friction coefficient of silicon nitride is not lowered by water when the speeds are too low for hydrodynamic lubrication [6].) This lowering of friction is attributed to the formation of stable aluminum hydroxides, which are modified to a layered structure (trihydroxide–bayerite) by the frictional shear stress.

C. Chemically Induced Fracture in Oxide Ceramics

Wallbridge et al. [16] have reported that wear of alumina sliding in water is higher than in air. With zirconia, water increases the wear rate by an order of magnitude over wear in dry nitrogen [4]. Even the humidity of room air causes an increase in the wear rate that is almost as large. This increase in wear occurs by intergranular fracture; we are in the presence of chemisorption embrittlement. According to Wiederhorn and Michalske and their coworkers [17,18], this phenomenon occurs by the attack of the bonds between neighboring metal and oxygen ions at a crack tip by water; it is related to the well-known tendency of oxide ceramics to form hydroxylic surfaces [12].

Adsorption-induced fracture occurs also in the presence of hydrocarbon lubricants [4]. In the case of zirconia, paraffin causes an increase in the wear rate of about 50% over the wear rate in dry nitrogen, where the friction coefficient is 0.7. When sliding occurs in paraffin with 0.5% stearic acid, which is a classic boundary lubricant, the friction coefficient decreases to 0.09 but wear increases by another factor of 3. Chemical attack of the grain boundaries and intergranular fracture are the causes of this increase in wear.

D. Boundary Lubrication by Paraffins

Polar molecules such as fatty acids, alcohols, and esters adsorb onto metallic surfaces and cause "boundary lubrication." In the case of metals, nonpolar hydrocarbons such as paraffins and the molecules of lubricant base stocks do not act as boundary lubricants. In the case of all ceramics investigated or known to us, paraffins act as boundary lubricants, with friction coefficients in the neighborhood of 0.12. We do not have a firm explanation for this difference; it probably has the same origin as the catalytic activity of these materials for the cracking and isomerization of hydrocarbon: the acid sites on the surface of ceramics, strong enough to break carbon–carbon bonds at elevated temperatures in catalytic cracking, are capable of absorbing the molecules at room temperature and provide boundary lubrication [19].

III. SURFACE CHEMISTRY OF CERAMICS

Ceramics are characterized by a large energy gap between their valence band (which constitutes the highest occupied molecular orbitals: HOMO) and their conduction band (or lowest unoccupied molecular orbital: LUMO) [20–24]. For this reason, chemical interactions with the environment are dominated by electron transfer, and most ceramics are considered as solid acids or bases according to the theory of Lewis.

Magnesia, alumina, and silica have similar electronic structures [20, 23], consisting of a split valence band; the lower band is in the nature of atomic oxygen 2*s* orbitals; the upper valence band consists of oxygen 2*p* orbitals with very slight cationic characteristics. The conduction band is nearly entirely cationic 3*s* and 3*p* in nature. The band gaps of these materials are among the largest measured, from 7.5 to 9.5 eV [24]. These characteristics classify these oxides as very hard acids and bases.

The electronic structure of water [25] is very similar to that of these oxides; it has valence and conduction band structures with a 7.5–8.0 eV energy gap. The reaction of water with these surfaces occurs with sufficient charge transfer to induce the adsorbed water molecule to dissociate into protons and hydroxyls. The surface charge of the oxide attracts the oppositely charged ions in solution, which forms a diffuse space charge region around the surface [26]. The charge double layer is affected by the pH of the solution. All these oxides also have stable hydroxide forms, and their pH-dependent solubilities in water range from 10^{-3} for silica and 10^{-4} for alumina.

Reactions with highly polar functional groups, such as —OH, —COOH, and —NH_2, which tend to dissociate or form polar hydrogen bonds [27–30], are similar to those one would expect to find with water; as a general rule of thumb, if water will dissolve in the compounds, alcohols, acids, and amines will tend to wet and spread onto the oxides discussed here.

The surfaces of these oxides are known to catalyze many ring-opening reactions when the ring is small and highly polar [19]. The general tendencies of alumina silicates to catalyze alkene production, cracking, and isomerizations are attributed to such reactions.

Silicon carbide and silicon nitride are narrow-band-gap materials (E_g = 2.8 eV for SiC and 5.0 eV for Si_3N_4 [24]); they can be considered semiconductors. In addition, both materials have strong tendencies toward nonstoichiometry, which further reduces the effective band gap.

Both these compounds react slowly with air or water [12,26] to form the more stable oxynitride or oxide surface coatings, which have the chemical properties mentioned above. Also, oxides are commonly added as tougheners and sintering aides to bulk silicon nitride and silicon carbide, which makes the grain boundary oxide phases even more prevalent. Consequently, these materials have surface chemistries in air that are nearly identical to that of silica.

The band gap of zirconia is 5.0 eV, much smaller than that of alumina and water, but the structure of the bands is similar. Because of this, ZrO_2 is a much softer acid base than is alumina or the oxide formations on Si_3N_4. The aqueous surface chemistry of zirconia is well known [31,32]; ZrO_2 is considered to be very weakly acidic in water. Measurements are complicated by the specific adsorption of softer complex anions onto the surface, such as NO_3 and ClO_4-, which produce considerable surface charge double layers even though no strong reaction with water has taken place.

Surface reactions with water tend to be a simple dissociation of the oxide to form Zr^{4+} and ZrO^{2+} in solution. As the concentrations of these increase to the order of 10^{-8}, complexations to $(Zr_4(OH)_8(H_2O^{16})^{8+}$ and eventually recrystallization of the oxide take place. No zirconium hydroxides are known to occur.

Unlike the aqueous chemistry, the organometallic chemistry of zirconia is very rich [30,31], with zirconium compounds acting as Ziegler–Natta catalysts for polymerization and causing olefin hydrozirconation reactions, which graft carbonyls onto alkenes to produce carboxylic acids. In tribological conditions, the potential exists for similar oxidation and catalysis reactions of organic lubricants with the surfaces or with wear debris to form complex oxidized oligomers. Analysis of the wear surfaces should easily reveal whether such reactions, probably accelerated by tribochemical mechanisms, are responsible for the wear found in lubricated zirconia.

IV. THEORETICAL PREDICTIONS

The surface chemistry of ceramics is determined by the electronic structure of the constituents and can be treated with the Lewis acid–base theory. The adsorption reactions are modeled as simple bonding problems between a single molecule and a surface, approximated by the linear combination of orbitals (LCO) [33]. We present estimations of the surface reactivity of ceramics with a model based on a theory originally formulated by Mulliken [34] and later extended by Salem [35] and Klopman [36] to describe acid–base interactions for individual molecules. The semiempirical extension to surface reactions used here has included the Fermi energy of the species as a parameter to predict average interactions for an ensemble [33].

During the surface reaction, a net charge is transferred from the base to the acid to form a net charge dipole at the adduct. Assuming a constant conformation for the water molecule and no effect of defect states in the substrate, Figure 1 shows a plot of the square of the calculated charge dipole formed as a function of band gap for water on a covalent ceramic surface. The dipole is seen to have a relative minimum of $E_g = 8.0$ eV that corresponds to zero charge transfer or a neutral reaction. Variation from this case increases the net charge transfer of the reaction.

Also plotted in Figure 1 are the reported solubilities of various covalent oxides [37] as a function of band gap [24]. With the scales suitably adjusted, a relationship between solubility and band gap is observed. Since water is highly

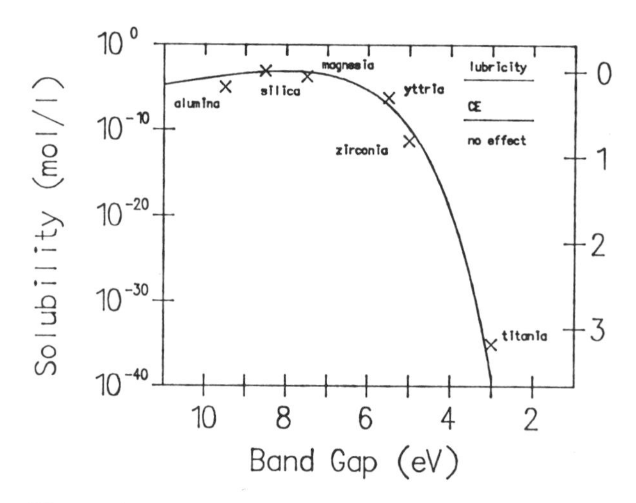

Figure 1 Relationship between the observed solubility of ceramics (left scale) and the calculated charge dipole formed in the reaction with water (right scale), as a function of the band gap of the ceramics. It predicts that silica and magnesia that undergo tribochemical reactions with water, yttria, alumina, and to a lesser extent zirconia, are susceptible to chemisorption embrittlement.

polar, the solubility of any species is related to the magnitude of this charge and the dielectric constant of the solvent [38].

A maximum chemical interaction is expected at the point of minimum net charge transfer. At this point, little electrostatic hindrance to a reaction is expected at surface sites. There is an equal likelihood for species dissociation, and since little complexation is required, the dissolution reaction is not hindered. At this point also, chemisorption embrittlement is predicted to be at a maximum. Changing the band gap of the substrate increases the magnitude of charge dipole formed by adduct formation. The polarity of the dipole is determined by the relative band gaps of the lubricant and the substrate if we neglect the effects of the Fermi energy, surface states, and doping. As a rule of thumb, the wider the band gap, the more acidic the material.

A maximum interaction is predicted for SiO_2 which has the highest solubility. The solubility of this material is sufficient to produce complete dissolution of the surface in a tribochemical wear situation, as observed [3]. Strong chemisorption is noted for alumina, magnesia, yttria, and potentially zirconia, but the magnitude of the adduct dipole limits the extent of any possible surface reaction. Sufficient reactivity is possible for chemisorption embrittlement [4] or the formation of stable hydroxides [15], but the solubility is too low for the smoothing of the surface. For titania, the adduct dipole is so large that no extensive reaction is predicted; neither chemisorption embrittlement nor surface smoothing are expected for titania.

The model interaction for a —CH_3 moiety as a function of band gap can be described in the same manner as for the water molecule. Figure 2 shows the

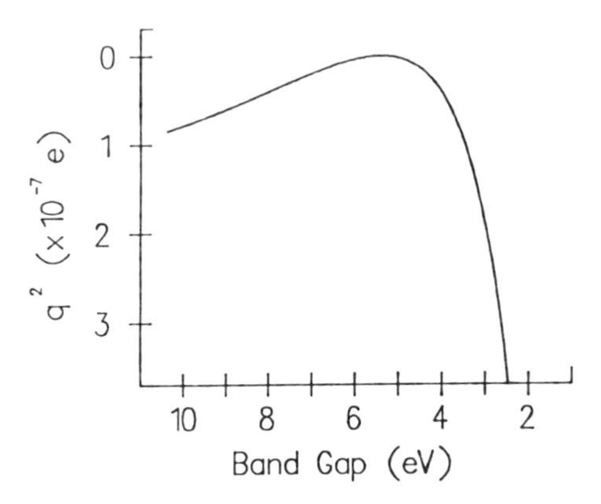

Figure 2 Calculated charge dipole formed in the adsorption for a –CH_3 group on an oxide ceramic surface as a function of the electronic band gap of the ceramic. It shows that yttria and zirconia are expected to be reactive towards hydrocarbons during sliding.

predicted relationship between the band gap and the surface adduct dipole. The relationship observed is similar to that of Figure 1, but it is shifted to a lower band gap as a result of the lower energy gap of the —CH_3 functional. Based on the results for water, significant adsorption is expected in the entire range of ceramics. A maximum for adsorption is expected for Y_2O_3, but solubility in a nonpolar medium is expected to be small because the dielectric constant is low. For this reason, chemisorption embrittlement is considered to be possible for Y_2O_3 and ZrO_2.

V. DISCUSSION

We have seen that the surface chemistry of ceramics plays a decisive role in their tribological behavior. In most cases, the chemical influence of the environmental lubricant is tribochemical, in that the reaction or attack is influenced by the mechanical phenomena that occur simultaneously. It appears that the tribochemical interaction takes four major forms, depending on the reactivity of the ceramic toward the environment.

At the highest reactivity, the environment attacks or corrodes the material generally. This attack can be accelerated by friction. We have encountered this example in the case of lubrication of zirconia by stearic acid [4]; attack on the grain boundaries leads to loss of individual grains everywhere on the material, but this loss is greatly magnified in the wear scar.

At intermediate reactivities, two different forms of tribochemical interaction take place. The interaction of water with silicon nitride, alumina, and zirconia is illustrative. With silicon nitride (Fig. 1), the reaction is not strong enough to cause general attack of the material, but simultaneous friction causes the tribochemical oxidation of the material to silicon oxide in humid environments and the dissolution of this oxide in liquid water [6,7]. The interaction of water with alumina is weaker and causes two different tribochemical interactions: the formation of stable hydroxides, which act as solid lubricants [15], and an increase in the wear rate, which probably takes the form of chemisorption embrittlement [16]. The weakest interaction of this type occurs with zirconia. In this case the reaction occurs only where it is mechanically stimulated by large mechanical stresses, which occur at crack tips. The chemisorption weakens the Zr—O bonds and lowers the threshold for crack propagation [17,18]. This form of chemisorption embrittlement is well known in silica glasses in the absence of friction. Silica thus presents the interesting case of chemisorption embrittlement in pure mechanical stress and tribochemical dissolution during friction [3].

In the case of hydrocarbon lubricants, our understanding is more limited. Yet precisely the tribochemical effects of hydrocarbon lubricants are the phenomena that require the most pressing development, since they determine the performance of ceramic elements in machinery that is lubricated by conventional oils and synthetic lubricants. Our observations on zirconia have shown the importance of these phenomena: certain hydrocarbons that provide low friction nevertheless increase the wear rate over that of unlubricated sliding.

The chemistry and tribochemistry of ceramics is a very rich area. Chemical interactions of these materials are strong and influence their tribological behavior in major ways. In contrast to the case of metals, the tribological utilization of ceramics is recent, and we have not accumulated a large amount of empirical experience. We feel that a systematic, scientific study of tribochemistry of ceramics could provide powerful support for the formulation of synthetic lubricants or lubricant additives for these materials.

ACKNOWLEDGMENTS

This work was supported by grants from the National Science Foundation, the Dow Chemical Foundation, and the Schaefer Fund for Excellence of Stevens Institute of Technology.

REFERENCES

1. H. Shimura, and Y. Tsuya, in *Wear of Materials—1977*, K. Ludema (ed.), American Society of Mechanical Engineers: New York, 1977, p. 452.
2. K. Suzuki and T. Sugita, in *Wear of Materials—1981*, S. K. Rhue, A. S. Ruff, and K. C. Ludema (Eds.) American Society of Mechanical Engineers: New York, 1981, p. 518.
3. T. E. Fischer and H. Tomizawa, *Wear, 105*, 29 (1985).
4. T. E. Fischer, M. P. Anderson, S. Jahanmir, and R. Salher, in *Wear of Materials—1987*, K. Ludema (Ed.), American Society of Mechanical Engineers: New York, 1987, p. 257.
5. S. Jahanmir and T. E. Fischer, *Tribol. Trans. 31*, 32 (1988).
6. H. Tomizawa and T. E. Fischer, *ASLE Trans. 29*, 481 (1986).
7. T. Sugita, K. Ueda, and Y. Kanemura, *Wear, 97*, 41 (1984).
8. C. Heinicke, *Tribochemistry*, Hanser: Munich, 1984.
9. T. E. Fischer, *Annu. Rev. Mater. Sci. 18*, 303 (1988).
10. A. J. Kiehle, L. K. Heung, P. J. Gielisse, and T. J. Rockett, *J. Am. Ceram. Soc. 58*, 17 (1975).
11. S. C. Singhai, *J. Am. Ceram. Soc. 59*, 82 (1976).
12. R. K. Iler, *The Chemistry of Silica*, Wiley: New York, 1979.
13. H. Tomizawa and T. E. Fischer, *ASLE Trans. 30*, 41 (1987).
14. T. E. Fischer, H. Liang, and W. M. Mullins, *Mater. Res. Soc. Symp. Proc. 140*, 339 (1989).
15. R. S. Gates, S. M. Hsu, and E. E. Klaus, *Tribol. Trans. 32*, 357 (1989).
16. N. Wallbridge, D. Dowson, and E. W. Roberts, in *Wear of Materials—1983*, K.

C. Ludema (Ed.), American Society of Mechanical Engineers: New York, 1983, p. 202.
17. S. M. Wiederhorn, S. W. Freiman, E. R. Fuller, and C. J. Simmons, *J. Mater. Sci. 27*, 3460 (1982).
18. T. A. Michalske and B. C. Bunker, *J. Appl. Phys. 56*, 2686 (1984).
19. H. Knozinger and P. Ratnasamy, *Catal. Rev.-Sci. Eng., 17*, 31 (1978).
20. J. A. Tossel, *J. Phys. Chem. Solids, 36*, 1273 (1975).
21. J. A. Tossel, *J. Am. Chem. Soc., 97*, 4840 (1975).
22. S. Ciraci and I. P. Batra, *Phys. Rev. B28*, 982 (1983).
23. I. P. Batra, *J. Phys. C.: Solid Phys. 15*, 5399 (1982).
24. W. H. Strehlow and E. L. Cook, *J. Phys. Chem.* 263 (1973).
25. R. M. Pitzner and D. P. Merfield, *J. Chem. Phys. 52*, 4782 (1970).
26. G. Y. Onoda, Jr., and J. A. Casey, in *Ultrastructures in Ceramics, Glasses and Composites*, L. L. Hench and D. R. Ulrich (Eds.), Wiley: New York, 1983, p. 374.
27. E. S. Tormey, Ph.D. thesis, Massachusetts Institute of Technology, 1982.
28. J. C. Bolger, *J. Electrochem. Soc. 128*, 82 (1981).
29. J. C. Bolger and A. S. Michaels, in *Interface Conversion for Polymer Coatings*, P. Weiss and P. G. Cheever (Eds.), American Elsevier: New York, 1968.
30. F. M. Fowkes (Ed.), *Industrial Applications of Surface Analysis*, ACS Symposium Series No. 199, American Chemical Society: Washington, DC, 1982.
31. K. C. Ray and S. Kahn, *Indian J. Chem. 13*, 577 (1975).
32. A. E. Regazzoni, M. A. Blesa, and A. J. G. Maroto, *J. Colloid Interface Sci. 91*, 569, (1982).
33. W. M. Mullins, *Surf. Sci. 217*, 459 (1989).
34. R. S. Mulliken, *J. Am. Chem. 74*, 811 (1952).
35. L. Salem, *J. Am. Chem. Soc. 90*, 543 (1968).
36. G. Klopman, *J. Am. Chem. Soc. 90*, 223 (1968).
37. W. F. Linke, *Solubilities: Inorganic and Metal–Organic Compounds*, Vol. 1 (1958) and Vol. 2 (1965), American Chemical Society: Washington, DC.
38. M. Maroncelli, H. MacInnis, and G. R. Fleming, *Science, 243*, 1674 (1989).

4

Friction and Wear of Silicon Nitride Exposed to Moisture at High Temperatures

Steven Danyluk
Georgia Institute of Technology
Atlanta, Georgia

M. McNallan
University of Illinois at Chicago
Chicago, Illinois

D. S. Park
Korea Institute of Machinery and Metals
Chang-Won, Korea

ABSTRACT

The friction and wear of polycrystalline silicon nitride has been measured in a sliding ball-on-flat apparatus in an environment containing water vapor at elevated temperatures. The friction coefficient and wear rate are typically high at the start of an experiment and then are significantly reduced at varying water partial pressures in the temperature range 300–800°C. The high friction and wear values are regained when the temperature is increased above 800°C. The high temperature transition depends on the water partial pressure.

The variability of the friction coefficient and wear rate is closely tied to the interaction of the water with the surface. Silicon nitride reacts with the water to form an amorphous layer of silica. The silica deforms as a result of the action of the ball and forms an unusual type of debris. Scanning electron microscopy of the wear tracks, formed when the friction coefficient was low, showed that the wear groove contained cylindrical particles (rolls) oriented perpendicular to the sliding direction. The "rolls" apparently support the load so that the friction is lowered and the wear rate is reduced. This chapter summarizes the experimental conditions that result in the formation of the rolls and the electron microscopy of the rolls.

I. INTRODUCTION

There is considerable interest in the automotive industry in replacing metallic components in internal combustion engines with structural ceramics, or in coating metallic components with ceramic films for high temperature oxidation and wear resistance [1]. Silicon nitride is particularly attractive for this application because of its high strength, low density, and good oxidation resistance. As a result, silicon nitride has been used in bearings and components of exhaust systems of rotary and gas turbine engines, seat guides of reciprocating engines, and coatings on tip seals. These applications require high temperature strength and stress rupture, as well as resistance to creep, fatigue, corrosion, and wear at temperatures up to 1200°C.

While the high temperature strength and creep resistance of silicon nitride, Si_3N_4, are known well enough to permit the use of this material for engine applications [2], other critical properties such as wear resistance, especially in reactive environments, are less well known. Wear of silicon nitride engine components will be expected to occur at elevated temperatures in the presence of reactive atmospheres containing combustion products, oxidants, and unburned fuels. Silicon nitride is not thermodynamically stable in such environments. Its resistance to oxidation and corrosion results from the formation of a protective oxidation product layer of SiO_2, which separates the ceramic from the reactive atmosphere [3]. This oxide layer will also be present on wear surfaces in reactive atmospheres and may influence the wear of the material in a positive or negative way. Water vapor, a major combustion product of hydrocarbon fuels, is known to have a significant effect on the physical and chemical properties of the SiO_2 layers that form on Si-based ceramics in oxidizing environments [4]. Thus, the lifetimes of ceramic components in engine applications may be limited by a complex wear–corrosion process in which both mechanical and chemical factors must be considered.

The research described in this chapter is a summary of earlier work in which the friction and wear of polycrystalline silicon nitride was measured in an environment containing Ar and $H_2O(g)$ gas mixtures at elevated temperatures [5], plus new results. In the earlier work, the friction and wear characteristics of a silicon nitride wear couple in a reciprocating ball-on-flat geometry were found to be related to the morphology of the debris generated in the wear track. Both the friction coefficient and wear rate were substantially lower in the water vapor containing environments at temperatures above 300°C than in dry environments at the same temperatures. This phenomenon was linked to the formation of "rolls" of wear debris oriented approximately perpendicular to the wear track. It was hypothesized that the debris particles may function like roller bearings to reduce the friction and material damage associated with the translation of the ball. Similar "rolls" and roll formation have been found on silicon nitride [6,7], alumina [8], and silicon carbide [9] in a variety of test geometries including room temperature, and the roll formation was always associated with water or a humid environment. None of these earlier studies, however, reported the details of the microstructure, nor were mechanisms for roll formation proposed.

Because "rolls" that can influence friction and wear behavior are formed on silicon nitride only at elevated temperatures and in the presence of $H_2O(g)$, it is unlikely that the rolls are the result of wear alone. Instead, they are the product of a combination of chemical and mechanical processes, where the properties of the oxidation product (amorphous SiO_2) are influenced by the presence of $H_2O(g)$ in the environment.

II. EXPERIMENTAL PROCEDURE

The ball-on-flat experiments on which the present experimental results are based were performed as described previously [5] with polycrystalline silicon nitride hot-pressed flats (NC 132) and hot isostatically pressed balls (NDB 100) obtained from the Norton Company. The average grain size of both types of material was 0.5 μm. The flats were prepared by cutting rectangular, thin sections of approximately $0.6 \times 2.5 \times 0.3$ cm^3 with a diamond-impregnated circular cutoff wheel using oil as a lubricant. One of the large flat surfaces was polished by standard metallographic techniques with a final polish consisting of an alumina powder slurry with a particle size of 0.03 μm. After polishing and before positioning on the specimen holder, the specimens were cleaned with acetone, trichloroethylene, and methanol.

The experimental apparatus (Fig. 1) consisted of a tube furnace with

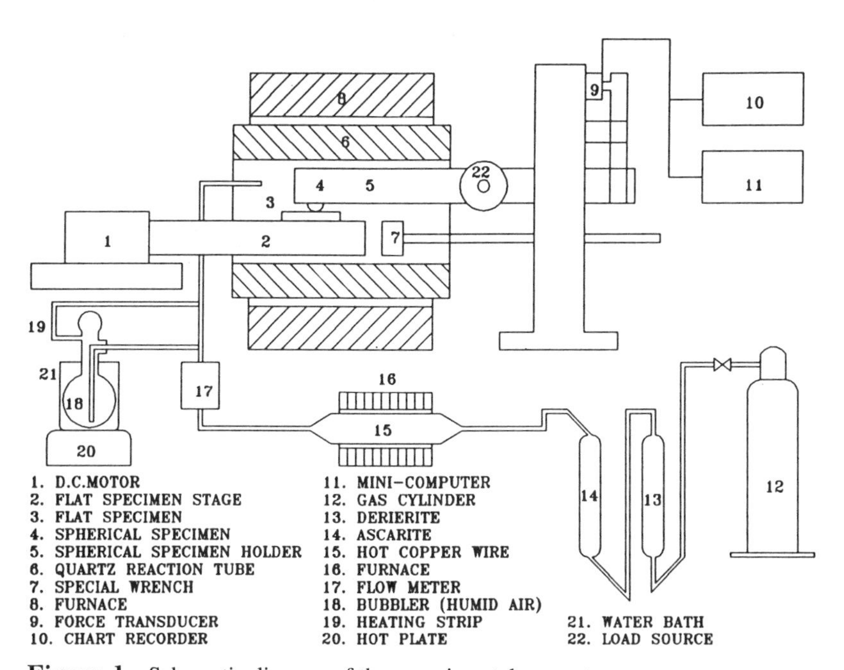

Figure 1 Schematic diagram of the experimental apparatus.

water-cooled endcaps through which two ceramic rods are inserted, one through each end. One rod holds the flat silicon nitride sample and the other is fitted with the 0.62 cm diameter ball. The rod with the ball is stationary and cantilevered so that the dead load (chosen to be 3.67 N) transmitted to the plate is fixed. The flat sample is moved in a straight line beneath the ball by the reciprocating motion of a stepper motor; the length of travel is 2.18 cm, and the velocity is 2 cm/s. The temperature of the furnace is set through the use of a controller. The friction coefficient was measured as a function of the number of reciprocating passes with a load cell mounted on the stationary ceramic rod.

The gas environments were produced from reagent grade argon, which was purified by passage through anhydrous calcium sulfate to remove residual water, ascarite to remove carbon dioxide, and copper turnings, which were inserted into a separate preheat furnace and heated to 500°C to to remove oxygen. The argon was then saturated with $H_2O(g)$ at a controlled temperature to fix the water vapor content of the gas mixture. For example, the water saturator was maintained at 73°C to produce a gas mixture containing 34% $H_2O(g)$ by volume. Because the water saturator was usually operated above room temperature, the gas lines between the water saturator and the furnace were maintained at temperature above 100°C to avoid condensation. All the experiments were performed at an argon flow rate of 370 cm^3/min at room temperature. A number of experiments were also conducted in a dry air environment. The data obtained in the humid environments were compared to those obtained in the air and argon environments.

The wear groove width and depth depended on the temperature, the water partial pressure, and the length of time required for the reciprocating motion. The wear rate was determined by interrupting the experiment at a predetermined number of strokes and measuring the groove width and depth. The wear scar on the balls was also examined. Since the velocitu of the reciprocating motion was constant, the number of strokes is related to the time of sliding. The friction coefficient was measured in situ during the forward and backward strokes, by a force transducer attached to the rod that contained the ball. The wear surfaces were examined by optical and scanning electron electron microscopy so that the wear mechanisms could be determined. Prior to electron microscopy, the sample surfaces were coated by a 2000 nm filament-evaporated gold film to enhance the topographic contrast.

Transmission electron microscopy was performed on extraction replica samples. A two-stage replica technique was used to extract the rolls onto an acetate sheet. Carbon was then evaporated on the side with the rolls so that the rolls were sandwiched between the acetate and the carbon film. The acetate was cut into smaller pieces, then inserted into an acetate bath to dissolve the acetate sheet. After approximately 3 days, the carbon film with the adhering rolls was then removed from the solution and mounted on 3 mm sample grids for examination in a 100 kV scanning transmission electron microscope (JEOL 100CX).

III. FRICTION COEFFICIENT AND WEAR RATE

Typical experimental data of the friction coefficient and the wear rate that reported in Reference 5 are shown in Figures 2 and 3. Figure 2 shows values of the steady state friction coefficient versus temperature after 10,000 strokes (equivalent to 218 m sliding distance) for samples exposed to argon, air, and 2, 8, and 34% partial pressure H_2O. Figure 2 shows that the friction coefficient in argon and air is approximately constant at a value of 0.65 over the entire temperature range tested, from room temperature to 1273 K. The error bars indicate the variation of the friction coefficient after the steady state has been reached. The samples exposed to the water vapor showed a reduced value of the friction coefficient, approximately 0.3 for the 8% water environment within the temperature range of 573–973 K. The higher value of friction coefficient was recovered when the temperature was raised beyond some critical value. For example, for the 8% water vapor environment experiments, the critical temperature is 973 K. The temperature at which friction returned to the high value depended on the partial pressure of the water.

Figure 3 shows the corresponding wear rate measurement for the same

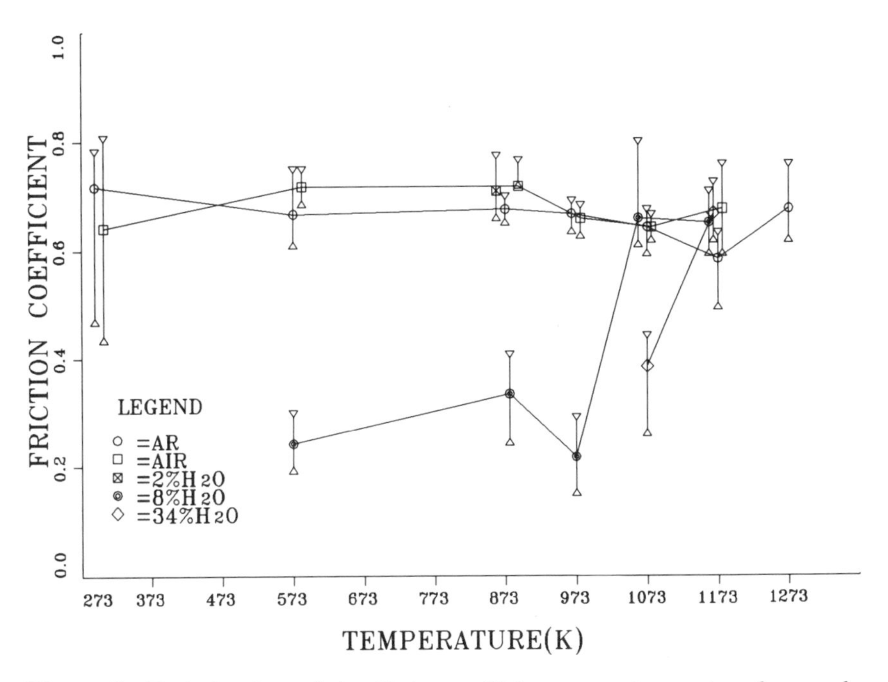

Figure 2 Typical values of the friction coefficient versus temperature for samples abraded in Ar, air, and 2, 8, and 34% water vapor pressure.

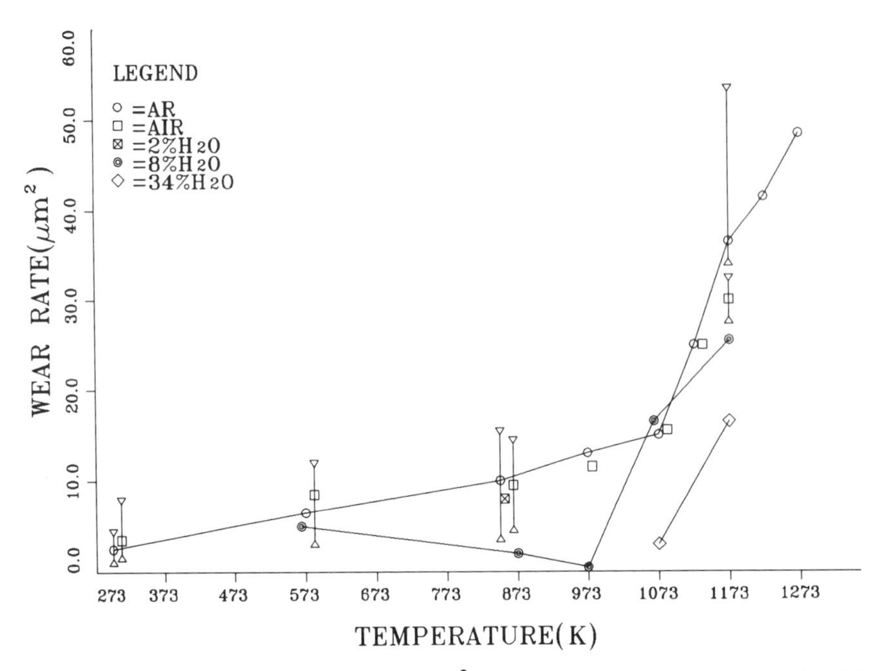

Figure 3 Typical values of wear rate (μm^2) versus temperature for samples abraded in Ar, air, and 2, 8, and 34% water vapor pressure.

environments displayed in Figure 2: the wear rate is seen to be correlated with the changes in friction coefficient. The wear rate is generally low at the low temperatures, increasing in an exponential manner as the temperature is increased. The wear rate is lower in the presence of water as compared with the argon and air environments, as can be seen for the 8% water vapor environment. The transition to the higher wear rate at the 8% H_2O can be seen at 973 K.

IV. OPTICAL AND SCANNING ELECTRON MICROSCOPY

Optical and scanning electron microscopy of the wear grooves showed that the surface morphology in the reduced friction and low wear rate regime were striated in the sliding direction, and the striations were composed of alternating smooth and rough regions. Figure 4 shows a typical optical micrograph of a sample surface from the center of the wear groove exposed to 8% H_2O water vapor at 973 K. The micrograph shows the striated region and some evidence of cylindrical particles oriented perpendicular to the sliding direction.

Selected samples of the striated grooves containing the cylindrically particles were examined by scanning electron microscopy. The geometry of the rolls varied with the temperature and water vapor content of the exposure. Figure 5 shows

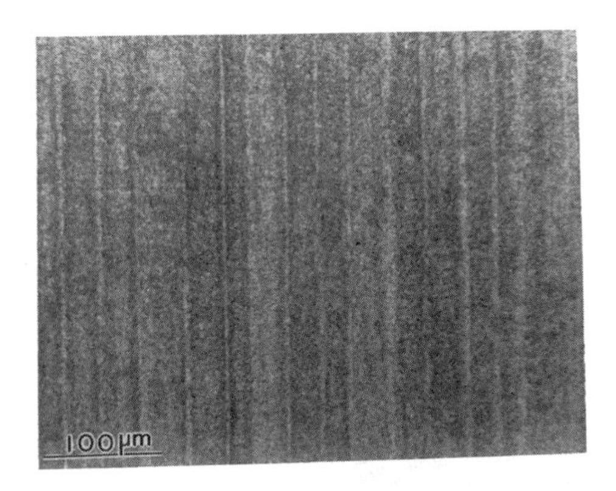

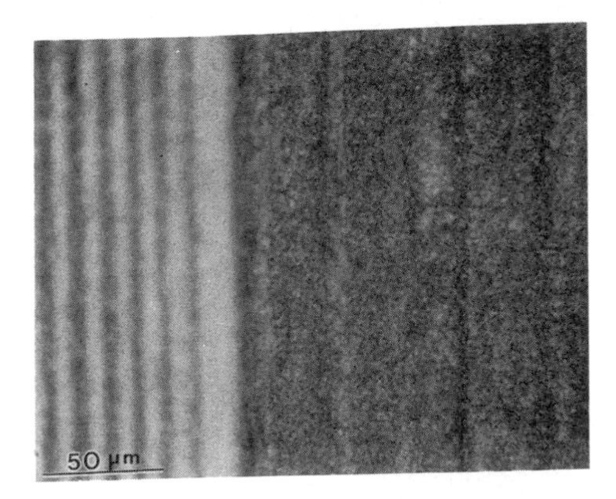

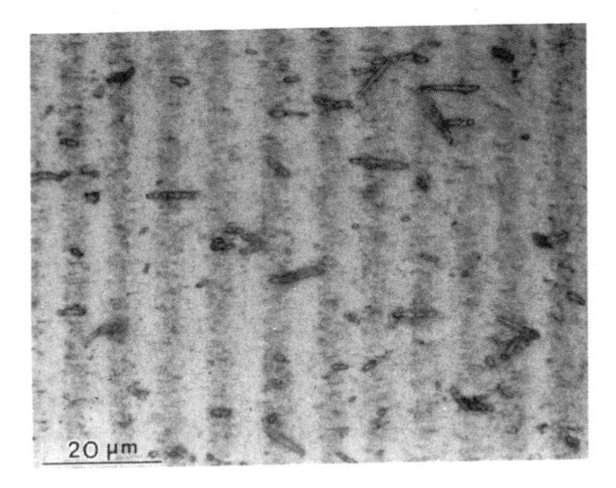

Figure 4 Optical micrographs in increasing magnifications of the wear groove. The wear groove contains parallel lines and cylindrical particles, which are oriented perpendicular to the scratching direction.

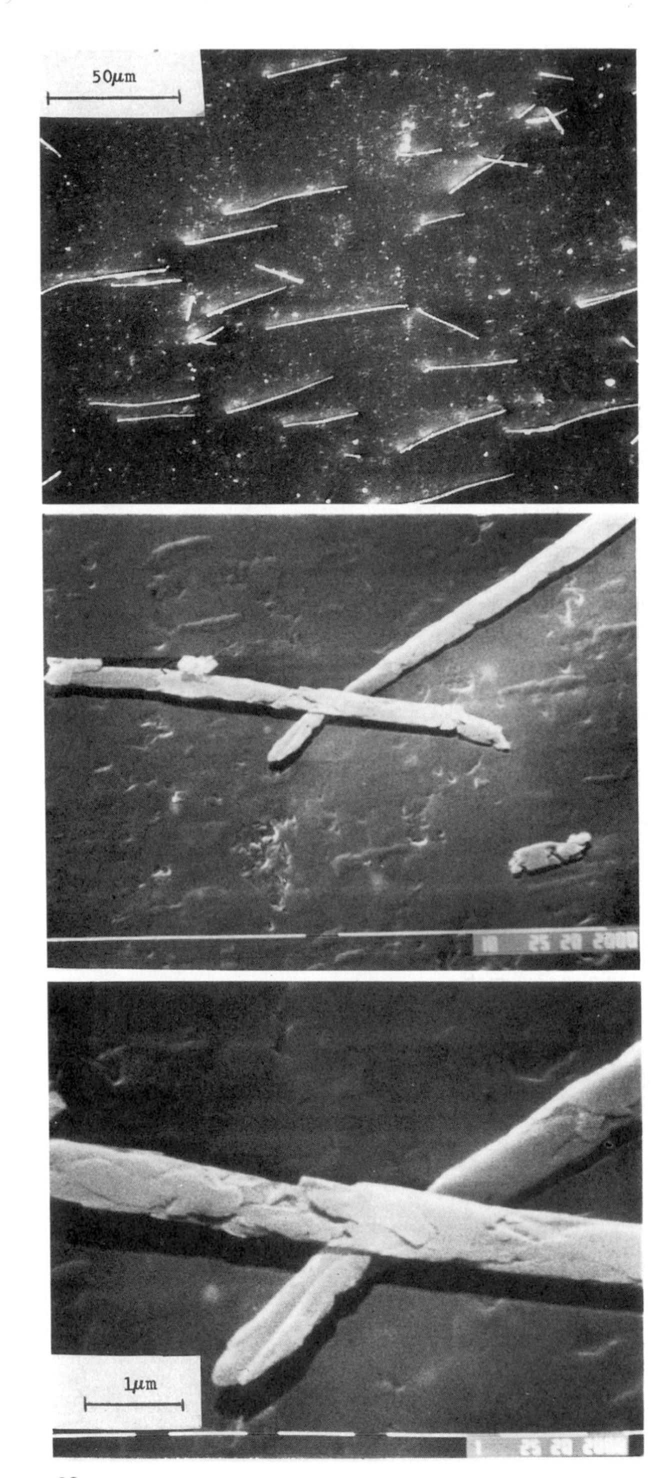
50μm
1μm

scanning electron micrographs of one such example of the wear groove produced in 34% water vapor at 973 K, the condition that produced the lowest friction coefficients and wear rates observed for this environment. These micrographs show that the debris is formed into "rolls," which are generally oriented perpendicular to the sliding direction. The rolls have cylindrical shapes and are straight, with uniform diameters in the range of 0.8–1.0 μm. The rolls in Figure 5 have a mean length of 26 μm with a standard deviation of 15 μm. The longest rolls are slightly more than 50 μm long. Some have broken into smaller pieces, accounting for the high standard deviation in the lengths. In some cases the rolls have been bent around each other, indicating their flexibility.

Figure 5 also indicates that the rolls are composed of compaction of thin layers. The higher magnification micrograph shows that the rolls are apparently formed from smaller wear particles, which adhere and form the cylindrical geometry by the sliding action of the ball.

Figure 6 shows another series of micrographs of a smooth region in the wear groove. These micrographs show that the grain boundaries have been delineated and in some cases, grain pullout may have occurred. The two higher magnification micrographs show evidence for the initiation of the rolls. The cxylindrical geometry appears to have been formed by the flattening of a debris particle, which curls into a roll, most likely as a result of internal residual stresses.

V. TRANSMISSION ELECTRON MICROSCOPY

The rolls were examined in a transmission electron microscope by bright field and electron diffraction techniques. Figure 7 shows two micrographs of a typical roll, at a midsection and near one of the ends. The microstructures are similar to those observed by scanning electron micrography. The surface structure is nonhomogeneous, appearing to be composed of smaller platelets attached to the main body of the roll. Figure 8 shows a bright field image of a fractured roll, with the corresponding electron diffraction pattern. Small debris particles are seen at the fractured section, and the diffraction pattern shows that the rolls are amorphous. In several cases, the diffraction spots were observed as shown in Figure 8. Another example of a fractured roll is shown in Figure 9. In this case, the roll appears to have fractured into smaller particles, and the diffraction pattern indicates that the structure is amorphous.

Figure 5 Scanning electron micrographs of the "rolls" formed in the sample abraded at 873 K in 34% water vapor pressures. The rolls are up to 100 μm long and 1 μm diameter.

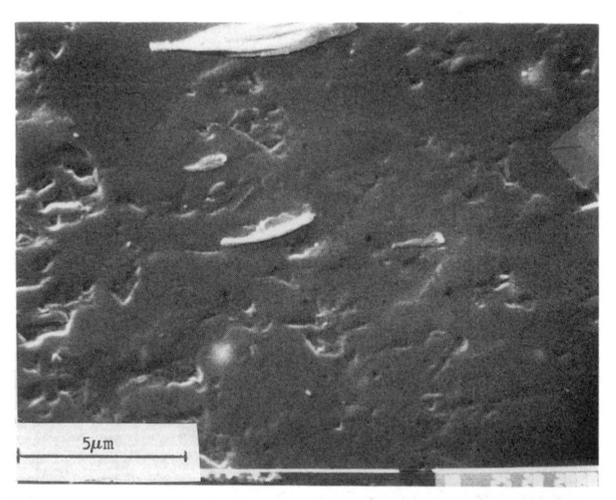
5μm

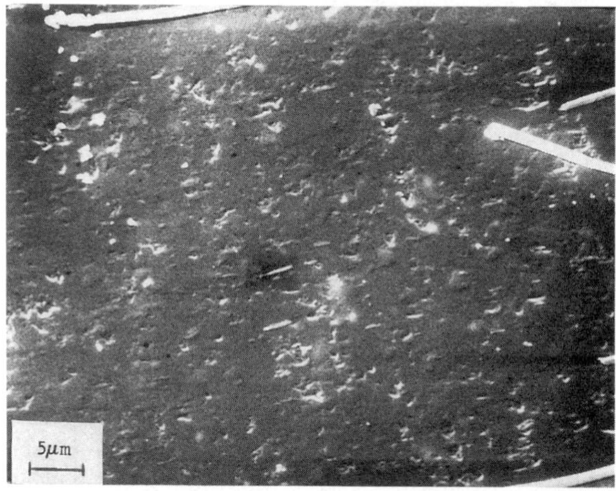
5μm

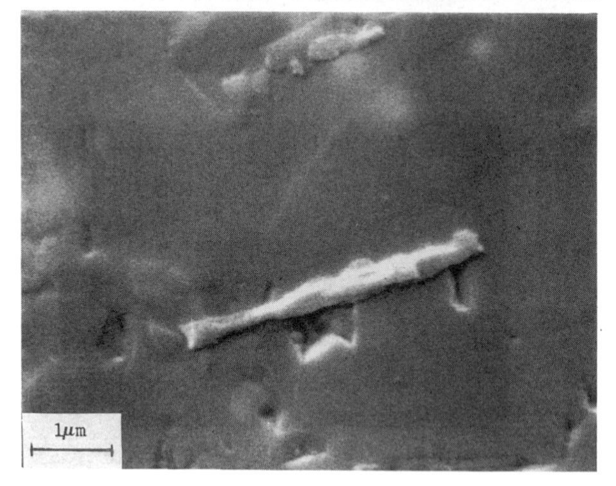
1μm

VI. DISCUSSION

The sliding of a silicon nitride ball on a silicon nitride flat in humid environments at elevated temperatures appears to result in two transitions in the friction and wear, namely a low and high temperature transition, as shown schematically in Figure 10. The friction coefficient has a transition at low temperatures that causes the friction to decrease to approximately half the room temperature value. As the temperature is increased beyond an upper critical temperature, the friction coefficient increases to the value obtained in the argon and air. The upper transition temperature depends on the relative percent water in the environment as shown in Figure 10. The wear behavior follows that of the friction values. Below the upper critical temperature, the wear in humid environments is lower than that in argon or air. As the humidity level increases, so does the transition temperature, and the wear rises to that of the inert gas and air.

The friction and wear are closely related to the chemical reactions that occur at the surfaces during the experiment. It is well known that the highly negative free energy of SiO_2 tends to oxidize silicon nitride by the reaction

$$Si_3N_4 + 3O_2 = 3SiO_2 + 2N_2$$

and the water vapor also modifies the structure of the SiO_2 to lower its viscosity and increase the diffusion coefficients of species within it [4]. The higher water vapor pressures would be expected to produce the most modified structures. The changes in the friction and wear are therefore related to the formation of the SiO_2 layer and its properties at the various temperatures.

At low temperatures the kinetics of the chemical reactions do not allow a sufficient amount of oxide to form, so the sliding of the hemispherical rider on the polycrystalline silicon nitride generates inter- and transgranular cracks, which initiate at the elastic–plastic interface [11] of the flat sample and propagate toward the surface. In addition, the oxides may be too hard and brittle to form rolls. The wear is then related to the propagation of lateral cracks and is not influenced by the chemical reactivity of the silicon nitride with the environment. The wear rate in this case would be described by

$$W = K\left(\frac{P}{H}\right)$$

where W is the wear rate, K is the wear coefficient, and P and H are the load on the hemispherical ball and the hardness of the silicon nitride flat, respectively.

Figure 6 Scanning electron micrographs of the smooth region of a sample abraded at 873 K in 34% water vapor pressure. The rolls appear to initiate from surface layers.

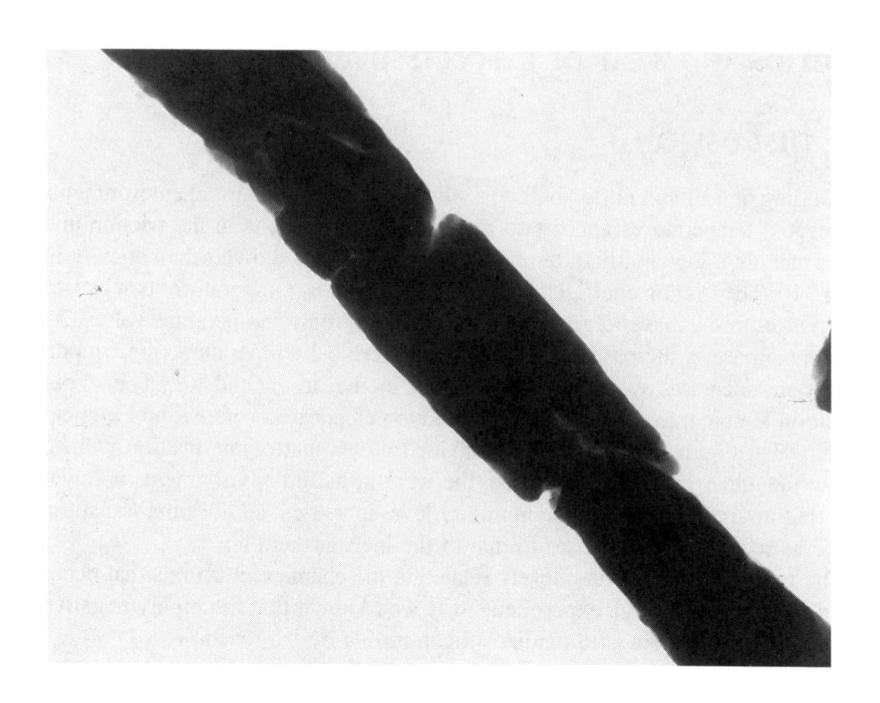

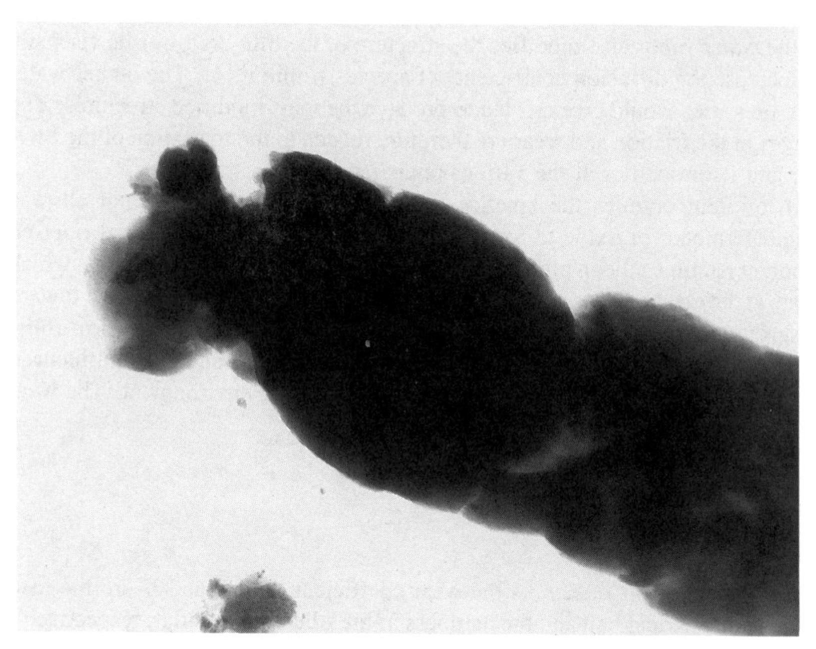

Figure 7 Transmission electron micrographs (bright field image) of a typical roll, at a midsection and at end.

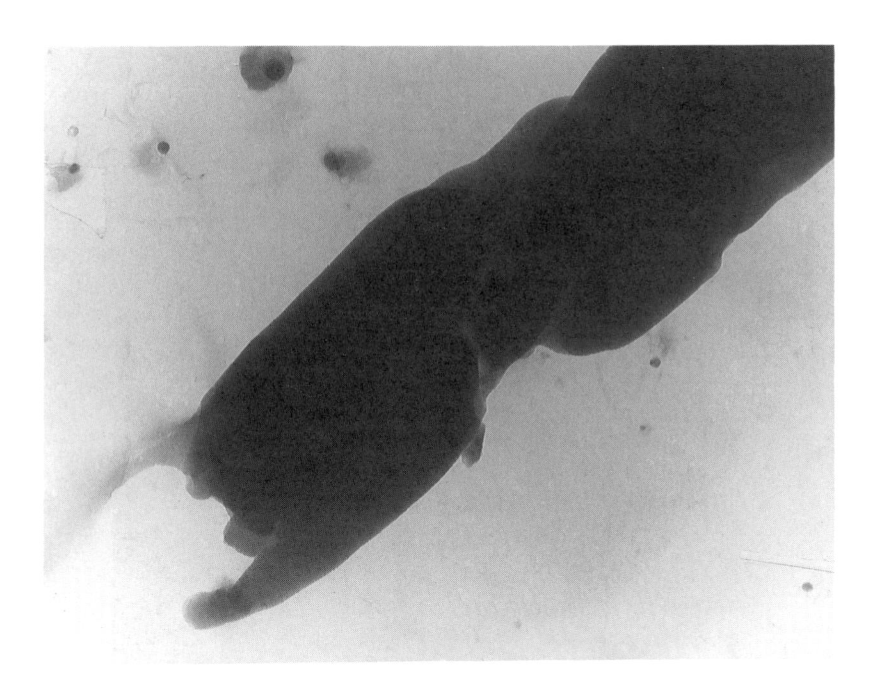

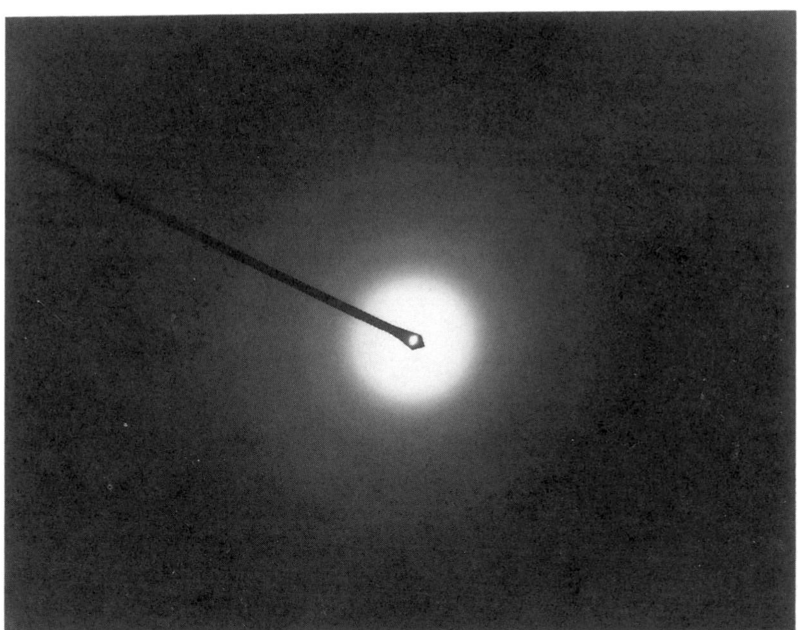

Figure 8 Transmission electron micrograph of a typical roll (bright field image) and the corresponding electron diffraction pattern.

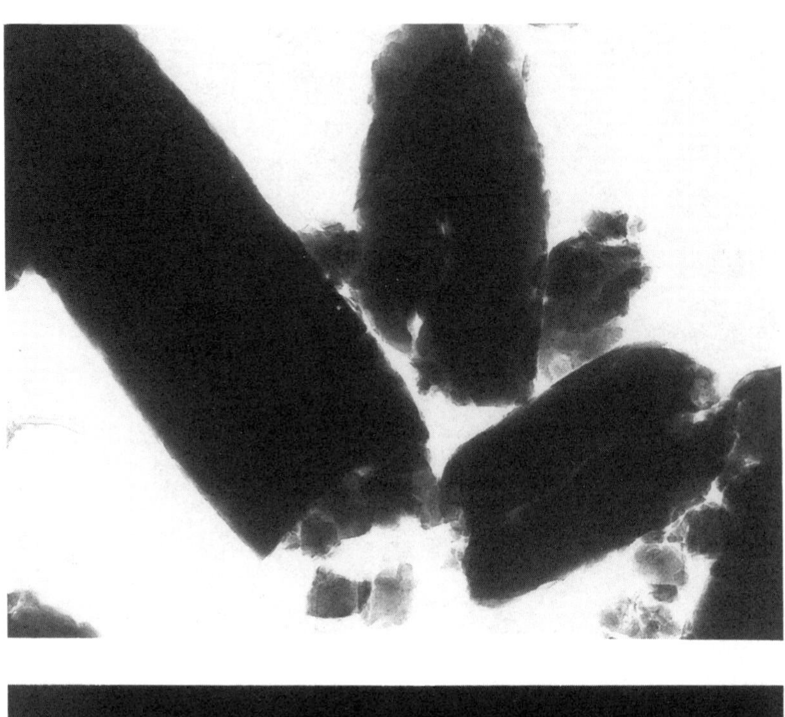

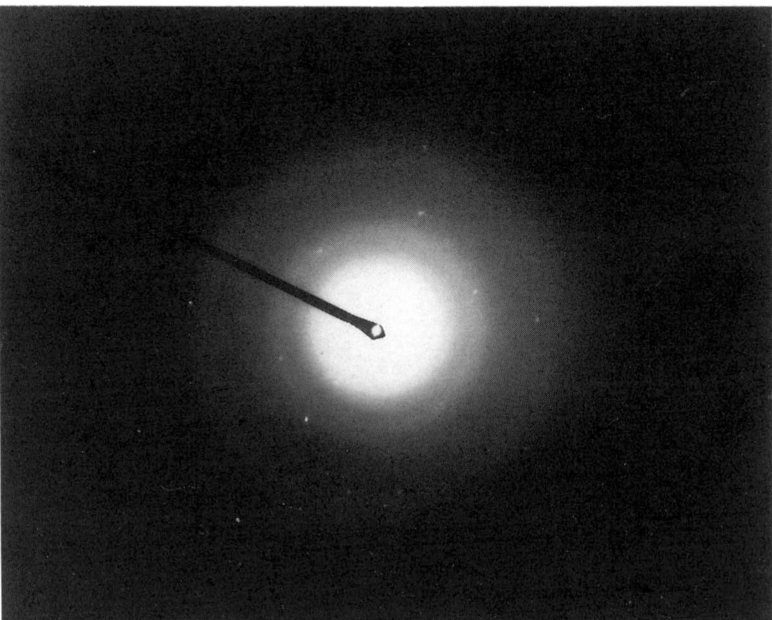

Figure 9 Transmission electron micrograph (bright field image) and the corresponding electron diffraction pattern of a typical roll.

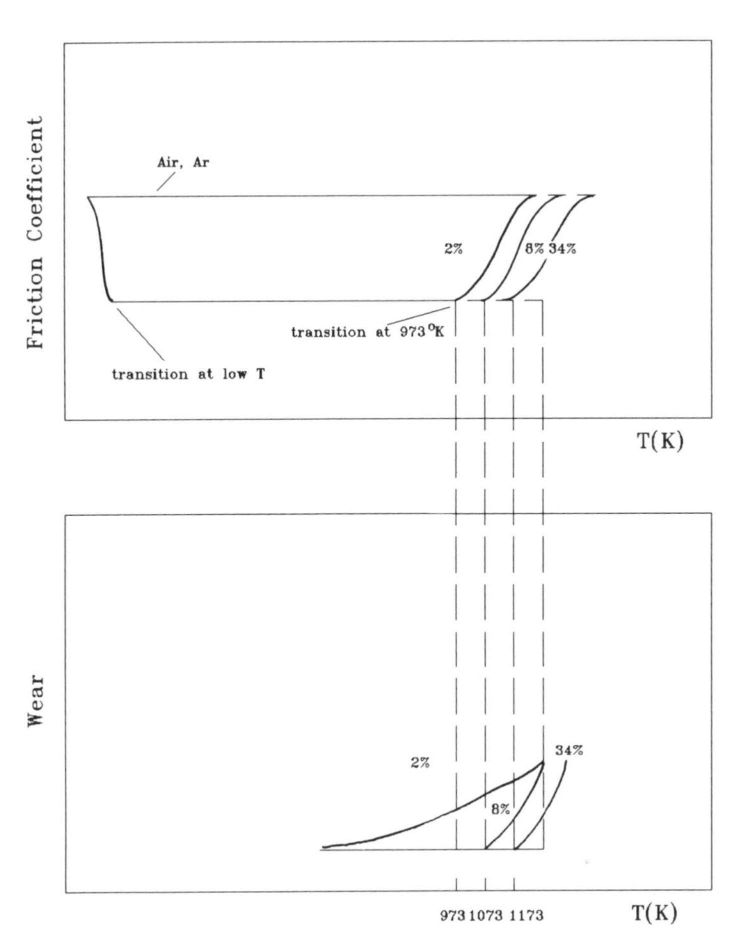

Figure 10 Summary of data for the friction coefficients and wear rates versus temperature.

decreases linearly with temperature at low temperatures and exponentially with temperature at elevated temperatures [12].

At the intermediate temperatures the silicon nitride reacts with the humid environment and a glassy silicon oxide film forms at the surface. The thickness of the film will depend on the water vapor pressure. If the film is sufficiently thick, the wear rate will be proportional to the viscosity of the glassy phase. As the sliding of the hemispherical ball causes the film to rupture, it curls as a result of the residual stresses. Long rolls form when small patches of film debris stick to roll nuclei. Additional oxide continues to grow in the newly exposed regions as the reaction between the silicon nitride and the water vapor proceeds.

As the temperature is increased beyond some critical value, the viscosity of the glassy film will decrease so that its lubrication properties will diminish. In effect, the film is squeezed out from underneath the ball so that the friction and the wear increase.

Our results indicate that this model can account for the low and high temperature transitions and predict that the low friction and wear could be modified by suitable doping of the oxide layer, as for example by the appropriate control of the gaseous environment.

VII. SUMMARY

Friction coefficients and wear rates of silicon nitride depend on temperature and humidity. Low and high temperature transitions are associated with the formation of elongated cylinders (rolls) in the wear track as a result of elevated temperatures and exposure to a humid environment. The transition temperature increases as the relative humidity increases. Electron diffraction analysis has shown that the debris is amorphous.

ACKNOWLEDGMENTS

This work was supported by the National Science Foundation grant MSM-874491 through the tribology program. We thank Dr. Jorn Larsen-Basse for his support. Thanks are also extended to J. Gramsas, S. Albertson, S. Dymek, and J. Tomei for technical assistance.

REFERENCES

1. *Proceedings of the Gas Research Institute Seminar on the Application for and Designing with High Temperature Materials in Natural Gas Usage*, J. R. Hellman, R. E. Tressler, and M. Lukasewicz (Eds.), Gas Research Institute: Chicago, and the Center for Advanced Materials, University of Pennsylvania: University Park, 1989.
2. V. J. Tennery, Ceramics in engines—An international status report, *Ceram. Bull. 68*, 362–365 (1989).
3. H. Du, R. E. Tressler, K. E. Spear, and C. G. Pantano, Oxidation studies of crystalline CVD silicon nitride, *J. Electrochem. Soc. 136*, 1527–1536 (1989).
4. S. C. Singhal, Effect of water vapor on the oxidation of hot pressed silicon nitride and silicon carbide, *J. Am. Ceram. Soc. 59*, 81–82 (1976).
5. D. S. Park, S. Danyluk, and M. J. McNallan, Friction and wear measurements of Si_3N_4 at elevated temperatures in air, Ar, and humid environments, in *Proceedings of a Symposium on Corrosion and Corrosive Degradation of Ceramics*, R. E. Tressler and M. J. McNallan (Eds.), American Ceramic Society: Westerville, OH, 1990.

6. T. E. Fischer and H. Tomizawa, Interaction of tribochemistry and microfracture in the friction and wear of silicon nitride, in *Proceedings of the International Conference on the Wear of Materials*, Vancouver, Canada, 1985.
7. C. S. Yust, J. T. Leitnaker, and C. E. Devore, Wear of an alumina–silicon carbide whisker composite, *Wear*, *122*, 151–164 (1988).
8. P. Block, F. Platon, and G. Kapelski, Tribological and interfacial phenomena in Al_2O_3/SiC and SiC/SiC couples at high temperature, *J. Eur. Ceram. Soc. 5*, 223–228 (1988).
9. C. DellaCorte, S. C. Farmer, and P. O. Book, Experimentally determined wear behavior of Al_2O_3-SiC Composite from 25 to 1200°C, NASA Technical Memorandum No. 102549 (1990).
10. S. Jahanmir and T. E. Fischer, in *Proceedings of the American Society of Lubrication Engineers, ASME/ASLE Tribology Conference*, Pittsburgh, 1986.
11. B. R. Lawn and T. R. Wilshaw, Indentation fracture: Principles and applications, *J. Mater. Sci. 10*(6), 1049–1081 (1975).
12. S. Guruswamy, J. P. Hirth, and K. T. Faber, *J. Appl. Phys. 60*, 4136 (1986).

5

Surface Damage and Mechanics of Fretting Wear in Ceramics

P. J. Kennedy and A. A. Conte, Jr.

Naval Air Development Center
Warminster, Pennsylvania

E. P. Whitenton, L. K. Ives, and Marshall B. Peterson

National Institute of Standards and Technology
Gaithersburg, Maryland

ABSTRACT

A study was conducted to determine the effect of low amplitude slip on surface damage for ceramic–ceramic and ceramic–metal combinations. The test arrangement consisted of a stationary ball loaded against a flat disk. An external drive imparted a twisting motion to the flat, producing slip amplitudes from 0.1 to 10 μm at the periphery of the contact. In some tests the influence of surface fatigue was examined by applying a rocking motion to the flat, to simulate oscillatory rolling without appreciable slip. Material combinations studied included silicon nitride–silicon nitride, silicon nitride–aluminum oxide, aluminum oxide–aluminum oxide, and silicon nitride–AISI 52100 steel. Damage characteristics are described and compared for each material combination.

I. INTRODUCTION

Monolithic ceramic bearings and other ceramic components will undoubtedly be incorporated in future military aircraft propulsion systems to promote the achievement of greater thrust-to-weight ratios. This transition from metals to ceramics is complicated by such factors as the unique tribology associated with ceramics, special ceramic lubrication requirements, the need to develop new high temperature lubricants and lubrication techniques, and the general lack of experience at the high operating temperatures required.

One important consideration in the use of ceramics for bearing applications is the effect of microslip on bearing life. During operation, a rolling element bearing experiences a certain amount of microslip at the ball–race interface which can result in surface damage. For ceramic bearings, damage associated with microslip can cause microfracture in the wear track, which because of the relatively low toughness of ceramics could lead to immediate bearing failure. The aim of this study was to determine the effect of microslip on damage in ceramic contacts and, in particular, to determine whether a critical slip amplitude exists that defines a lower limit below which no surface damage occurs. This value would represent a parameter that could be incorporated in future ceramic bearing designs.

Relatively little work has been done on the effect on fretting of extremely small slip amplitudes (i.e., <0.1–$10\ \mu m$). The upper limit of motion for fretting was estimated by Ohmae and Tsukizoe [1] and Vingsbo and Soderberg [2] as approximately 300 μm. Beyond this length, the fretting wear process changes to one that is more typical of cyclic sliding. The lower limit for the minimum amount of slip necessary to produce fretting is more difficult to establish. When a tangential force is applied to two contacting bodies, resulting in a shear traction, the displacement is initially elastic. With increasing force, some amount of the displacement will occur within the contact by plastic flow and slip. Sliding at the contact will require a tangential force sufficiently large to overcome the static friction force. Johnson [3], in very accurate work, measured elastic microdisplacement values of 1–3 μm preceding large-scale slip for steel on steel. Courtney-Pratt and Eisner [4] reported similar data and also showed that the displacement necessary to produce slip was decreased in the presence of a lubricant. The exact value for the amount of elastic deformation that precedes slip would, of course, depend on experimental conditions. In an earlier study [5] we reported a value of 0.06 μm for the minimum slip amplitude associated with oxidative damage for fretting couples of AISI 52100 steel. The value of 0.06 μm was measured at the low pressure, outer edge of the wear scar where elastic deformation would be minimal; thus it indicates the actual amount of slip necessary to produce the first evidence of fretting. This value is significantly greater than the value of 0.002 μm reported by Tomlinson [6]. The work of Sato et al. [7] can be cited as another case of damage observed at a low amplitude of slip. Using a steel ball and glass flat configuration, these investigators reported visual evidence of surface damage at a slip amplitude of 4 μm. However, the minimum slip amplitude associated with surface damage was not evaluated in the study.

Silicon nitride, one of the most promising ceramics for use in high temperature bearing applications, was the primary material employed in this study. The other materials, alumina and AISI 52100 steel, were selected principally for comparative purposes. Alumina was chosen because it was believed that its wear characteristics would differ significantly from those of silicon nitride. Tests using bearing steel–ceramic couples were conducted, since this type of bearing construction is presently being used in several applications. Finally, to compare damage associated with fatigue occurring with little or no slip, some tests were conducted in which microrolling motion was applied to the contacting specimens using the same test apparatus.

II. EXPERIMENTAL

Fretting tests are generally conducted by imparting a linear oscillatory motion to one of two contacting specimens. Extremely small linear displacements are, however, relatively difficult to reproduce accurately and to measure. Fretting motion can also be produced by twisting one specimen with respect to the other. The advantages in the second case are that slip amplitudes much less than 1 μm can be generated easily, and the magnitude of the displacement is easily measured, since it is proportional to the angle of rotation. The latter approach was taken in this investigation. The specimens consisted of a ball, which was held stationary, and a flat, which was given an oscillatory rotational or twisting motion. Rotation was about an axis perpendicular to the flat and through the center of the contact with the ball. The resulting movement of the two specimens produced fretting damage in the contact area, which could be correlated with the slip amplitude. The test device was also capable of rocking the flat against the ball to simulate an oscillating rolling motion without appreciable slip. An advantage of the specimen configuration used is that precision-ground ceramic balls of a variety of different materials are available from several suppliers, and flat specimens can be easily machined from stock materials.

The test rig employed in this investigation had been used in an earlier study of microslip at concentrated contacts [8] and in other studies of fretting [5,9]. Schematic drawings of side and top views of the test rig are shown in Figure 1. The major components are a pneumatic bellows arrangement for application of a load to the ball, a gas bearing stage to support the flat, electromagnetic drivers to rotate or rock the gas bearing stage, and capacitance probes to measure the stage displacement. Additional details of the rig and components for mounting the specimens can be found in Reference 5.

Important characteristics of the stress distribution at a ball-on-flat contact are illustrated schematically in Figure 2. Assuming elastic conditions, a circular contact region is produced with an applied normal stress τ_N that is zero at the periphery of the contact and rises to a maximum at the center according to the classical Hertz analysis. Application of a torque to rotate the flat relative to the ball about the z-axis without slip will result in the shear stress τ_S distribution, also shown in Figure 2. Slip will occur in the outermost annular region if $\tau_S/\tau_N > f_S$, where f_S is the static coefficient of friction. This slip region, of course,

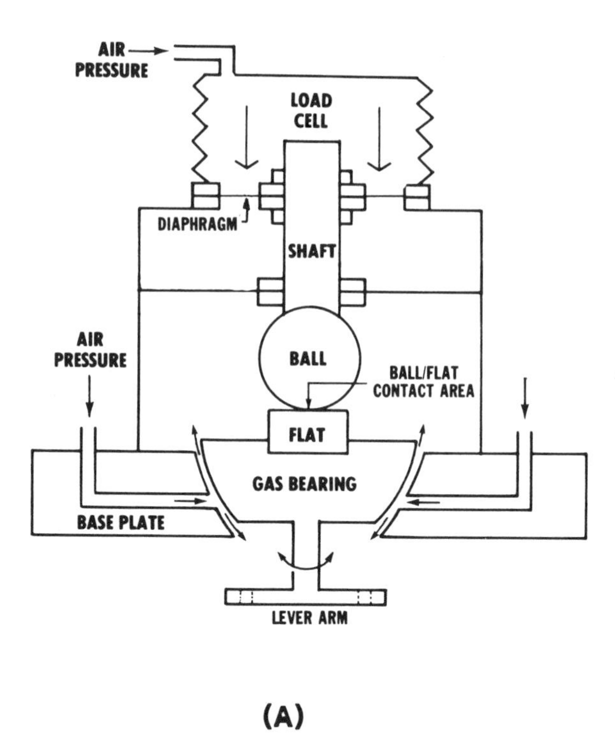

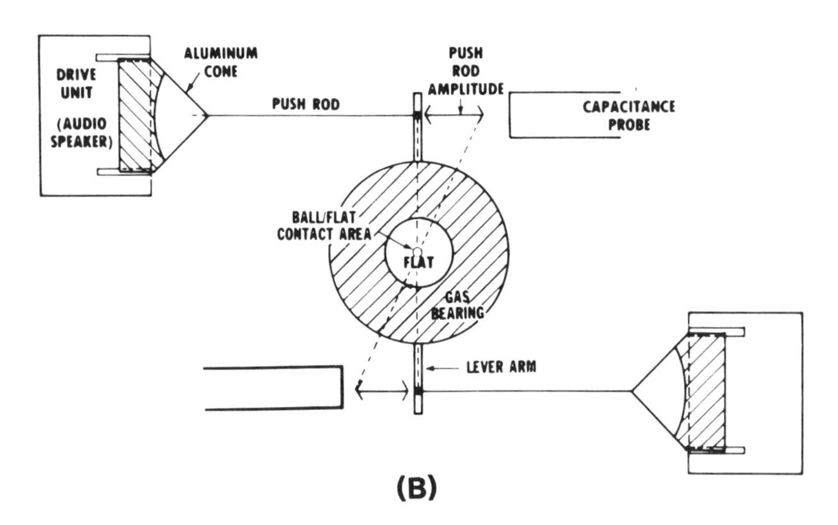

Figure 1 Schematic drawing of test rig showing (A) side view and (B) top view looking down on the specimens displaying the arrangement of the drive components.

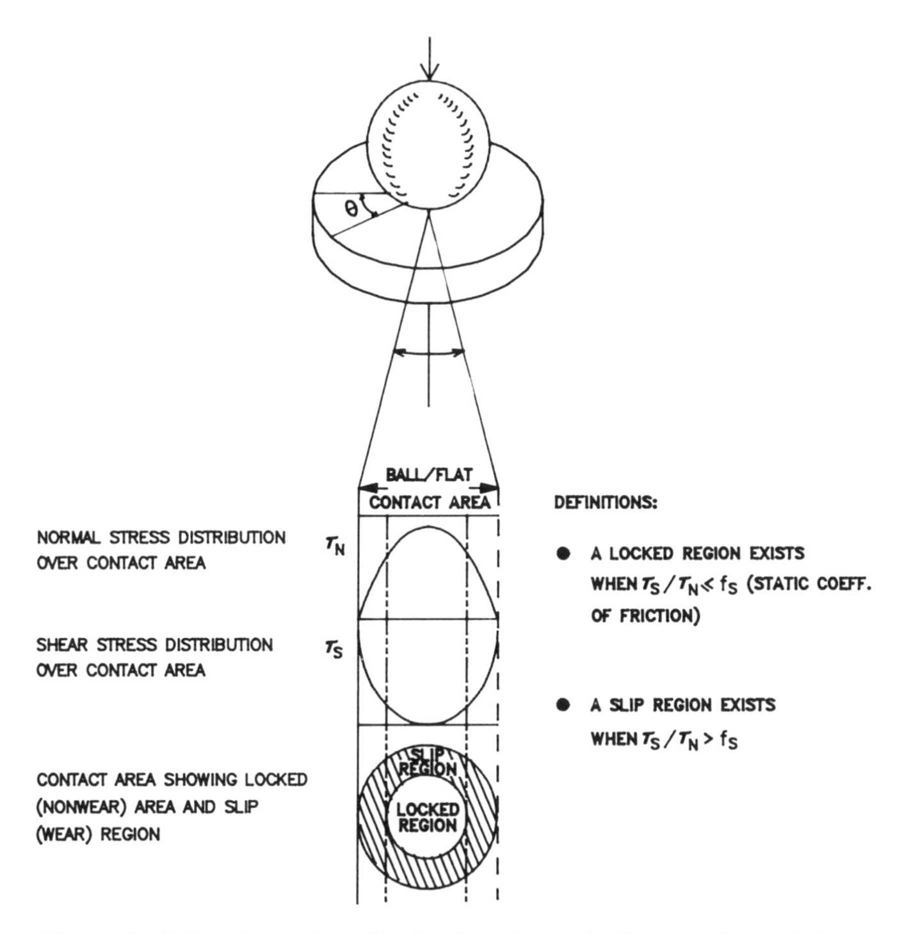

Figure 2 Ball-on-flat configuration showing relationship of shear and normal stresses to scar features.

corresponds to the region that is subject to fretting wear. The slip amplitude at any radial position in the slip region is determined by the magnitude of the applied angular amplitude of rotation corrected for elastic displacement. The slip amplitude will vary from zero at the periphery of the locked region to a maximum value at the outer boundary of the contact area. In this investigation, slip amplitude is reported as the displacement calculated at the outer diameter of the contact area based on the applied angular amplitude of rotation.

Testing was normally done at room temperature using an applied load of 88 N and a frequency of 210 Hz. The primary experimental variables were the slip amplitude, the normal load, and the duration of the test.

The silicon nitride ball specimens were grade 5, 12.7 mm diameter bearing balls made from either Norton uniaxially hot-pressed NC-132 or hot isostatically pressed NBD-100 material. The flats were in the form of small cylinders 8 mm in diameter and 3 mm in height. They were machined from an NC-132 silicon nitride tile. Other materials include 12.7 mm diameter alumina balls (Coors AD995) and standard grade 25 EP (extra polish) AISI 52100 bearing steel balls, sapphire flats, and hardened AISI 52100 steel flats. The surface roughness of the test specimens was less than 1 μm rms.

III. RESULTS AND DISCUSSION

A. Nature of Surface Damage

A topographic map obtained by stylus profilometry of a typical wear scar is shown in Figure 3. The scar consists of a central locked region, where there is no visible damage, and an annular slip region worn to various depths, with the greatest depth located just outside the locked area. The material appears to have been extruded out of the worn area to form a lip beyond the outer contact edge. This extrusion might be expected, since the normal pressure (Fig. 2) decreases rapidly as the edge of the scar is approached. The maximum depth of wear in this case was about 1 μm.

Details of the surface damage can be seen in Figure 4, which presents scanning electron micrographs of the specimen in Figure 3 taken at various magnifications. Figure 4A shows the flat specimen with the small wear scar at its center. Figure

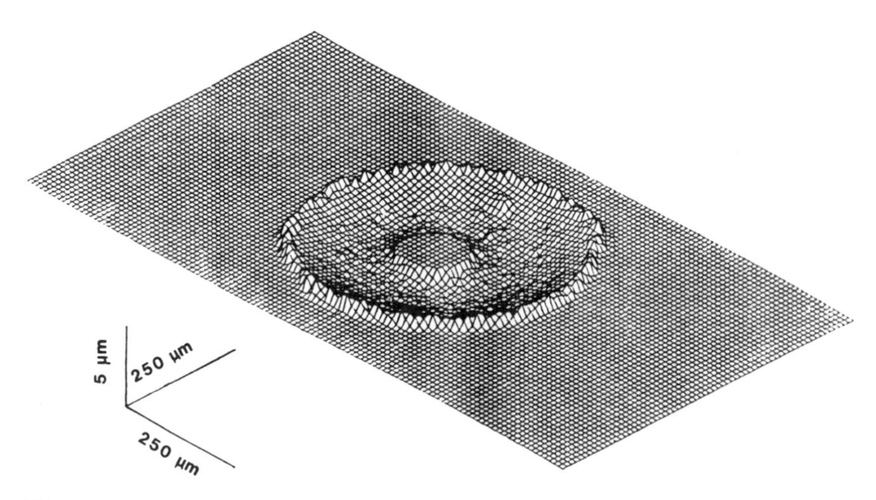

Figure 3 Map of wear scar topography generated by stylus profilometry.

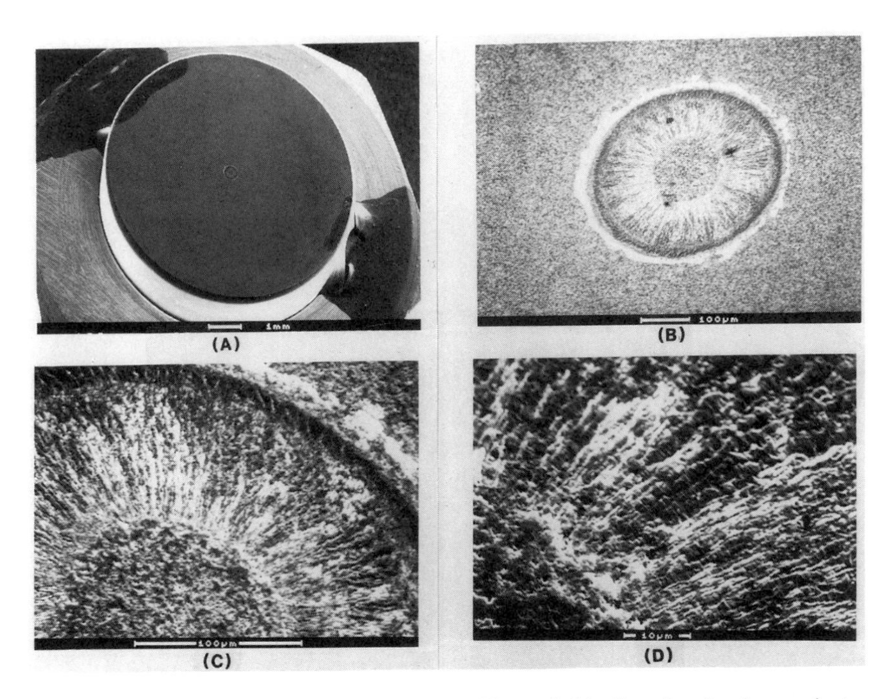

Figure 4 Typical wear scar formed on a silicon nitride flat after fretting against a silicon nitride ball: (A) flat with wear scar at its center, (B) wear scar at a higher magnification, and (C, D) details of damage in slip region.

4B shows the scar at a higher magnification, revealing its general features. Details of the damage in the annular slip region are shown in Figure 4C; the locked region is at the bottom of the photograph. Two important characteristics of the slip region were the presence of debris and closely spaced grooves extending radially from the locked region to the outer periphery (Fig. 4C, D). The outer lip consisted of a ring of debris. The section that follows describes changes in the damage pattern as a function of amplitude and sliding time.

B. Effect of Slip Amplitude and Sliding Time on Damage

A series of fretting tests were run with slip amplitudes ranging from 0.1 to 6 µm at 210 Hz and sliding times from 10 to 360 minutes at a load of 88 N. Microscopic examination and surface profile analyses of the resulting wear scars were carried out.

At a slip amplitude of 0.5 µm (Fig. 5), no damage was found after tests of 10 and 30 minutes duration. At 180 minutes a very faint ring of stain was

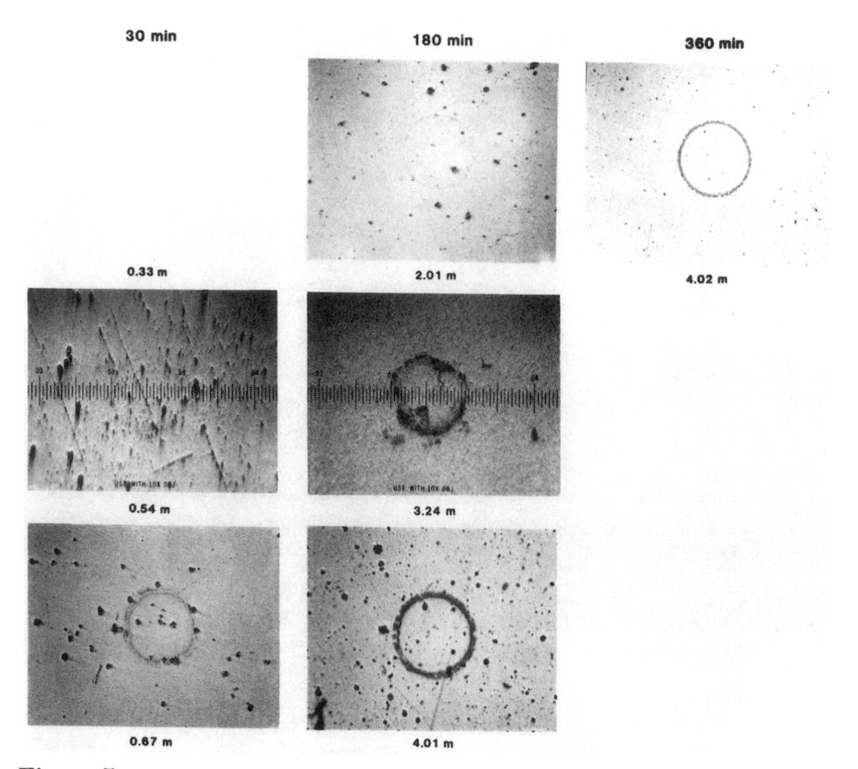

Figure 5 Light optical micrographs showing the relationship of increasing slip amplitude and time to wear scar formation.

observed, which became more obvious at 360 minutes. At an amplitude of 0.8 μm, staining was well advanced at 180 minutes; at an amplitude of 1.0 μm, staining was readily observed at 30 minutes. Thus, it was found that the first indication of wear occurred at shorter times at higher amplitudes.

Additional micrographs of wear scars formed at progressively higher slip amplitudes are presented in Figure 6 to show the transition from what begins as a faint ring of stain to fairly severe wear. Although debris buildup is evident in all three specimens, profile traces indicated that what is seen in micrographs A and B is a thin layer of accumulated debris and that little material has been removed from the surface of the silicon nitride flat. Micrograph C clearly shows a significant reduction in the size of the locked region, a slipped region with considerable depth of wear, and a pile-up of extruded debris around the wear track.

Examination of the surfaces indicated that significant damage did not occur below a slip amplitude of 0.2 μm regardless of the time involved in testing.

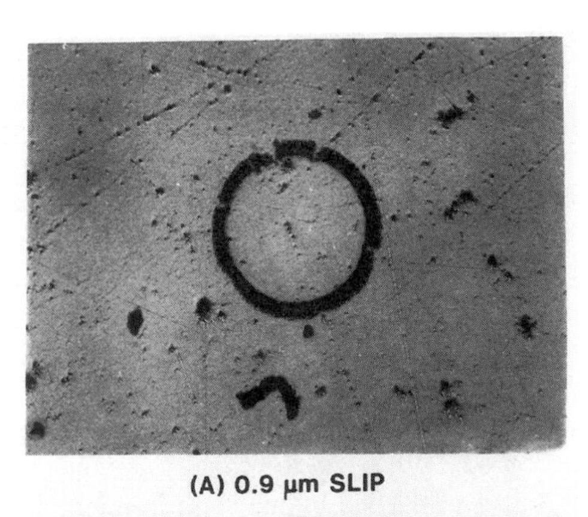

(A) 0.9 μm SLIP

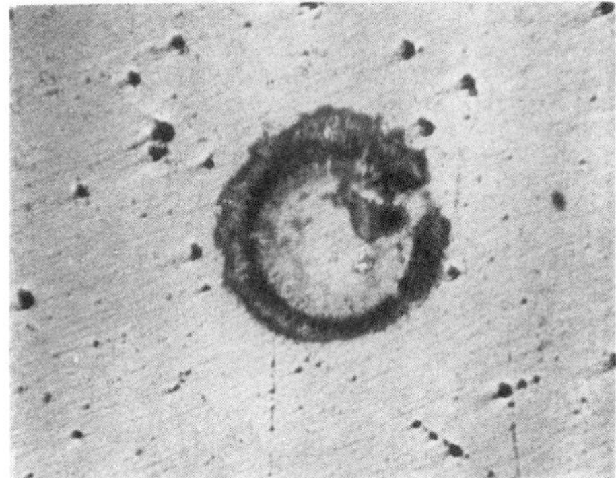

(B) 1.8 μm SLIP

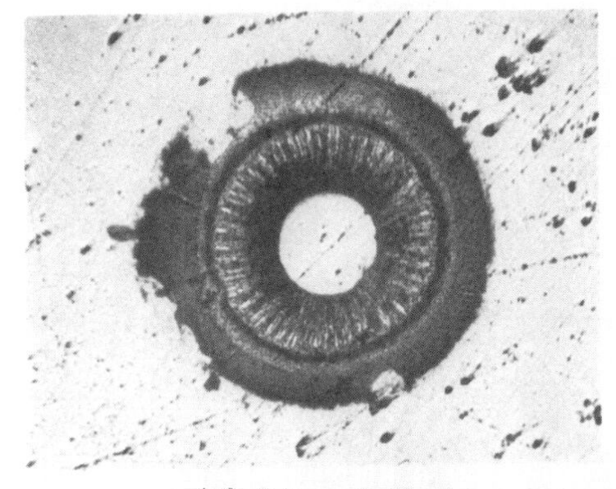

(C) 6.2 μm SLIP

Figure 6 Wear scars showing degree of surface damage as a function of slip amplitude: (A, B) increasing buildup of a tribochemical film; (C) significant surface damage in the slip region.

Above a slip amplitude of 3 μm, both wear and surface damage increased significantly. Wear was not linear with time. This relation is illustrated in Figure 7, which plots wear volume, determined from surface profilometer traces, against slip amplitude for 126,000 and 378,000 cycles. The similarity in wear volumes for the two sets of data indicates that most of the surface damage had occurred before 126,000 cycles, with the rate of further material removal drastically decreased after that. Data obtained by measuring wear scar dimensions as a function of time at a given slip amplitude also displays this effect (Fig. 8).

Wear rates for the silicon nitride combination, slip amplitude 3.25 μm, calculated from surface profile traces were 2×10^{-16} $m^3/N \cdot m$ at 1 minute and 10 minutes and decreased to 6×10^{-17} $m^3/N \cdot m$ at 180 minutes. This decrease in wear rate is the result of the reduction due to wear in the contact pressure in the slip region. Eventually all the load will be carried by the locked region.

Surface profiles of the silicon nitride ball and flat were measured after testing. The profiles are compared in Figure 9 to show the corresponding scar topographies. The difference in magnification for the vertical and horizontal scales causes the curve that defines the original undamaged ball surface in Figure 9 to appear elliptical rather than spherical. The volume of material removed from the slip region of the flat was approximately equal to that removed from the ball, 5×10^{-5} mm^3. The volume of debris material in the

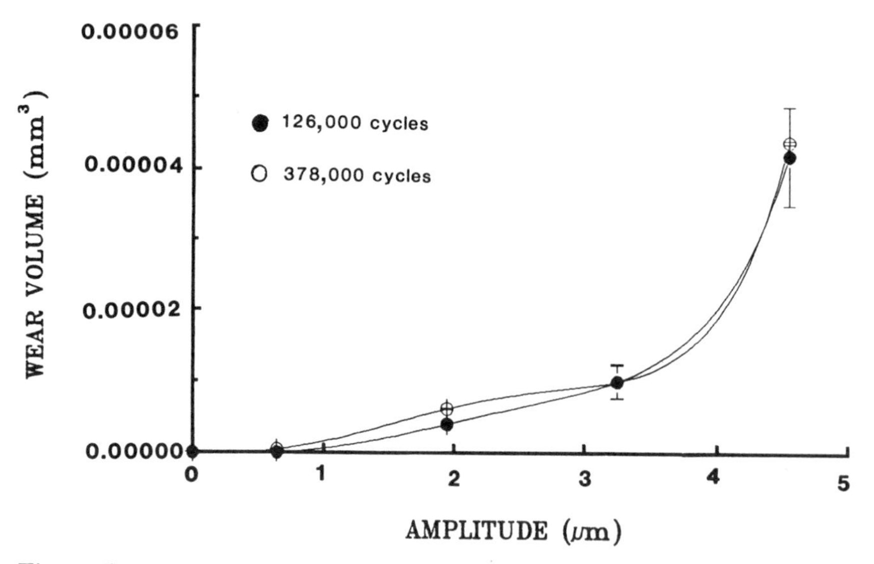

Figure 7 Relationship of slip amplitude to surface damage determined after 126,000 and 378,000 fretting cycles.

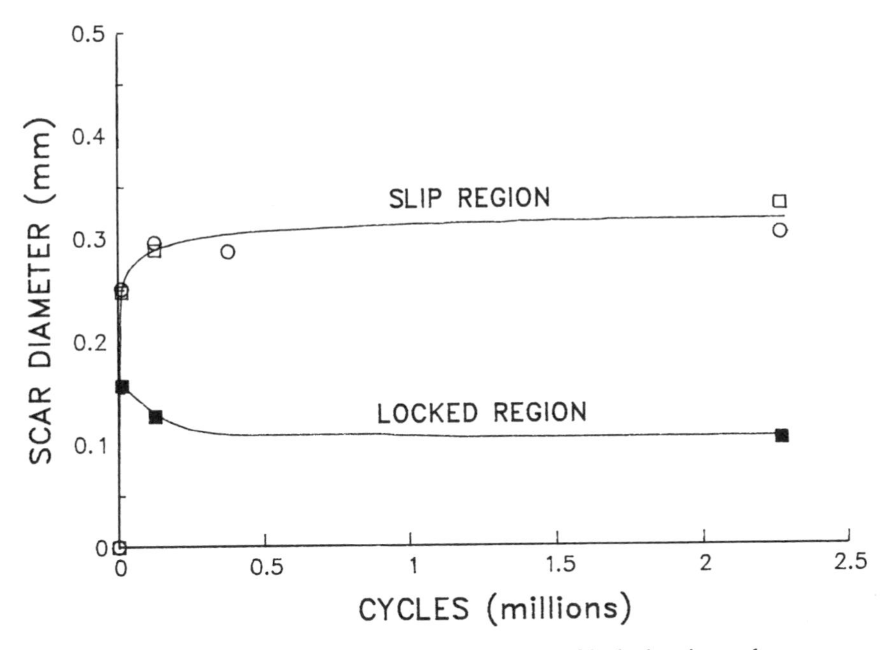

Figure 8 The effect of number of cycles on diameter of locked region and wear scar.

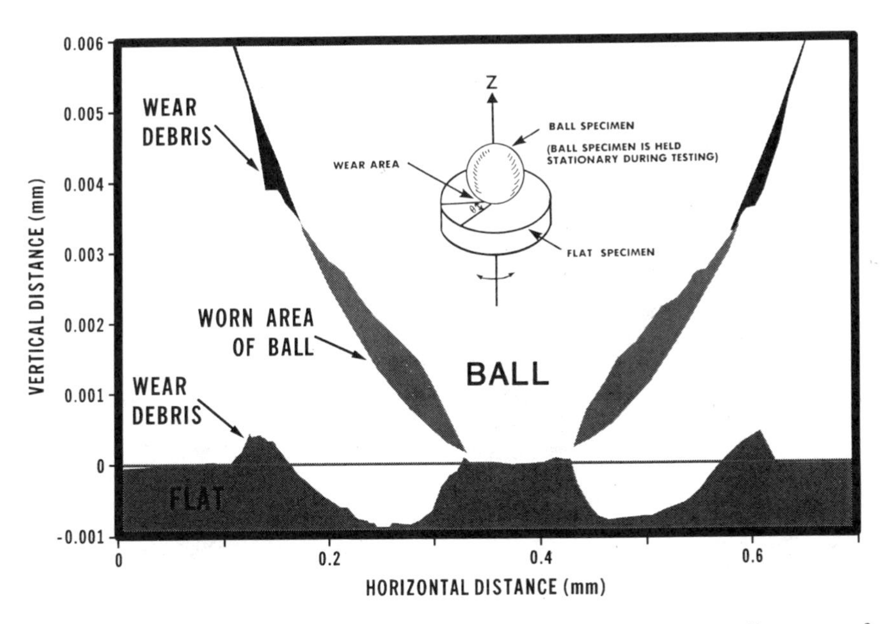

Figure 9 Matched profiles of scars on ball and flat showing corresponding areas of wear and debris buildup after completion of the test. Elliptical shape of ball is a result of larger vertical compared to horizontal magnification.

built-up regions at the outer edge of the wear scars was 1×10^{-5} mm^3. However, much of the extruded debris was not firmly attached to the surface and had been removed before the specimens were traced. Note that the specimens in Figure 9 were traced in the unloaded condition. Application of the 88 N load employed in the test would result in elastic deformation to bring the surfaces more closely into contact.

C. Wear Mechanism for Silicon Nitride

A characteristic type of sunburst pattern was found in the slip region, as shown in Figure 10A, when the total slip distance was in excess of 1 m. Compacted debris was present within the grooves in the slip region, as well as at the periphery of the scar. The relative softness of this layer is illustrated by Knoop indentation marks (Fig. 10B, C). For comparison, an indentation in the silicon nitride surface well outside the scar area is shown in Figure 10D. It can be seen that the

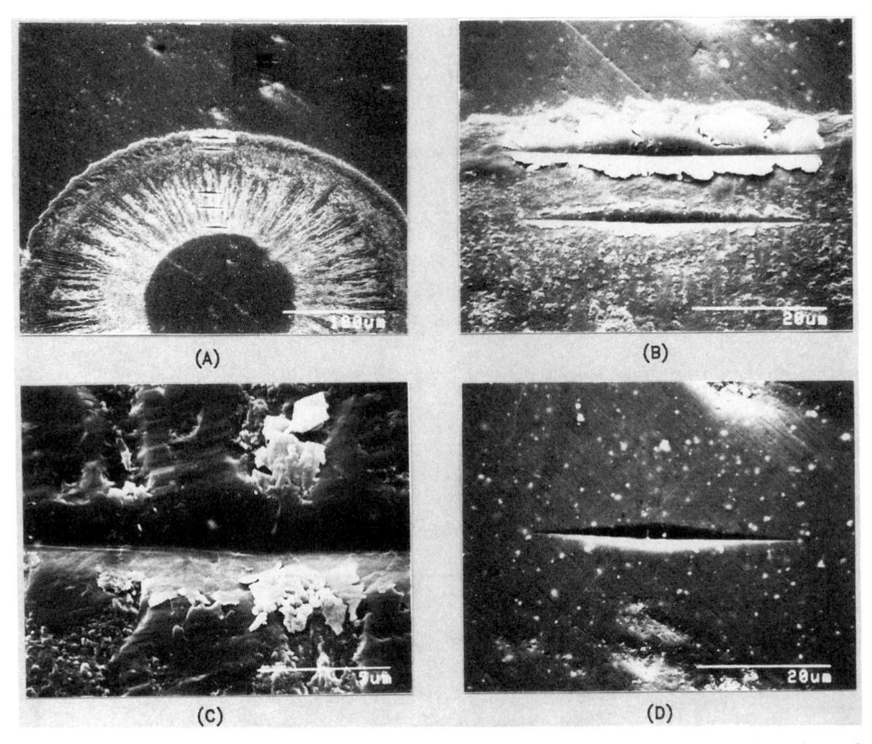

Figure 10 Knoop indentations illustrating plasticity of debris layer: (A) location of indentations, (B) indentation in debris layer at periphery of scar, (C) indentation in slip region, and (D) indentation in unworn surface distant from scar.

indentations in the layer at the periphery of the scar in Figure 10B and within the slip region in Figure 10C are quite different from that in Figure 10D and are characterized by the plastic extrusion of thin lips of material. In Figure 10C it is interesting to note the presence of loose debris in the grooves. This debris material was examined in the transmission electron microscope (TEM) and found to be amorphous. Moreover, electron energy loss spectroscopy carried out in the TEM indicated the presence of silicon and oxygen but not nitrogen. This result suggests that the film material is an oxide or hydrated oxide of silicon. It appears that an oxidized film is initially formed on the surface in the slip region. This relatively soft film is then sheared from the surface and extruded at the outer periphery of the contact. In addition to this oxidative wear process in the slip region, fragments may fracture from the edge of the locked region because of fatigue. These fragments may be "milled" and oxidized in the slip region to contribute to the extruded debris.

At very low amplitudes of slip, some evidence was observed for a fracture type of wear mechanism within the slip region. Figure 11 is a micrograph of an

Figure 11 Feature indicating delamination of thin flake from slip region of scar on silicon nitride flat.

area in the slip region of a scar on a flat subjected to a slip amplitude of less than 1 μm. Delamination of a thin flake of material approximately 10 μm in length and width and less than 1 μm thick has occurred. Similar damage was observed on the ball as well. This spalling damage may be due to the restricted access of oxygen and water vapor into the concentrated contact, limiting tribochemical film formation and the associated lubrication which would prevent such damage.

The most striking characteristic of the silicon nitride–silicon nitride combination was the development of the sunburst wear pattern, mentioned earlier. The micrographs in Figure 12 show the pattern in detail. These micrographs indicate that the sunburst pattern is the result of many periodically spaced, smooth, radial grooves. Figure 12C, taken at an angle and thus accompanied by considerable foreshortening, displays the topography of the grooves. In general, the grooves started near the edge of the locked region. The spacing increased with increasing radial distance from the center of the scar. This general trend in spacing suggests that groove spacing might be a measure of actual slip as a function of radial distance. For example, the wear scar shown in Figure 12 represents the effect of a slip amplitude of 3.25 μm applied for 180 minutes. The average spacing in Figure 12B is approximately 2.3 μm, or somewhat less than the applied slip amplitude.

A more detailed study of this relationship of groove spacing to applied slip amplitude was carried out by plotting groove spacing as a function of radial distance for two other wear scars that were formed at two different slip amplitudes. The results indicated that the groove spacing was equal to approximately half the applied slip amplitude at the outer contact radius. It has been suggested that a significant amount of applied slip is lost as a result of the compliance of the mechanical components of the test rig when applied slip amplitudes are extremely small [10]. This possibility was investigated in the present study by attaching a flag in the position where the ball would normally touch the flat specimen, in order to measure actual movement at that point. Good agreement was found between the slip amplitude calculated from the applied angular displacement and the measured slip amplitude. Thus, the difference in groove spacing and slip amplitude could not be explained by an error in the determination of the slip amplitude.

The precise mechanism for the formation of the grooves was not determined. Since the silicon nitride is harder than the film that forms on the surface, the grooves are not worn into the surface by direct abrasion during the extrusion of film debris. Furthermore, an extensive examination of many specimens showed no groove making in progress by silicon nitride debris fragments. If this had occurred, one would expect to see partial grooves containing silicon nitride debris. The current hypothesis is that the grooves are associated with vibration that occurs during sliding.

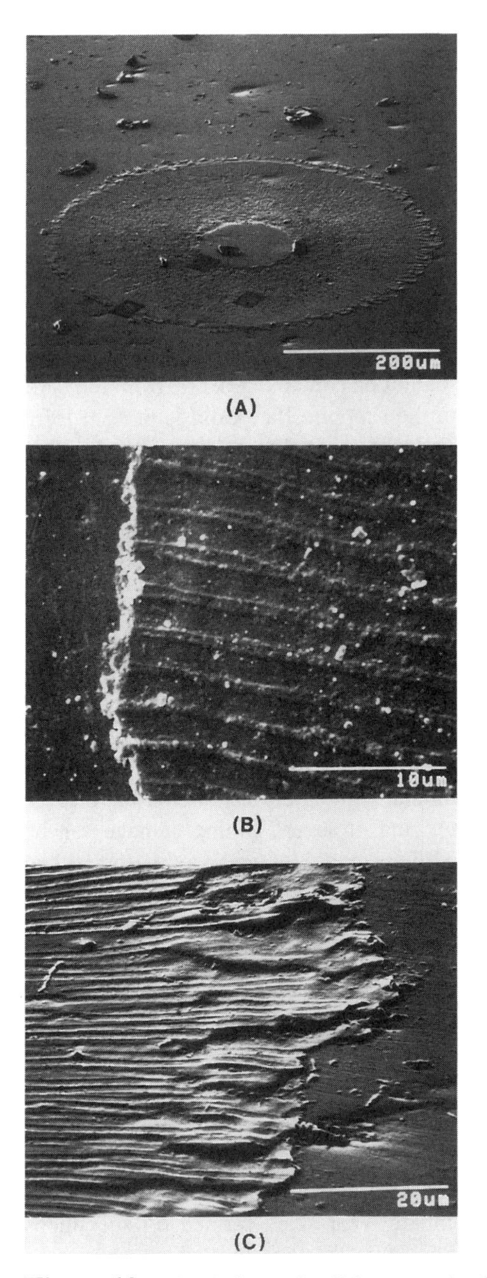

Figure 12 Morphology of radial grooves in slip region of scar in a silicon nitride flat: (A) scanning electron micrograph of entire scar, (B) grooves viewed in normal direction, and (C) grooves viewed at an angle to show topography.

D. Other Material Combinations

Fretting wear couples consisting of combinations of silicon nitride, alumina (polycrystalline and sapphire), and AISI 52100 steel were also studied. With a silicon nitride ball against a sapphire flat at slip amplitudes less than 3 μm, a fine granular wear debris was produced, suggesting that microfracture of the sapphire surface had occurred. At higher slip amplitudes, cracks in the surface were observed. Figure 13 shows micrographs of a typical wear scar formed on a sapphire flat: the groove pattern and a large crack in the slip region in Figure 13A and fracture at the edge of the locked region in Figure 13B. An energy-dispersive X-ray spectrum from the grooved area is shown in Figure 13B. The silicon peak indicates the presence of silicon containing debris from the ball. The aluminum peak is, of course, from the sapphire flat. Gold is present from the sputtered gold coating that was applied to prevent charging. Surface cracking similar to that shown in Figure 13 was not observed with any of the silicon nitride flats.

Results obtained by using a polycrystalline alumina ceramic ball against a sapphire flat were entirely different. In this case, the surface of the sapphire flat was gently worn away, producing fine debris and exposing what appeared to be a crystallographically related surface structure. Surface cracking was not observed.

For the silicon nitride–AISI 52100 steel combination, the wear mechanism involved the transfer of metal to the ceramic, as seen in Figure 14. Figure 14A is an overall view of the wear scar formed on an AISI 52100 steel ball, showing the locked region at the center and the annular slip region. Within the slip region, an inner ring of mild wear and an outer ring of severe galling damage can be distinguished. The original surface finish is essentially unchanged in the locked region (Fig. 14B). Details of the inner, mild wear ring in the slip region appear in Figure 14C. Figure 14D is a micrograph of the wear scar formed on the silicon nitride flat; transferred metal can be seen on the silicon nitride flat corresponding to the galled region on the steel ball. Damage to the silicon nitride flat was less than that to the steel ball.

IV. TYPE OF MOTION

The effect of type of relative motion on surface damage is illustrated in Figure 15. In this example the couple consisted of a silicon nitride ball and an AISI 52100 steel flat. The micrographs compare wear scars formed on specimens subjected to a rolling motion (Fig. 15A, B) with scars formed on specimens subjected to the twisting motion studied earlier (Fig. 15C, D). The rolling motion was achieved by rocking the flat specimen back and forth at 210 Hz against the stationary ball. It can be seen that the damage to the flat in rolling is very different

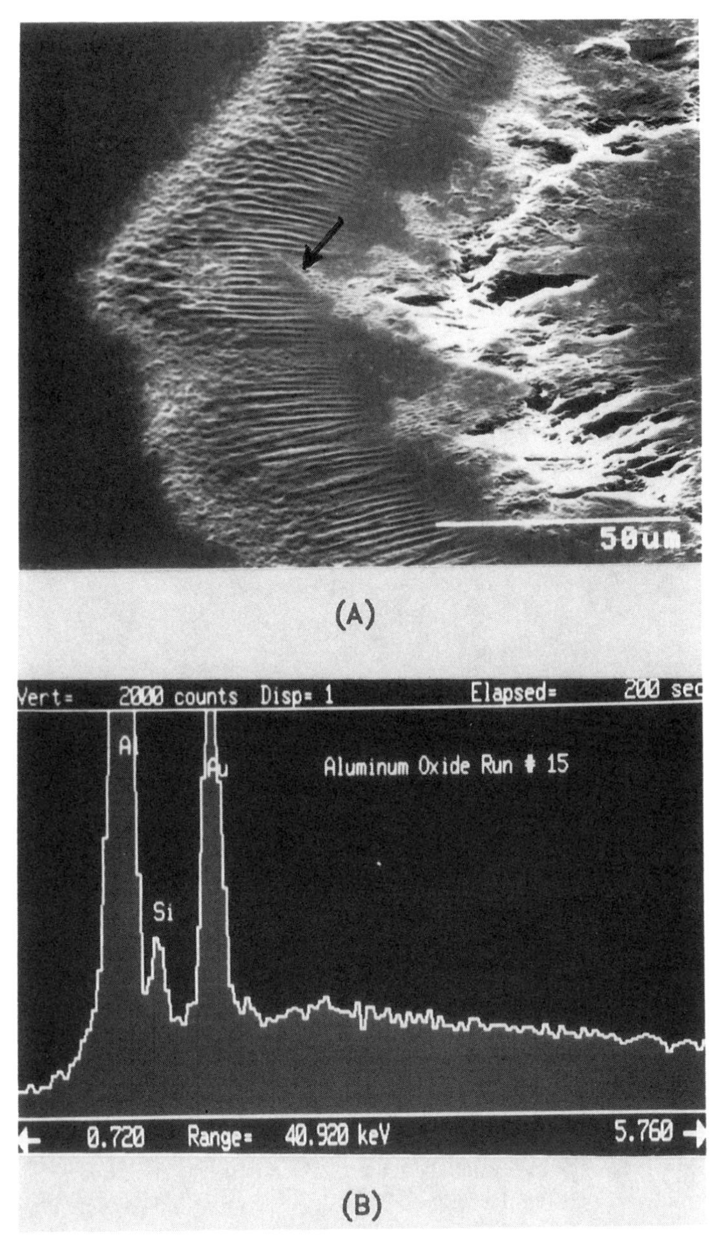

Figure 13 Wear scar formed on a sapphire flat after fretting by silicon nitride ball: (A) large crack (arrow) and groove pattern in slip region, and (B) energy-dispersive X-ray spectrum from debris layer in grooved region.

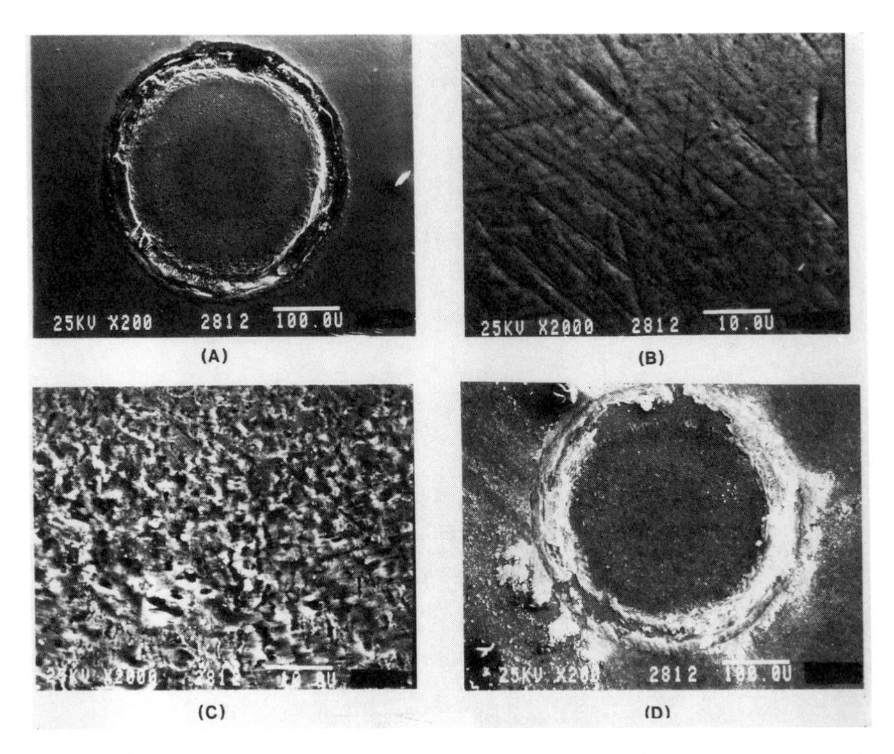

Figure 14 Mating wear scars formed on an AISI 52100 steel ball and silicon nitride flat: (A) scar on 52100 steel ball showing severe damage in outer part of slip region, (B) locked region on 52100 steel ball showing absence of damage, (C) the inner part of the slip region exhibiting mild wear, and (D) scar on silicon nitride flat with ring of metal transfer.

from the damage caused by slip. In rolling, the damage region consists of a multitude of pits (Fig. 15B). Twisting, as seen above, results in galling damage (Fig. 15C, D). The pits associated with rolling are probably due to fatigue. The same type of pitting occurs in rolling contact bearings and gears.

V. CONCLUSIONS

This study of fretting damage in ceramic contacts showed that the type of damage depends on the material and the material combination, and that the damage is a nonlinear function of both slip amplitude and number of cycles. Above a certain amplitude, damage occurred regardless of the time in contact. Below this critical amplitude, the slip region exhibited a thin oxide film without significant surface damage.

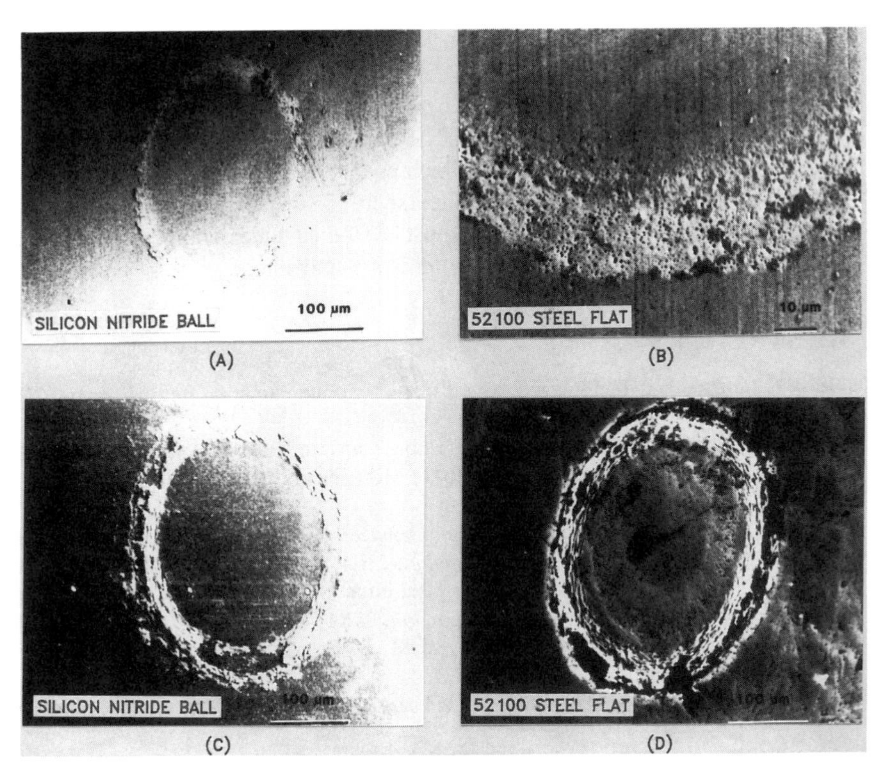

Figure 15 Damage due to rolling (A and B) compared to that caused by twisting motion (C and D) for silicon nitride ball against AISI 52100 steel flat.

Silicon nitride–silicon nitride formed a surface film in the slip region which, under the fretting action, was extruded to the outer edge of the contact. Significant wear began at slip amplitudes of 2–3 μm with a wear rate of 2×10^{-16} $m^3/N \cdot m$. This result is about two orders of magnitude *less* than that reported for sliding wear [11]. The discrepancy is probably due to the protection provided by the film trapped within the contact area. An unusual type of damage pattern was found which consisted of small radial grooves extending from the locked region to the edge of the contact. The groove spacing was a function of the slip amplitude.

For silicon nitride balls on sapphire flats, brittle fracture and cracking of the flat occurred at large slip amplitudes. At smaller slip amplitudes, wear occurred by microfracture within the contact area, which produced powdered wear debris. When a polycrystalline alumina ball was used, only powdered debris was produced, with no evidence of cracking of the sapphire flat.

For steel against ceramic, metal transfer to the ceramic dominated the wear process. Transfer led to a situation characterized by contact that was the same as it would have been if a steel–steel combination had been used.

Unfortunately, generalizations are hard to make. Each material combination behaves in its own individual manner. The events that occur, and their sequence, are a function of preceding events, which change the nature of the surface and the operating conditions at the interface such as stress and temperature. This means that each material combination must be studied under a range of conditions to ensure that its wear and damage patterns are understood.

REFERENCES

1. N. Ohmae and T. Tsukizoe, The effect of slip amplitude on fretting, *Wear*, *27*, 281–294 (1974).
2. O. Vingsbo and S. Soderberg, On fretting maps, in *Wear of Materials—1987*, K. C. Ludema (Ed.), American Society of Mechanical Engineers: New York, 1987, pp. 885–894.
3. K. L. Johnson, Surface interaction between elastically loaded bodies under tangential forces, *Proc. R. Soc. London*, *230A*, 531–548 (1955).
4. J. S. Courtney-Pratt and E. Eisner, The effect of tangential force on the contact of metallic bodies, *Proc. R. Soc. London*, *238A*, 529–550 (1957).
5. P. J. Kennedy, S., J. Calabrese, and M. B. Peterson, Microdamage in sliding contacts, Wear, *121*, 223–238 (1988).
6. G. A. Tomlinson, The rusting of steel surfaces in contact, *Proc. R. Soc. London*, *115A*, 472–483 (1927).
7. J. Sato, M. Shima, J. Igarashi, M. Tanaka, and R. B. Waterhouse, Studies of fretting, Part I, *J. JSLE, Int. Ed. 3*, 15–19 (1982).
8. M. B. Peterson, B. F. Geren, E. B. Arnas, S. Gray, S. F. Murray, J. W. Lund, and F. F. Ling, Analytical and experimental investigation of gas bearing tilting pad pivots, Mechanical Technology Incorporated, NASA CR 72609 (MTI TR-32) (1969).
9. P. J. Kennedy, M. B. Peterson, and L. Stallings, A study of surface damage at low amplitude slip, *ASLE Trans.* *27*(4), 305–312 (1984).
10. Y. Berthier, C. Colombie, L. Vincent, and M. Godet, Fretting wear mechanisms and their effects on fretting fatigue, *Trans. ASME*, *J. Tribol.* *110*(3), 517–524 (1988).
11. T. E. Fischer and H. Tomizawa, Interaction of tribochemistry and microfracture in the friction and wear of silicon nitride, *Wear*, *105*, 29–45 (1985).

6

Abrasive Wear of Ceramics

Jorn Larsen-Basse

National Science Foundation
Washington, DC

ABSTRACT

Abrasive wear of ceramics is controlled by several different mechanisms, depending on the relative mechanical properties of the abrasive and the run-in ceramic surface. These mechanisms are inferred from analogy with cemented carbides and cermets, which have been studied more extensively. They include plastic deformation, surface and subsurface brittle cracking, and "pothole" damage at edges, defects, and other points of stress concentration. Experimental results from abrasion tests on a number of different alumina and silicon nitride compositions with 100 μm silicon carbide and quartz abrasives and with 1 μm diamond particles are used to illustrate qualitatively the effects of hardness, fracture toughness, and microstructure of the material being abraded, and the effect of the type of abrasive used. Results are compared with data for cemented carbides and cermets.

I. INTRODUCTION

A. Definitions and Classifications

Abrasion has been defined as a process in which hard particles or protuberances are forced against a solid surface and move along it [1]; the resulting removal of material from the surface is termed "abrasive wear." The hard particles are the "abrasives." The word comes from Latin *ab-* (= off) + *radere* (= to scrape).

The definition does not specify that the hard particles be harder than the solid, but that seems to be an implicit assumption. One may wonder, therefore, if abrasion could be a problem in the normal use of advanced ceramics, since these materials are much harder than almost all the naturally occurring abrasives. In the following brief discussion we hope to show that this traditional view, which has been derived from studies of abrasion of ductile metals, does not fully hold for hard, fairly brittle materials, for which abrasion can, indeed, be a significant cause of wear. It should be pointed out that in many practical situations that are classified as abrasive conditions, the ceramic material may actually be exposed to a combination of impact and abrasion. An example is a tool insert used in interrupted cuts. In these cases the ultimate failure is usually by gross fracture, and toughness often is as important as abrasion resistance. We do not address those conditions here; we deal only with abrasive interactions without gross impact.

Abrasion processes can be classified in a number of different ways. One often distinguishes between two-body abrasion and three-body abrasion. In two-body abrasion, the particles are fixed like the teeth in a file or the grit on an abrasive paper, and they attack the solid surface like a number of minute cutting tools. In three-body abrasion, the abrasives are loose third bodies caught between two surfaces that are moving relative to each other. For metals, it is generally found that wear rates in two-body abrasion are significantly greater than in three-body abrasion [2]. However, the mechanisms of material removal, to the extent that they are known, seem to be quite similar in both cases, involving heavy plastic deformation of the surface layers, formation of minute machining chips, and extrusion of metal into ridges along the edges of grooves and craters [3].

Very little research has been published on the abrasive wear of ceramics. It is helpful, therefore, to refer to relevant findings for similar materials, especially cemented carbides and cermets which, because of their use in rock drilling and similar abrasive applications, have been the subjects of considerable study.

B. "Hard" and "Soft" Abrasives

According to conventional wisdom, the harder a material is, the less it will wear in abrasive situations. Kruschov and Babichev [4] showed in pin-on-disk (two-body) tests against silicon carbide (SiC) abrasive papers that in most cases the abrasion resistance—defined as the inverse volumetric wear rate—is directly proportional to hardness (Fig. 1). The proportionality factor is greater for metals than for minerals, and this difference was attributed to the partially brittle behavior of the minerals. When the abrasive and the solid are similar in hardness, there is a more pronounced change in wear rate. Nathan and Jones [5] confirmed this long ago when testing a wide range of abrasives and metals; they consistently found a large drop in wear rate as the ratio of the hardness of the abrasive H_a to the hardness of the surface being abraded H_m fell below about 1.2. Tabor [6] noticed that the same value is the ratio between indentation hardnesses of adjacent minerals on the Moh scratch hardness scale, on which each of the 10 minerals on the scale is able to scratch those below it and to be scratched by those above it. It was concluded that hard abrasives cause high wear rates because they can indent the surface and cut grooves in it. Softer abrasives are unable to indent or cut and must remove material by much less efficient mechanisms. The transition between the high wear caused by "hard" abrasives and the low wear caused by "soft" abrasives is, then, a relative effect, and the terms "relatively hard" and "relatively soft" abrasives have been used. The transition generally takes place in a band of H_a/H_m values centered at or near the above-mentioned ratio of approximately 1.2.

Figure 2 shows typical hardness values for some common abrasives and structural materials. In this simple presentation the bulk hardness values have been used, while it would be more correct to use the surface hardness or, rather, the average hardness that an indenting abrasive will see. The two scales are adjusted to the 1.2 ratio such that a given abrasive is hard relative to any structural material to the left of it and soft relative to those to the right. Thus, for example, quartz, the most abundant and ubiquitous naturally occurring abrasive, is hard relative to most metals, including hardened steels, but it is soft relative to most advanced tool insert materials and ceramics. SiC is a hard abrasive for tungsten

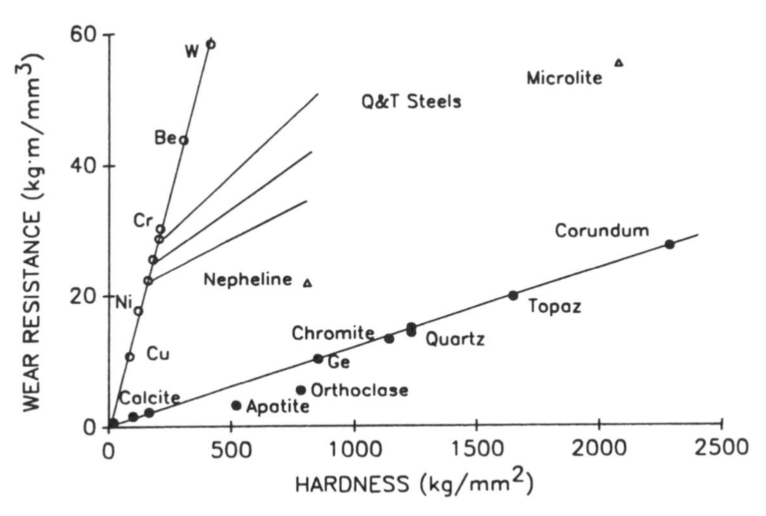

Figure 1 Abrasion resistance versus bulk hardness for pin-on-disk abrasion by SiC papers [4].

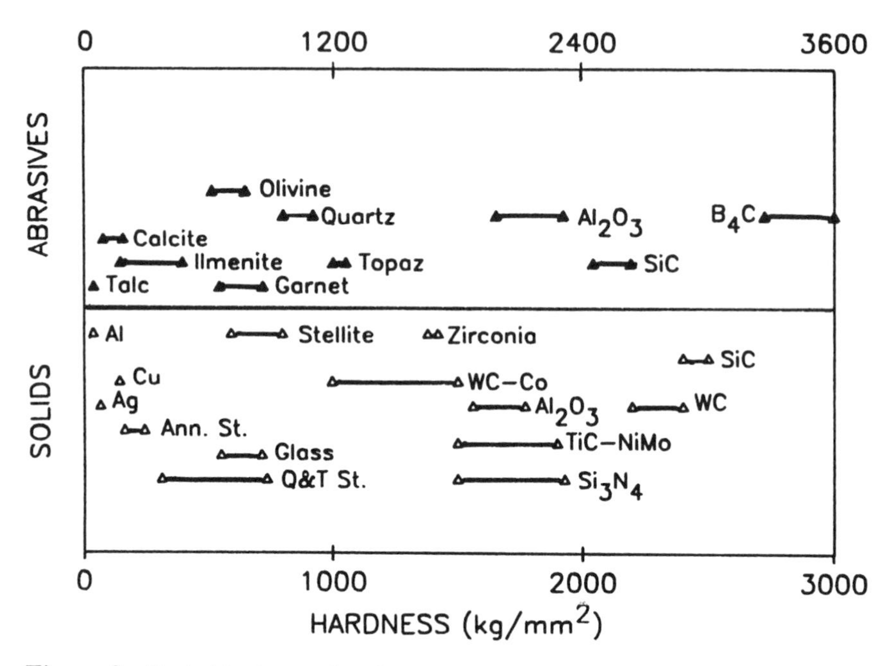

Figure 2 Typical hardness values for some common abrasives and structural materials.

carbide–cobalt (WC-Co) composites, but not for the individual WC grains; and boron carbide is a hard abrasive for almost all materials except diamond. At the other end of the scale, calcite is a relatively hard abrasive for metals such as silver, but a relatively soft one for cold worked copper and anything harder.

One may conclude from Figure 2 that abrasive wear of structural ceramics is not significant because these materials are much harder than most common abrasives, certainly the naturally occurring ones. This conclusion assumes that the wear of ceramics by soft abrasives is very low, as it is for metals. The assumption is not entirely correct, however, as can be inferred from the results shown in Figure 3 [7]. It is noticed that all the various cemented carbide compositions and hardened and annealed steels show the above-mentioned transition at $H_a/H_m = 1.2$. However, while the wear rate below this point is near zero for metals, it is significant for the WC-Co composites, increasing as the cobalt content decreases and the toughness consequently drops. The primary reason is that the surface forces associated with even relatively soft abrasives may be sufficient to cause brittle microfracture at points of stress concentration. For ceramics, this mechanism of wear may be quite significant. The wear mechanisms for the two different hardness regions, to the extent that they are known at this stage, are discussed briefly below.

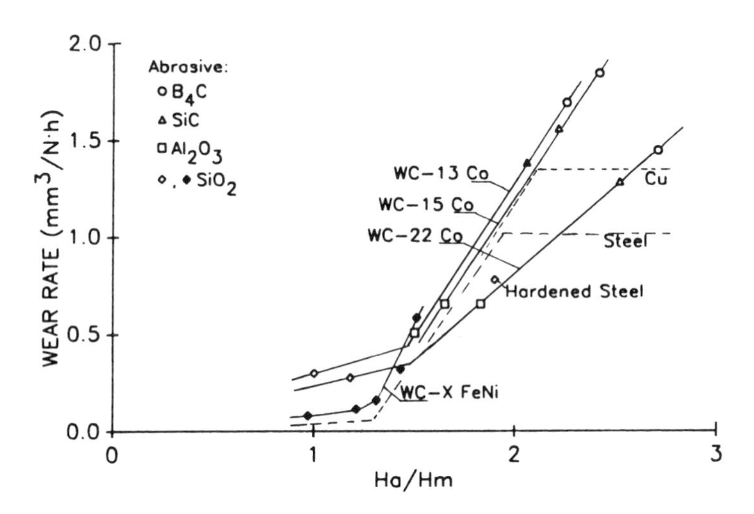

Figure 3 Wear rate of some cemented carbides in three-body abrasion by various abrasives of 100 μm diameter [7]. Extrapolated values for copper and annealed steel included for comparison.

II. ABRASIVE WEAR MECHANISMS

A. The Abrasive–Surface–Environment System

In discussing the various removal mechanisms that operate during abrasion, it is useful to review the general aspects of the abrasive grit's interaction with the surface (Fig. 4). An early model of the wear by hard abrasives compares the grit to a hardness indenter, which forces a groove to form [2], as in Figure 4A. If the material is ductile, it will form small machining chips at the front edge of the grit and will also be extruded to the edges of the grooves. The surface layers eventually become severely work hardened.

For materials with relatively limited ductility, such as cemented carbides of the WC-Co family, very similar wear mechanisms operate, except that the removed machining chips usually are in the form of agglomerates of small fragments of material [8]; there also may be some instances of brittle fracture removing small spalls of material in regions near the indenting abrasive grit [9]. In addition, the surface damage may include tensile cracks, which form under the sliding indenter (Fig. 4B).

When the hardness of the abraded material increases and moves closer to the hardness of the abrasive, it becomes necessary to consider the behavior of the system as a whole, not just as an invariant indenter pushing into a softer surface. Now, the indenter also deforms as a result of the load, and if the ratio of groove depth to grit diameter is below a certain value, there will be no material removed

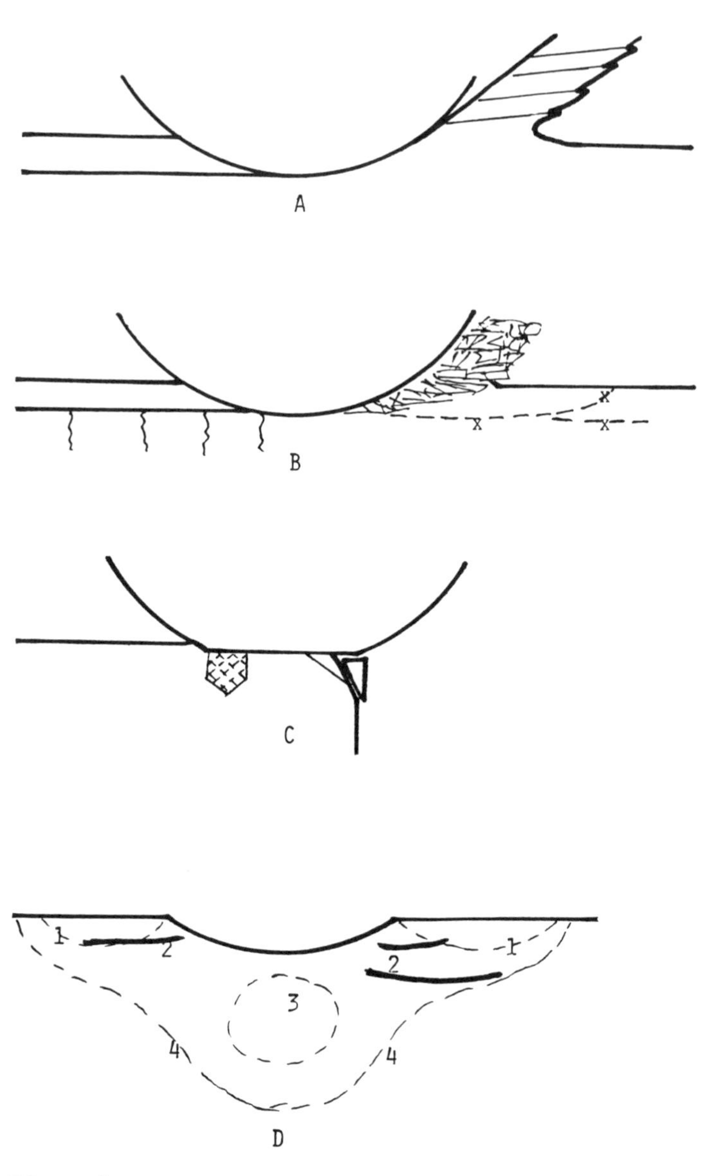

Figure 4 Sketches of various abrasive–surface interactions: (A) ductile metal–hard abrasive, (B) low ductility composite–hard abrasive (x = additional intergranular subsurface crack possibly expected for some ceramic materials), (C) low ductility composite–soft abrasive, near-surface irregularity, and (D) cracks formed at indentation in brittle material: (1) radial cracks, (2) lateral cracks, (3) median crack, and (4) crack front after unloading [15].

by plastic deformation. Then the abrasive, while considered to be hard according to the simple $H_a/H_m = 1.2$ rule discussed above, may in fact behave as a soft abrasive. This type of behavior may be seen, for example, when the operating conditions result in low loads on large abrasives [7].

When the abrasive is "relatively soft," it can no longer indent the surface (Fig. 4C). It slides over the surface as an asperity, which exerts frictional forces on the surface material. These forces result in surface deformation. For metals they result in a slow accumulation of strain and in the displacement of hard second-phase particles, such as carbides in steels. Accumulation of strain can lead to opening of low cycle fatigue cracks [10]. For metals, the wear in this region is low, especially in comparison to the wear due to hard abrasives, but it is nevertheless real: see Figure 3. For less ductile metals, the frictional forces can result in removal of fragments by brittle fracture. For example, for cemented carbides abraded by quartz, it was found [11] that the soft abrasives tend to pull off whole carbide grains at edges and defects in a "pothole" type of damage, resulting in gradual lateral growth of these defects in the "upstream" direction (Fig. 4C). It also appears that individual carbide grains may fragment into very small particles similar in size to the expected dislocation cell structure in the material [3], possibly spontaneously as a form of elastic strain energy release. From this, one would expect brittle materials to wear relatively more than metals in this region, as is generally confirmed by the available results: see, for example, Figure 3.

The H_a/H_m ratio is, as mentioned, only a general indicator of the abrasive grit's ability to indent a surface. Directional properties of both solid and abrasive can result in quite different hardness levels for different crystallographic directions that are not included in the general hardness values. Abrasive shape also plays an important role: sharper abrasives are obviously able to indent surfaces more readily than a more rounded grit. In this connection it is also worth recalling that abrasives, like most other ceramic materials, have mechanical properties that are very sensitive to the environment, especially to the presence of polar molecules, specifically water vapor or liquid. For SiC and alumina (Al_2O_3) abrasives, water vapor facilitates crack propagation which, under the right conditions, may result in self-sharpening of the abrasives during their contact with the surface [12]. The result can be increased wear. On the other hand, for an abrasive such as quartz, which is extremely sensitive to water-assisted crack propagation, the presence of moisture may also facilitate crushing of the abrasives, in which case the wear rate would decrease.

B. Expected Removal Mechanisms in Abrasion of Ceramics

Some qualitative ideas of the possible abrasive wear mechanisms for ceramics can be derived from the foregoing general discussion. At present, insufficient information is available to permit development of more quantitative relations. Clearly, however, hardness and toughness are important properties in this connection.

Figure 2 shows that all naturally occurring abrasives are soft relative to advanced ceramics and, therefore, will abrade them by the "soft" mechanisms. That includes particles of the abundant quartz, which account for 60% of the Earth's crust and is present in sand, dust, ash, etc. Therefore, most of the anticipated abrasive wear problems in real life will fall in the "soft" region for advanced ceramics. "Hard" abrasion wear mechanisms could be due to particles of industrial abrasives left over from manufacturing processes, if they are harder than the component's surface. Also, wear particles from upstream components that are harder or contain hard particles (e.g., WC grains from WC-Co alloys) can cause "hard" abrasion, as can same-material wear particles if they have significant directional properties. So, while most abrasion probably will be in the soft mode, the hard mode will not be completely insignificant. One might also expect the mechanisms controlling material removal in diamond finish grinding to be some of the same hard mode mechanisms when the effects of high temperatures and velocities and of cooling fluids are allowed for.

The toughness effect will determine how great a role brittle fracture will play in the material removal. Toughness is important both in the hard mode, where cracks may form below and around indentations and at the trailing edge of a sliding indenter, and for the soft mode, where frictional forces may generate cracks in material at edges and defects. Attempts to quantify the effects of fracture toughness have not been successful, but the effects of toughness may be understood qualitatively, as further discussed below.

C. "Hard Abrasion" Mechanisms

The size and shape of an indentation and the local toughness determine the deformation pattern, obviously. For very shallow indentations, no cracks form but there is plastic deformation under the indenter. This regime is the same as the "ductile grinding" regime used in finishing operations of ceramics [13] and quartz wafers [14]. In this region, flow stress and/or low cycle fatigue strength at high strain and strain rates should control the wear resistance, as they do for metals [3]. For larger indentations and/or more brittle materials, cracks of various types form: first the radial cracks to the side of the indentation, then the lateral cracks running under the surface; finally, for large indentations of brittle materials, the deep median cracks (Fig. 4D).

A number of wear models have been developed based on the assumption that all the material subjected to cracking is removed and also based on the implicit assumption that the abrasives are so hard that their deformation is negligible (see, e.g., Refs. 15, 16). These models tend to be complicated expressions. For example, in

$$\dot{V} \propto K_c^{-3/4} H^{-1/2} P^{5/4} \tag{1}$$

from Reference 15, $\dot{V}$ is the volume loss per unit distance of sliding, K_c is the fracture toughness, H is the indentation hardness, and P is the applied load. A somewhat similar expression is proposed in Reference 16:

$$\dot{V} \propto P^{1.125} K_c^{-0.5} H^{-0.625} \left(\frac{E}{H}\right)^{0.8} \tag{2}$$

where E is the elastic modulus. Unfortunately, it has not been possible to validate these models by experiments, as discussed by Ajayi and Ludema [17]. The effects of local toughness are real and can be seen qualitatively in abrasion of series of WC-Co alloys [7] and cast, wear-resistant alloys [18]. However, Equations (1) and (2) clearly overstate the effects of brittle fracture. Possible reasons are that not all the cracked material is removed; that the cracks in the surface material can relieve some stresses and, thereby, make the surface more resistant to the damage from subsequent indenters; and that the macroscopic properties of undamaged material used in modeling may not well represent the microscopic events in severely damaged material that constitute wear.

Many structural ceramics are fine grained and have weak grain boundaries because, for example, of the presence of sintering aids or microporosity. Therefore, the cracks around sliding indenters may not be as neatly linear as sketched in Figure 4D, but may follow grain boundaries, instead. For certain conditions, one might expect the stress wave ahead of a moving hard indenter to be able to force subsurface cracks to form parallel to the surface (Fig. 4D) and thereby create an additional submechanism of wear.

D. "Soft Abrasion" Mechanisms

One expects in-surface forces—the friction forces from nonindenting abrasive asperities—to dominate in the "soft abrasion" mode. Most of the damage will occur at weak spots in the surface, such as the partially supported material at edges, grinding scratches, and pores. This "pothole" effect will spread in a counterflow or upstream direction, as mentioned above. The phenomenon has not been studied in much detail, but one would expect the state of the initial surface to have a very large role in determining the onset of damage. In addition to the edge effects, there might be damage from contact fatigue. In addition, for WC grains in WC-Co composites it was noticed [3] that the individual grains tended to fracture in small fragments (the size of expected dislocation cells), and it was suggested that brittle grains might fragment spontaneously as a strain energy release mechanism once a certain threshold of stored elastic strain energy has been reached. If this mechanism is real for WC-Co, it probably would be found for ceramics also.

E. Other Considerations

It is important to consider the whole system. Not only does the environment affect the behavior, as discussed above, but it should be recalled that abrasives themselves wear. Thus, for example, in tests of long samples of hard WC-Co alloys, SiC abrasives under two-body conditions would act as hard abrasives at the leading edge but, because they wore and were blunted during the passage, they could begin to behave as soft abrasives near the trailing edge [7].

III. SOME EXPERIMENTAL RESULTS

To underscore some of the discussion above, it is informative to examine some experimental results, which illustrate some of the preceding points. The results are combined from several studies [11,19–22].

A. Experimental Details

The tests were three-body abrasion of alloys in the WC-Co family by SiC [19] and by SiO_2 [11]; abrasion of WC-FeNi by SiC [20]; abrasion of (TiC-*X*% TiN)/12Ni 10Mo 10VC by SiC, SiO_2, and diamond [21]; and abrasion of various alumina and silicon nitride (Si_3N_4) materials also by SiC, SiO_2, and diamond polish [22]. The abrasives were 80 μm SiC, 100 μm sharp particles SiO_2, and 1 μm diamond polish. The coarse abrasives were supplied to the contact area between a 55 mm diameter, 1.5 mm wide steel wheel (annealed SAE 1020) and the 2 mm wide specimen surface. The wheel operated with a surface velocity of 0.085 m/s. For the SiC abrasives the applied load was 10.9 N; 10 minutes running time was sufficient to generate significant wear. For the quartz abrasives, most tests were done at 4 or 5 N applied load because higher loads tended to crush the abrasives. It was found that abrasion by quartz would not readily initiate on a smooth surface. Therefore, the quartz tests were conducted in the wear scars from the SiC tests. The weight loss was measured and converted to volumetric wear rates.

In the tests with the diamond polish, a 19 mm hardened steel ball was supported between the vertical sample surface and the rotating conical tips of two adjoining coaxial shafts. The ball was covered with 1 μm diamond paste before each test and the paste was replenished at intervals. The scar size was measured by optical microscopy after 30 minutes of running.

The three different abrasives were chosen to represent expected different abrasion mechanisms: SiC as a hard abrasive, SiO_2 as a relatively soft one, and fine diamond particles as possibly giving ductile regime deformation by a hard abrasive.

Before each test the sample surface was polished with 9 μm diamond polish. Table 1 shows typical ranges of hardness and fracture toughness K_{Ic} for the various materials groups, and Table 2 gives some more details on the ceramic materials.

Table 1 Typical Property Ranges for the Materials Studied

Material	Vickers Hardness (kg/mm^2)	K_{Ic} ($MPa \cdot m^{1/2}$)
WC-Co/WC-FeNi	850–1600	16–10
TiC/NiMo; (TiC + TiN)NiMoVC	1500–1870	11.3–7
Si_3N_4	1530–1900	5.4–4.3
Al_2O_3	1550–1760	5–3

Table 2 Processing and/or Composition Details of Ceramic Materials

Material	Vickers Hardness (kg/mm^2)	K_{Ic} ($MPa \cdot m^{1/2}$)
(TiC + X% TiN)/(12.5 Ni + 10 Mo + 10 VC)		
X = 11%	1630	9.2
X = 20%	1685	9.2
X = 30%	1735	8.4
X = 40%	1813	7.9
X = 60%	1870	6.8
TiC-NiMo (X = 0)	1505	11.3
Al_2O_3		
High purity, sintered	1562	
Transformation strengthened	1680 and 1755	
Si_3N_4		
a. sintered; Binder: 8 Y_2O_3 + 5 W + 4 Al_2O_3	1895	
b. HP-RBSN; binder: 8 Y_2O_3 + 1 Al_2O_3	1806	
c. HP; binder: 1 MgO + 1 Al_2O_3	1700	
d. HP; binder: 5 CeO_2 + Al_2O_3	1705	
e. HP; binder: 8 Y_2O_3, amorphous	1525	
f. HP; binder: 8 Y_2O_3, crystalline	1625	
g. HP; binder: 1 MgO + 1 Al_2O_3	1680	

B. Results

Table 3 shows typical ranges of wear rates, normalized for load. It is seen that the SiC abrasive gives wear rates 10–30 times greater than the wear by SiO_2, at low load and 100–150 times greater than the wear by the diamond polish. The wear rates by SiO_2 at 4 N are 50–70 times greater than those by the same abrasive at 26 N load because of the above-mentioned crushing of the quartz at the higher loads.

The results are plotted in Figures 5–8. The results for the hard abrasive SiC (Fig. 5) indicate that wear rates generally decrease as the hardness increases and tend toward zero around the expected point of about 85% of the hardness value for SiC. The curve for each family of alloys generally lies at lower wear rates for higher K_{Ic} values. This is not quite the case for the Si_3N_4 family compared to the titanium carbide alloys. The latter have K_{Ic} values in the range 7–11.3 $MPa \cdot m^{1/2}$ and the former in the range 4.3–5.4. Yet, the Si_3N_4 materials have somewhat better wear resistance than the tougher TiC alloys at the same hardness. Also, the alumina samples have only slightly lower toughness than the silicon nitrides but significantly lower wear resistance under these conditions. Clearly, the silicon nitrides have quite good resistance to abrasion by hard abrasives in comparison to other materials of similar hardness. It can also be concluded that abrasion resistance in this region is not determined by hardness and fracture toughness alone. Additional factors, probably related to the fracture path in the microstructure, must be included.

Figure 6 shows results for abrasion by the fine diamond particles. The general ranking of Si_3N_4 versus the TiC alloys is the same as for SiC, but the aluminas now have wear resistance similar to the silicon nitrides for the pure alumina, and similar to the TiC alloys for the transformation-strengthened material. It is unclear why the curves show a peak in wear rate at a hardness of about 1700 DPN and tend to zero near 2000. Perhaps the interaction between the backing ball and the diamond particles must also be taken into account in this region. In comparison to the SiC data, the hardness has a much more significant influence on wear rate, while fracture toughness seems to play a very minor role. This is as expected, since the 1 μm diamond particles most probably remove material by

Table 3 Typical Abrasive Wear Rates

Abrasive	Wear rate ($mm^3/N \cdot h$)
80 μm SiC	1.5–4.5
100 μm quartz, low load (4 N)	0.05–0.5
100 μm quartz, high load (26 N)	0.001–0.0015
1 μm diamond	0.01–0.04

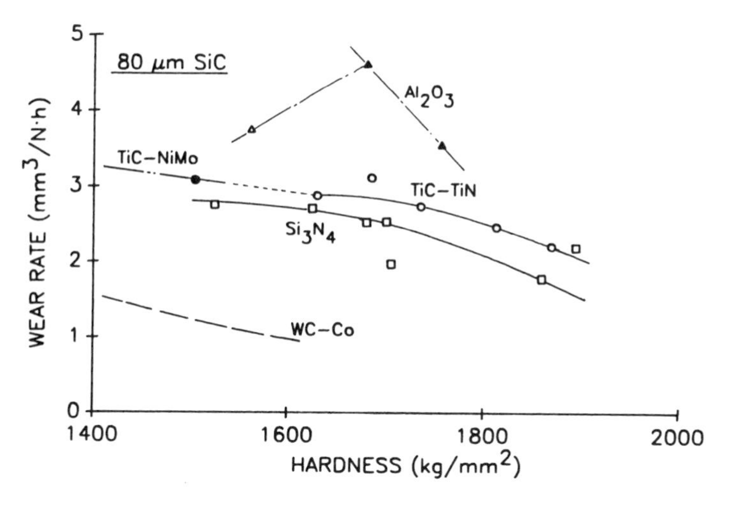

Figure 5 Wear rates versus bulk hardness in three-body abrasion by SiC.

a ductile regime mechanism. The values for abrasion by SiO_2 at high load are included for comparison. They are much lower than the wear rates by the diamond polish.

The data for SiO_2 abrasives are shown in Figure 7 for a broad range of materials and on an expanded scale for the ceramic materials alone in Figure 8. The cemented carbides of the WC-Co and WC-FeNi families have lower wear rates, probably

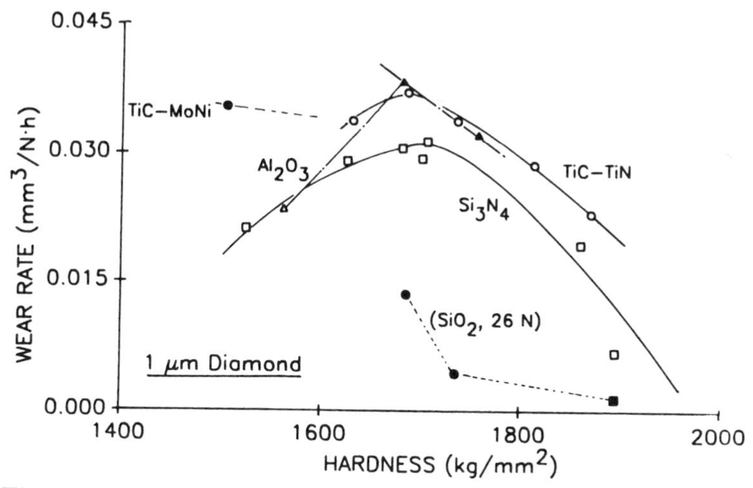

Figure 6 Wear rates versus bulk hardness in three-body abrasion by 1 μm diamond polish. Results from abrasion of three of the compositions by 100 μm quartz at high load (26 N) included for comparison.

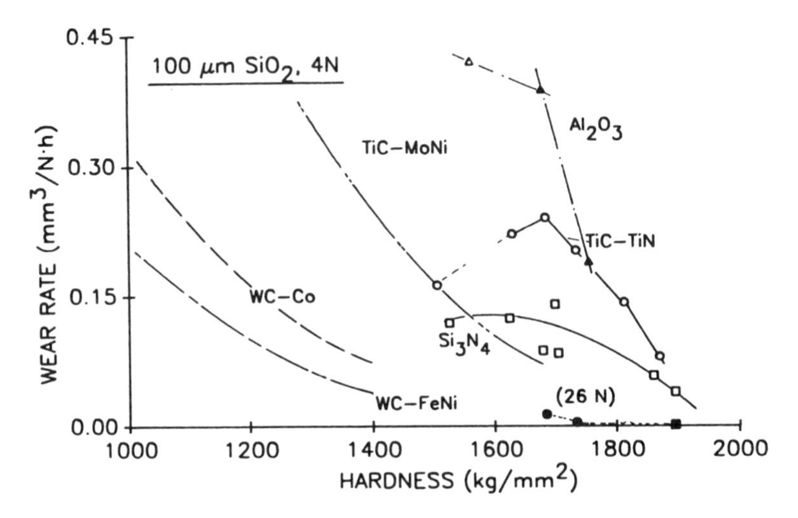

Figure 7 Wear rates versus bulk hardness in three-body abrasion by SiO_2, low load (4 N). A few values obtained at high load (26 N) included for comparison.

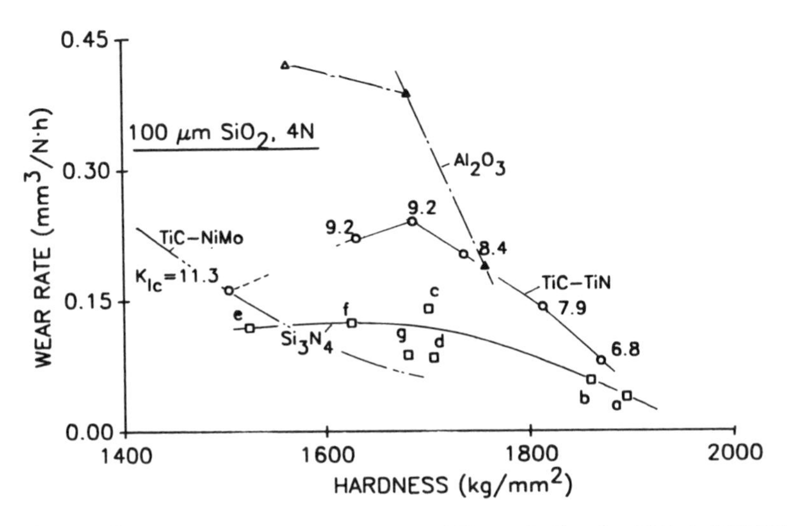

Figure 8 Expanded high hardness part of Figure 7. For the (TiC + *X*%TiN)/12.5Ni 10Mo10VC compositions the values of K_{Ic} are indicated. For the Si_3N_4 compositions, the specimen notations from Table 2 are indicated.

because of their greater ductility, but their hardness range does not extend as far. For the harder materials the curves all tend toward zero around 1800–1950 kg/mm^2. Near this hardness value the influence of K_{Ic} decreases, probably because the hardness difference between abrasive and solid becomes so great that the contact stresses, whose maximum values are dictated by the strength of the quartz abrasive, become so low relative to the fracture stress of the solid that the contact zone generates only a few points of significant stress concentration in the solid, with local stresses sufficient to cause material removal. The TiC-MoNi family has better wear resistance than the other materials. Increasing the hardness of this family of alloys by TiN substitutions in the matrix (to form titanium carbonitrides) and VC in the binder material results in greater hardness values but also increased wear rates. The alumina with the lowest hardness, which is the sintered material, has much lower wear resistance than the transformation-strengthened material, more so than expected from their differences in typical K_{Ic} values. A possible explanation is that the strengthened material may act to stop the microcracks responsible for material removal by the quartz abrasives.

The expanded results in Figure 8 show the complexity of the combined effect of hardness and fracture toughness on wear resistance. On one hand, the effect of fracture toughness is seen for the TiC alloys, where going from the straight TiC-NiMo alloy to (TiC-11TiN)/NiMoVC alloy increases hardness from about 1500 to about 1620 and decreases K_{Ic} from 11.3 to 9.2. At the same time, the wear rate goes up by about 20%, indicating that in this case, it is controlled primarily by fracture toughness. On the other hand, if the hardness of the TiC-TiN alloys is increased by adding TiN, both fracture toughness and wear rate decrease, indicating that in this instance hardness has the greater effect on wear rate. For the silicon nitrides it is noted how much the type of binder and processing technique affect hardness, although these variables have relatively little influence on wear rate, certainly in comparison to the aluminas. This result is most probably due to the general absence of porosity in the Si_3N_4 materials. The highest hardness and best wear resistance in this group are shown by the materials with the greatest amount of binder material, compositions a and b. Compositions c and g are nominally identical and have the same hardness. Yet, the wear rate of c is about 60% greater than the wear rate of g, showing the influence of unspecified processing parameters or unspecified impurities. Compositions e and f differ only by e having amorphous binder material and f having been heat treated to cause the binder phase to crystallize. The treatment increased the hardness from 1510 to 1620 kg/mm^2 but made no difference in wear rate.

IV. CONCLUSIONS

The abrasive wear resistance of ceramic materials is determined by a combination of factors:

1. The properties of the abrasive, primarily its hardness relative to the ceramic surface and its crushing strength, or friability.
2. The size of the abrasive indentation, which, in combination with the local fracture toughness of the surface material, determines whether material removal is in the ductile regime or includes some microcracking.
3. The local properties of the microstructural components of the ceramic surface and near-surface material, including the presence of pores, weak grain boundaries, and transformation-strengthened regions.

Hardness and fracture toughness give only very general indications of a material's resistance to abrasive wear but can be used, in combination with information about the microstructure, in the qualitative prediction of general levels of abrasion resistance.

ACKNOWLEDGMENTS

The experimental work discussed in this chapter was performed while the author was affiliated with the University of Hawaii. It was supported, in part, by the National Science Foundation. The opinions presented are solely those of the author, not necessarily those of the NSF.

REFERENCES

1. Organization for Economic Cooperation and Development, Research Group on Wear of Engineering Materials, *Glossary of Terms and Definitions in the Field of Friction, Wear and Lubrication (Tribology)*, OECD: Paris, 1969.
2. E. Rabinowicz, *Friction and Wear of Materials*, Wiley: New York 1965, Chapter 7.
3. J. Larsen-Basse, Role of microstructure and mechanical properties in abrasion, *Scripta Metal. Mater. 24*, 821–826 (1990).
4. M. M. Kruschov and M. A. Babichev, *Issledovania Iznashivania Metallov*, Akademia Nauk USSR: Moscow, 1960, p. 226.
5. G. K. Nathan and W. J. D. Jones, Influence of hardness of abrasives on the abrasive wear of metals, in *Lubrication and Wear, Fifth Convention, Proc. Inst. Mech. Eng., 181*, Pt. 30, 215–221 (1966–1967).
6. D. Tabor, The physical meaning of indentation and scratch hardness, *Br. J. Appl. Phys. 7*, 159–166 (1956).
7. J. Larsen-Basse, Some mechanisms of abrasive wear of cemented carbide composites, *Metaux, Corros. Ind.* No. 653, 1–8 (January 1980).
8. J. Larsen-Basse, Effect of composition, microstructure, and service conditions on the wear of cemented carbides, *J. Met., 35*(11), 35–42 (1983).
9. J. Larsen-Basse, Effect of hardness and local fracture toughness on the abrasive wear of WC-Co alloys, in *Tribology—Friction, Lubrication and Wear, Fifty Years*, Vol. I, Institution of Mechanical Engineers: London, 1987, pp. 277–282.

10. J. Larsen-Basse and B. Premaratne, Effect of relative hardness on transitions in abrasive wear mechanisms, in *Wear of Materials—1983*, K.-C. Ludema (Ed.), American Society of Mechanical Engineers: New York, 1983, pp. 161–166.
11. J. Larsen-Basse and E. T. Koyanagi, Abrasion of WC-Co alloys by quartz, *Trans. ASME, J. Lubr. Technol., 101*, 208–211 (1989).
12. J. Larsen-Basse, Effect of atmospheric humidity on the dynamic fracture strength of SiC abrasives, *Wear* (in press).
13. K. Subramanian and S. Ramanath, Mechanism of material removal in the precision grinding of ceramics, in *Precision Machining: Technology and Machine Development and Improvement*. M. Jouaneh and S. S. Rangawala (Eds.), American Society of Mechanical Engineers: New York, 1992, pp. 1–19.
14. T. G. Bifano and J. B. Hooker, Precision grinding of ultra-thin quartz wafers", in *Precision Machining: Technology and Machine Development and Improvement*, M. Jouaneh and S. S. Rangawala (Eds.), American Society of Mechanical Engineers: New York, 1992, pp. 21–28.
15. A. G. Evans and T. R. Wilshaw, Quasi-static solid particle damage in brittle solids, I. Observations, analysis and implications, *Acta Metal, 24*, 939–956 (1976).
16. A. G. Evans and B. D. Marshall, Wear mechanisms in ceramics, in *Fundamentals of Friction and Wear of Materials*, D. A. Rigney (Ed.), American Society for Metals: Metals Park, OH, 1981, pp. 439–452.
17. O. O. Ajayi and K. C. Ludema, Surface damage of structural ceramics: Implications for wear modeling, *Wear, 124*, 237–256 (1988).
18. K.-H. Zum Gahr, How microstructure affects abrasive wear resistance, *Met. Progress*, September 1979, pp. 46–52.
19. J. Larsen-Basse and P. A. Tanouye, Abrasion of WC-Co alloys by loose hard abrasives, in *Advances in Hard Material Tool Technology*, R. Komanduri (Ed.), Carnegie-Mellon University: Pittsburgh, 1976, pp. 188–199.
20. J. Larsen-Basse, Abrasive wear resistance of tungsten carbide composites with iron–nickel binder, in *Wear of Materials—1981*, S. K. Rhee, A. W. Ruff, and K. C. Ludema (Eds.), American Society of Mechanical Engineers: New York, 1981, pp. 534–538.
21. J. Larsen-Basse, Abrasive wear of some titanium carbonitride-based cermets, *Mater. Sci. Eng. A, 105/106*, 395–400 (1988).
22. J. Larsen-Basse, Abrasive wear behavior of some hard tool materials, *Eurotrib* (1993) (in press).

III

SOLID LUBRICATION AND COMPOSITES

7

A Review of the Lubrication of Ceramics with Thin Solid Films

Ali Erdemir

Argonne National Laboratory
Argonne, Illinois

ABSTRACT

Advanced structural ceramics are being considered for use in a large variety of mechanical systems that involve high loads, speeds, and temperatures. Development of new solid and liquid lubricants and lubrication concepts is essential for achieving satisfactory tribological performance under these stringent application conditions. This chapter provides an update of the state of the art in lubrication of ceramics, with particular emphasis on solid lubricants and lubrication concepts. A summary of the wear mechanisms of ceramics is followed by a discussion of current approaches to and advanced methods for solid lubrication of ceramics. Finally, recent tribological experiences with solid-lubricated ceramics are presented in greater detail.

I. INTRODUCTION

Increasing demand for higher performance energy conversion and utilization systems has led to growing interest in the friction and wear characteristics of such advanced structural ceramics as alumina (Al_2O_3), silicon nitride (Si_3N_4), silicon

---

Work supported by U.S. Department of Energy, Office of Transportation Materials, under Contract W-31-109-Eng-38.

carbide (SiC), and zirconia (ZrO_2) [1–5]. A major reason for this interest is that conventional tribomaterials often lack the capability to operate efficiently under the stringent tribological conditions encountered by these energy conversion and utilization devices. As a true reflection of the growing interest, the number of publications addressing both the tribology and the application aspects of ceramics has increased dramatically in recent years. In addition to numerous journal articles, several special technical publications [6,7], conference proceedings [8,9], books, and book chapters [10–13] have been devoted to the tribology of ceramics.

Most of the earlier tribological studies concentrated on four structural ceramics: Al_2O_3, Si_3N_4, SiC, and ZrO_2. The friction and wear characteristics of these ceramics have been investigated quite considerably under a wide range of tribological conditions. These studies have resulted in a better understanding of the friction and wear mechanisms of these materials. Such mechanistic understanding is essential for the successful development of large-scale energy conversion and utilization systems.

Besides the structural ceramics, ceramic composites and coatings have been the subjects of increasing interest in recent years. This trend was reflected in a recent conference on the tribological and application aspects of ceramic matrix and metal matrix composites [14]. These high performance materials combine attractive mechanical, thermal, and chemical properties that are critically important under the demanding operating conditions of advanced tribological devices. Ceramic fiber reinforced, metal matrix composites have also been developed in recent years, and their potentials are currently being evaluated for various tribological applications [14,15]. The use of sialon (SiAlON), aluminum nitride (AlN), boron carbide (B_4C), and boron nitride (BN) for tribological purposes was suggested in the past, and some experimental data pertaining to their tribological behavior have recently been presented in the literature [16–18].

Over the years, great strides have been made in the processing, fabrication, and diverse utilization of ceramic and composite materials for a wide range of engineering applications. The state of the art has reached a point that permits the fabrication of intricate machine parts from fine ceramic powders with near-net-shape precision. In keeping with these developments, ceramic prototypes of several tribocomponents have also been successfully manufactured in recent years and are currently being tested in actual tribological environments. While some tribocomponents are still in the exploratory stage and require further refinement, others have been developed successfully and are offered on a commercial scale.

Unique properties that make ceramics very attractive for future tribological applications include high hardness and strength, low bulk density, excellent chemical inertness, and good thermal insulation. Moreover, these properties are largely retained at fairly elevated temperatures.

Low thermal conductivity of ceramics forms the basis for low heat rejection engine concepts currently being explored in various laboratories [4,8,9,11,19–

23]. Some recent feasibility studies indicate that with the realization of ceramic-based, low heat rejection engine systems, billions of dollars worth of fuel will be saved per year in the United States alone [5,24]. Because of their excellent insulating capabilities, partially stabilized ZrO_2 ceramics appear very promising for these applications, which are being explored by major engine manufacturers.

Because of their light weight, high hardness, high elastic modulus, good chemical inertness, and excellent mechanical strength, Si_3N_4 and SiC ceramics are being considered for ball- and roller-bearing applications [5]. Various bearings have already been manufactured from these ceramics and are commercially available. For applications that involve high loads, high speeds, and high temperatures, Si_3N_4 bearings have the potential for use in applications under more extreme conditions (especially higher temperatures) than those which are feasible for steel bearings [25,26]. At present, of course, it must be said that steel bearings are much more reliable in their range of applicability. Indeed, reliability of ceramics in general is a serious problem, which requires much more research and development. Si_3N_4 and SiC ceramics are being considered for use in piston rings and liners, piston pins [4,20–22], intake and exhaust valves, camshafts, cam lobes [4,5,27–30], high temperature seals, and turbocharger rotors [4,5,30]. Prototypes of these components have been fabricated, and their long-term reliability in real application environments is being assessed [30].

Apart from structural and composite ceramics, ceramic coatings have received increasing attention in recent years. Great strides have been made in controlling the thickness, morphology, and adhesion of these coatings. Among others, ZrO_2 and Cr_2O_3 coatings have attracted particular attention for high temperature tribological applications [19]. ZrO_2, because of its excellent insulating properties, is employed as a thermal barrier coating in advanced heat engine systems. The results of recent field tests are quite encouraging. These coatings have been successfully applied to combustion chamber walls of metal-based engine systems; as a result, a substantial increase in fuel economy was achieved [4,5,23]. Cr_2O_3 coatings are primarily applied to control friction and wear at elevated temperatures [19,31].

The primary purpose of this chapter is to provide an overview of the state of the art in lubrication of ceramics with thin solid films. A brief summary of the wear mechanisms of advanced ceramics is followed by a discussion of current approaches and advanced methods for lubrication of ceramics, with special emphasis on solid lubricant films applied on ceramic surfaces by various advanced deposition methods.

II. WEAR MECHANISMS IN CERAMICS

Despite the intense technological interest in using ceramics for diverse tribological applications, their widespread utilization is not expected until the year 2000

[32]. Until now, only ceramic bearings and cutting tool inserts have been developed successfully and made available commercially. Although prototypes of other tribocomponents (e.g., valve trains, cams and followers, and piston rings and liners) have also been fabricated from ceramic materials [4], their large-scale utilization has not yet been realized. Some of the reasons for this are high fabrication cost, insufficient reliability and knowledge base, and, most important, poor friction and wear performance of ceramics under dry sliding conditions.

The reliability of most ceramics for many tribological applications is not yet well established. The results of recent field and laboratory tests are somewhat inconsistent and often discouraging. One exclusive finding from these tests is that under dry sliding conditions, both the friction and wear coefficients of ceramics are too high to be practical for most tribological applications [2,33–39]. The present consensus is that without the development of new liquid and/or solid lubricants and advanced lubrication concepts, large-scale use of ceramics in future tribosystems appears highly unlikely. Accordingly, new lubricants and lubrication concepts are being explored in various research laboratories and tested on ceramic surfaces. Before describing the new lubricants and lubrication methods, a brief summary of the wear mechanisms of ceramics is presented, because the lubricants and lubrication methods currently being explored are largely based on mechanistic understandings of the friction and wear of ceramics.

A. Microfracture

Based on some recent experimental studies, several investigators have concluded that microfracture is by far the greatest source of wear in ceramic materials [22,33–35]. This is because, unlike metallic materials, most ceramics are inherently brittle. When pulled in tension at room temperature, they show essentially no sign of plastic elongation. The yield and fracture strengths of most ceramics are found to be almost identical [40]. Nevertheless, microscale plastic deformation occurrs in some ceramics under highly localized compression (e.g., diamond indentation), or in slow-strain-rate creep situations [41]. The inherent brittleness of these materials is thought to stem from their usually limited number of slip systems available for plastic flow and the inherent difficulty of dislocation glide in their structures [40,41].

A major cause of microfracture in ceramics is related to internal volume defects. Despite all the technological advances made in their fabrication, ceramics still contain many internal defects (e.g., flaws, voids, inclusions, weak grain boundaries). During sliding contact, these defects act as stress concentration points. Under the influence of normal and tangential forces, microcracks can initiate from these defects, where dislocations may sometimes pile up and thus block slip-band propagation [42,43].

For static contact conditions, it is known that real contact occurs at asperity contacts and that normal forces acting on the asperities can easily translate into extremely high contact pressures. Metals accommodate such stresses by undergoing plastic deformation. However, most ceramics cannot easily yield in a plastic mode; instead, they may undergo brittle fracture.

For conditions of dynamic sliding contact, in addition to high contact stresses acting in a normal direction, shear stresses acting tangentially are developed at the contact interfaces. The magnitude and location of the maximum shear stresses are dependent on the extent of frictional traction between sliding surfaces. According to published data [2,21,33–35], the friction coefficients of ceramic couples may range from 0.4 to 1.2. As explained in Reference 44, if the friction coefficient of a sliding couple exceeds 0.3, the maximum orthogonal shear stresses are shifted from the subsurface to the sliding surface. As a result, high tensile stresses are generated behind the trailing edges of the moving surface asperities. It has been found that beyond a critical threshold, tensile stresses can initiate surface cracks with an alignment often perpendicular to the direction of relative motion. Because of the cyclic nature of the sliding contact, fatigue may be instrumental in the initiation of microcracks, especially from those defects that act as stress concentration points.

Hsu et al. [45] found that with increasing load and velocity, the wear rate of Al_2O_3 may undergo distinct transitions. To illustrate the critical conditions under which wear transitions occur, these investigators constructed a series of three-dimensional wear maps for both dry and oil-lubricated sliding contacts. Skopp et al. [46] and Woydt et al. [34,47] also reported steep transitions in the wear rates of ZrO_2, SiC, and Si_3N_4 test pairs with increasing velocity. Similar observations were made on the wear rates of Al_2O_3 and ZrO_2 tested over a wide range of sliding velocities [48,49].

The scanning electron micrograph in Figure 1a shows the morphology of a typical wear track formed on a flat specimen of Al_2O_3 during oscillatory sliding against an Al_2O_3 ball. According to the technical data provided by the manufacturer (Morton Thiokol, Inc., Alfa Oxide Ceramics Division) these ceramics were fabricated from high purity Al_2O_3 powders (99.7 wt %) by means of sintering at 1800°C. It is evident that the worn surface consists of smooth plateaus divided by deep microcracks. Figure 1b is a close-up of a microcrack that was initiated from a pore located at a grain boundary.

B. Tribochemical Wear

Despite their relatively high chemical inertness under static conditions, both nonoxide and oxide ceramics have been shown to interact with their surroundings during dynamic contact and to undergo tribochemical degradation. For example, Fischer and Tomizawa [37,38] have shown that sliding interfaces of Si_3N_4 can

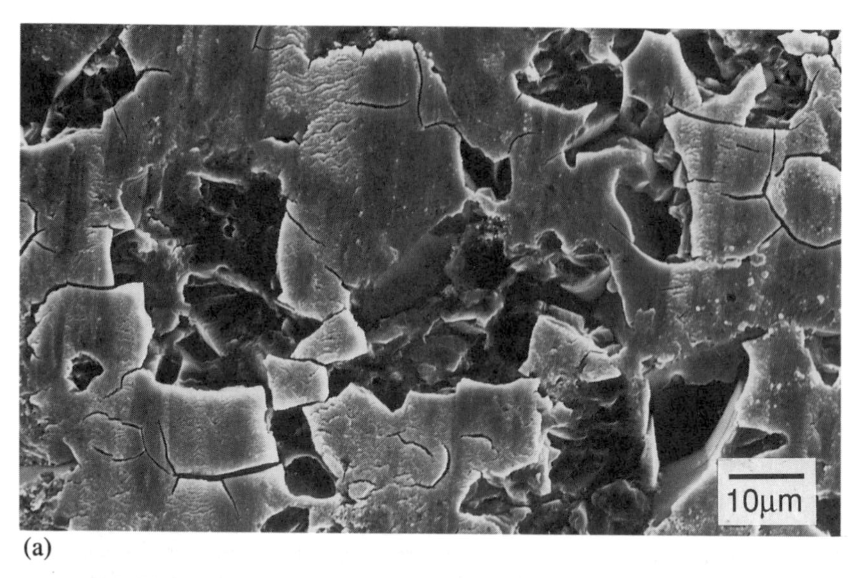

(a)

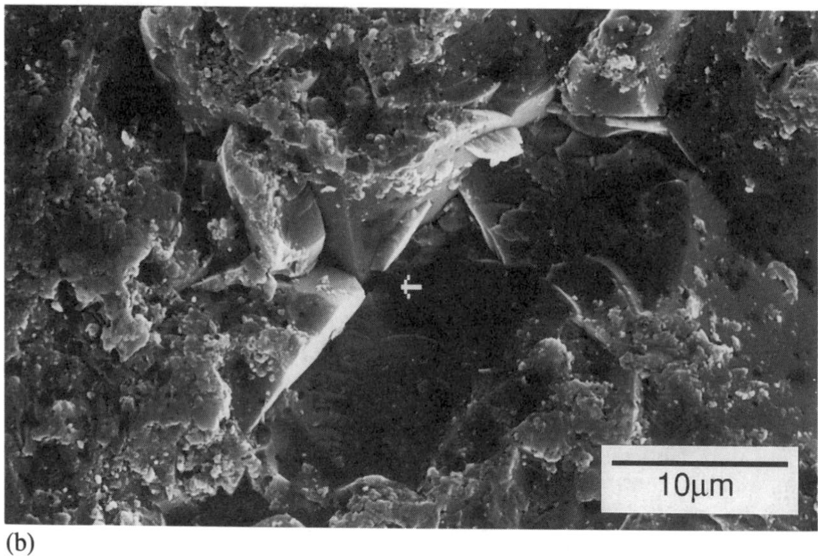

(b)

Figure 1 (a) General and (b) detailed scanning electron micrographs of a wear track formed on a flat Al_2O_3 specimen during sliding against an Al_2O_3 ball. Test conditions: load, 10 N; temperature, 23°C; sliding velocity, 0.05 m/s; relative humidity, 50%; radius of ball, 6.35 mm. (Before electron microscopy, loose wear debris particles were removed from the surface by ultrasonic agitation.)

react with oxygen and/or water molecules in their surroundings to form oxygen-rich films. The degree of oxidation is greater in humid environments and at elevated temperatures. Under certain tribological conditions, these films profoundly affect the wear and friction behavior of Si_3N_4. Scanning electron micrographs (Figs. 2a,b) show the shape and morphology of such oxide films that had formed on the wear scar of an Si_3N_4 ball (Cerbec Ceramic Bearing Co., NBD-100) that was slid against an Si_3N_4 flat (Kyocera Co., SN-220). As is evident from Figure 2a, these surface films are rather thick, somewhat fragile, and easily detachable from the underlying ceramic. The X-ray map in Figure 2c verifies that the film shown in Figure 2b is indeed rich in oxygen. In dry air or inert gases, the wear of Si_3N_4 was deduced to be largely controlled by microfracture [37].

Under the dynamic action of sliding contact, even oxide ceramics have been observed to interact with the chemical species in their surroundings and to undergo distinct transitions in their friction and wear behavior. For example, some investigators reported that Al_2O_3 wears at a much higher rate in water than in normal air, mainly because of stress–corrosion cracking [50,51]. In contrast, Gates et al. [52] reported low friction and low wear rates for Al_2O_3 in water-containing environments and attributed these findings to the formation of lubricious aluminum hydroxide films on the sliding interfaces.

Few tribological data are available on the friction and wear behavior of nonoxide structural ceramics at elevated temperatures. Available wear data suggest that most ceramics wear at significantly higher rates at elevated temperatures than at room temperature [38,46,48,53]. Occasionally, on some ceramics, a transition from low to high and back to low wear rates with increasing temperature has been observed [46,53]. Such transitions have been attributed to the changes in the mechanical and chemical characteristics of the sliding interfaces [46]. For example, low friction and wear of Si_3N_4 at high temperatures have been attributed to the formation of silicon oxide films, which soften considerably at high temperatures [38,46]. SiC, which undergoes oxidation in open air and graphitization under inert atmosphere during sliding contacts at elevated temperatures, also enjoys low friction and wear [53].

C. Thermomechanical Wear and Melting

It is known that essentially all the mechanical work done to overcome friction is converted into heat and that this heat is generated in the immediate vicinity of the sliding interfaces. The amount of frictional heat flux q is proportional to the friction coefficient μ, the normal force F, and the sliding velocity v but inversely proportional to the nominal contact area A_n, as given by $q = (\mu F v)/A_n$ [36,54,55]. The real area of contact, being much smaller than the nominal contact area, gives rise to much higher local heat fluxes in the vicinity of asperity

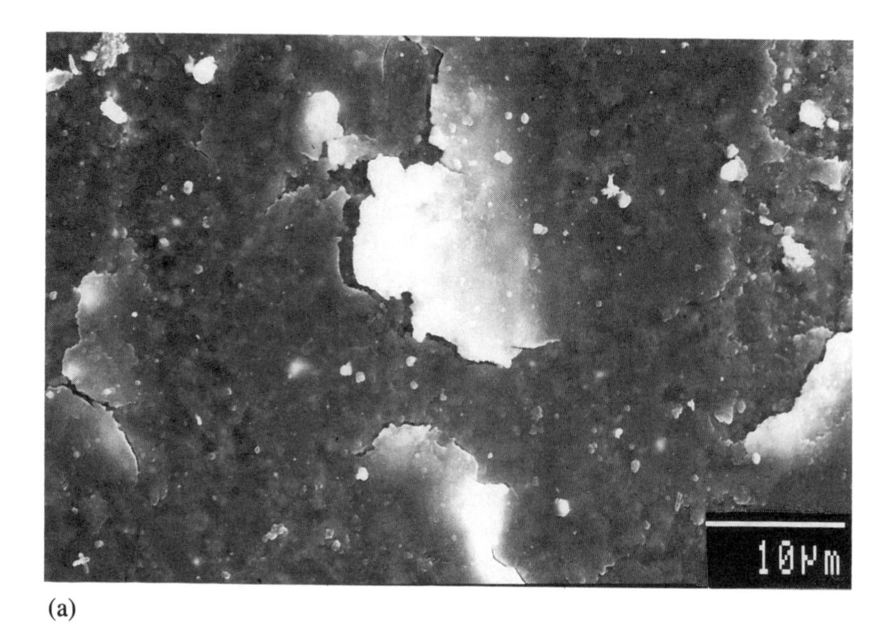

(a)

Figure 2 (a) and (b) Typical scanning electron micrographs of oxygen-rich surface films formed on an Si_3N_4 ball during reciprocating against a flat Si_3N_4 specimen in air of about 40% relative humidity. (c) Oxygen X-ray map of the area covered in (b). Test conditions: load, 5 N; temperature, 23°C; average sliding velocity, 0.05 m/s; pin radius, 6.35 mm.

contacts. Because the frictional heat flux enters the contacting bodies through these contact spots and/or junctions (known as "hot spots"), their local temperatures, referred to as "flash temperatures," can be much higher than the overall or "bulk" surface temperature.

In recent years, several theoretical and phenomenological models have been proposed to explain the detrimental effect of sliding velocity on wear. For example, Winer and Ting [36] have proposed a thermomechanical wear theory for ceramics that emphasizes the effect of sliding velocity on wear. Except for SiC and AlN, most ceramics possess significantly lower thermal conductivity than metals do. As a result, they cannot dissipate frictional heat generated at sliding interfaces as effectively as most metallic alloys. Large temperature gradients often develop between areas of real contact and surrounding regions, thus creating high thermal stresses. When these thermal stresses are combined with normal and tangential stresses (due to applied load and frictional traction), plastic yielding may occur and result in wear [36]; alternatively, these brittle ceramics may fracture, hence may suffer severe wear losses [22,49]. Taking into

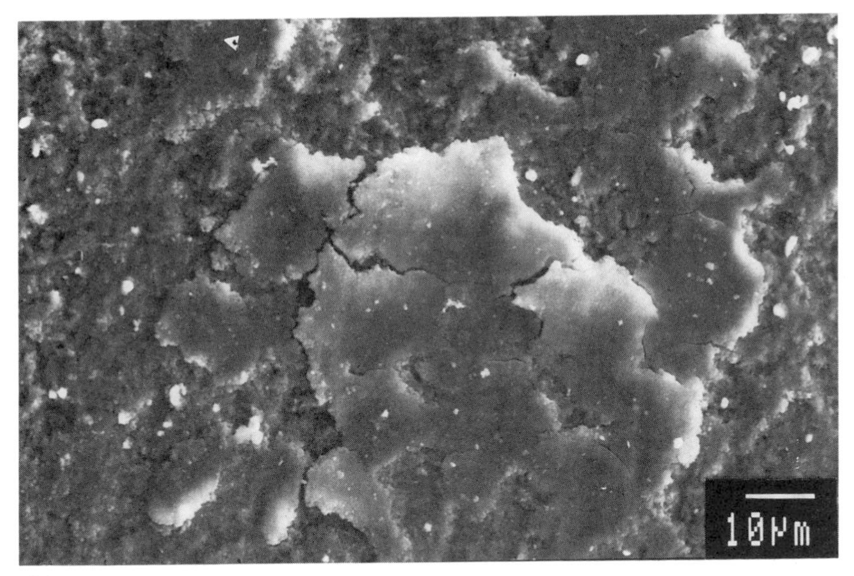

(b)

(c)

account the state of thermal and contact stresses, as well as material properties, Ting and Winer [55] constructed wear maps for partially stabilized ZrO_2 and identified the regions of wear, conditional wear, and no wear.

Using the surface temperature model of Ashby et al. [54], Erdemir et al. [49] estimated the bulk and flash temperatures of partially stabilized ZrO_2 sliding against ZrO_2 at different velocities. For the material properties shown in Table 1 and test conditions selected, a flash temperature comparable to the melting point of ZrO_2 was calculated at 2 m/s. To provide a better representation of the estimated flash temperatures, a temperature map (Fig. 3) was constructed for the test pair by using the T-MAPS software of Ashby et al. [56]. Experimental test conditions (1 and 2 m/s, 5 N) are located in this map, which shows the variation of bulk and flash temperatures of a zirconia couple over a wide range of load/pressure and sliding velocities. It is predicted that for the nominal pressures employed in these tests, local asperity melting of ZrO_2 is feasible at sliding velocities greater than about 1.7 m/s, which is designated as the critical velocity v_c. Table 1 gives the source and properties of ZrO_2 used in these calculations.

Although such temperature estimates are notoriously variable and depend on

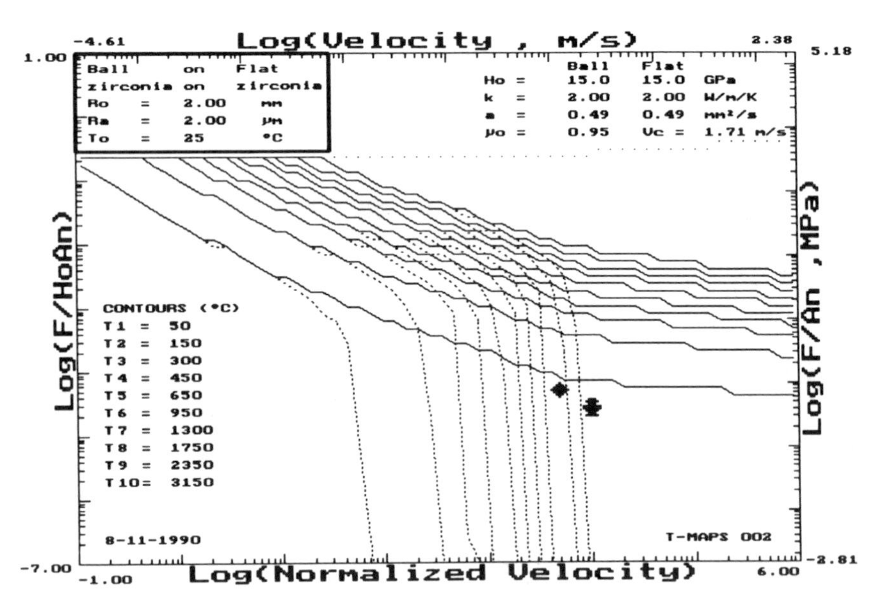

Figure 3 Temperature map of a sliding pair consisting of ZrO_2 ball and ZrO_2 disk. Solid lines represent bulk temperature contours; broken lines represent flash temperature contours; ♦, flash temperature at 1 m/s; ♣, flash temperature at 2 m/s. R_0 is nominal contact radius, R_a, radius of single isolated asperity junction; T_0, ambient temperature; H_0, hardness of ZrO_2 at room temperature; k, thermal conductivity; a, thermal diffusivity; μ_0, friction coefficient; v_c, critical velocity (material properties are given in Table 1).

Table 1 Properties of ZrO_2 and Al_2O_3 Used in the T-MAPS Software [a]

Property	Units	ZrO_2 [b]	Al_2O_3 [c]
Density	kg/m^3	5400	3800
Young's modulus (E)	GPa	160	343
Temperature dependence of modulus (T_m/E) (dE/dT)[d]	—	0.5	0.35
Yield strength	GPa	3	2.4
Temperature dependence of yield strength (T_m/σ_y) ($d\sigma_y/dT$)[d]	—	0.5	0.35
Hardness at room temperature	GPa	15	16
Melting point	K	2893	2323
Thermal conductivity	W m^{-1} K^{-1}	2	25
Thermal capacity	J kg^{-1} K^{-1}	750	794
Latent heat of melting	kJ/kg	706	551

[a]Definitions of temperature-dependent material properties can be found in Reference 56.
[b]Data from the material data sheet of the Morton Thiokol, Inc., Alfa Ceramics Division, 152 Andover Street, Danvers, MA 01923, and Reference 56.
[c]Data from the material data sheet of the Kyocera International, Inc., 8611 Balboa Ave, San Diego, CA 92123, and Reference 56.
[d]Melting temperature T_m in kelvins; σ_y, yield strength at room temperature.

the particular choices of important parameters, it should be noted that selected parameters are similar to those deduced by Ashby et al. [56] from quantitative matching of theory and available experimental data for a similar ZrO_2 material. Thus, it is reasonable to say that these estimates have at least some semiquantitative reliability.

In support of the predicted flash temperature Erdemir et al. [49] presented scanning electron micrographs that revealed evidence of local melting on rubbing surfaces. The micrographs presented in Figure 4 show features that suggest local melting on the rubbing surfaces of ZrO_2/ZrO_2 at 2 m/s and Al_2O_3/Al_2O_3 at 4 m/s. Figure 5 is a temperature map for Al_2O_3/Al_2O_3 test pairs obtained with the T-MAPS software [56]. The material data sheet for Al_2O_3/Al_2O_3 and $PSZ/ZrO_2/ZrO_2$ test pairs are given in Table 1. It is apparent from Figure 5 that asperity melting becomes feasible for Al_2O_3 test pairs at sliding velocities greater than about 3.4 m/s, the so-called critical velocity v_c. The critical velocity of the Al_2O_3 test pairs is higher by a factor of 2 than that of ZrO_2 pairs. This difference can be attributed to the significant differences in material and thermal properties of these two ceramics. Again, because Ashby and coworkers have chosen key parameters through quantitative matching of theory and experiment, we feel some degree of confidence in these admittedly different temperature estimates.

How can local melting produce wear? As shown in Figure 6, some of the ZrO_2 and Al_2O_3 materials may become molten at asperity contacts, whereupon

(a)
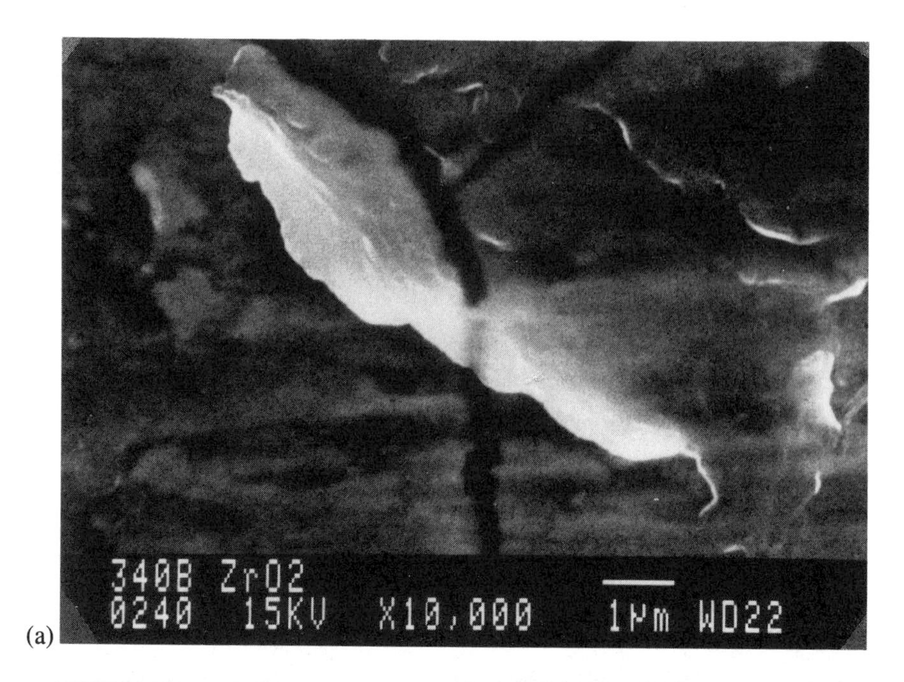

(b)
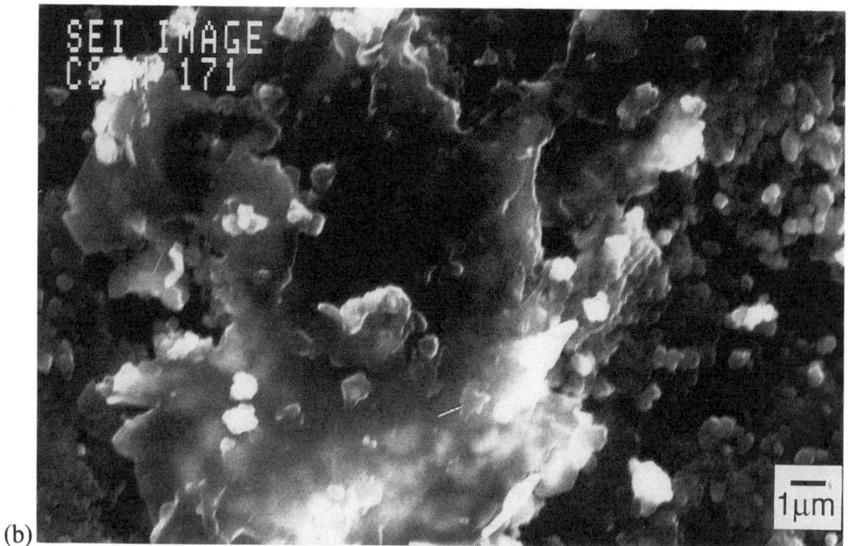

Figure 4 Scanning electron micrographs of wear scars of (a) a ZrO_2 ball slid against a ZrO_2 disk at 2 m/s and (b) an Al_2O_3 ball slid against an Al_2O_3 disk at 4 m/s. Microfeatures shown are indicative of local melting. Test conditions: load, 5 N; relative humidity, > 1%; temperature, 23°C; pin radius, 4.77 mm.

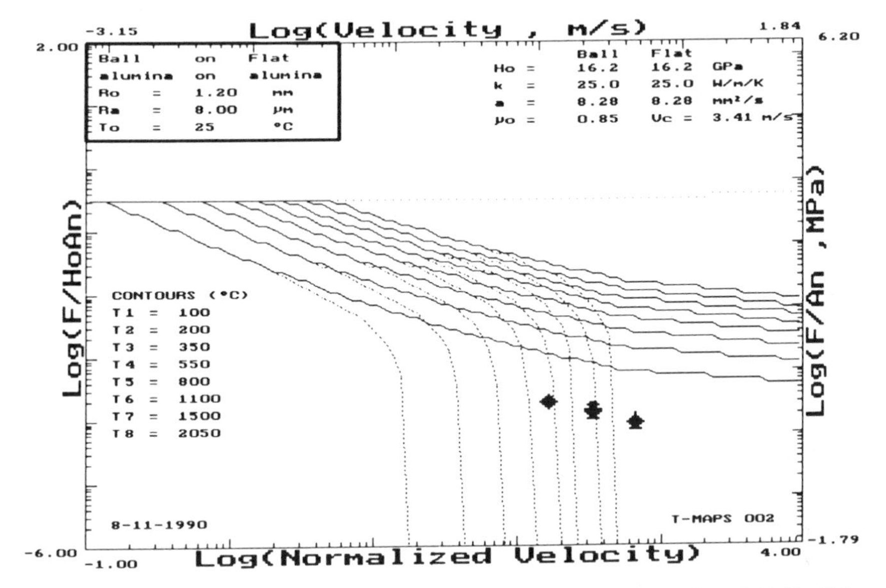

Figure 5 Temperature map of a sliding pair consisting of Al_2O_3 ball and Al_2O_3 disk. Solid lines show bulk temperature contours; broken lines show flash temperature contours. Data points: ♦, ♣, and ♠ represent the flash temperatures of interfaces during sliding at 1, 2, and 4 m/s, respectively. R_0 is nominal contact radius; R_a, radius of single isolated asperity junction; T_0, ambient temperature; H_0, hardness of ZrO_2; k, thermal conductivity; a, thermal diffusivity; μ_0, friction coefficient; v_c, critical velocity (material properties are given in Table 1).

they are squeezed out of the sliding interface in the molten state and then solidify and accumulate around the trailing edges of the rubbing balls. Whereas some fragments are broken and become loose wear fragments, others remain with the balls. Figure 7 shows that the wear rate of ZrO_2 balls increases markedly with increasing velocity, suggesting that the thermal and/or thermomechanical instabilities due to high frictional heating are detrimental to the wear behavior of low thermal conductivity ceramics. Note that the ZrO_2 balls experience especially high wear at 2 m/s, which had caused melting of asperities (as indicated in Fig. 4).

D. Plastic Flow

One must be very cautious when judging a microfeature resembling plastic flow. For example, looking at a micrograph taken from the wear track of an Al_2O_3 sample formed during sliding at 200°C (see Fig. 8a), one might think that the flat plateaus are the result of some gross plastic flow that occurred on the sliding

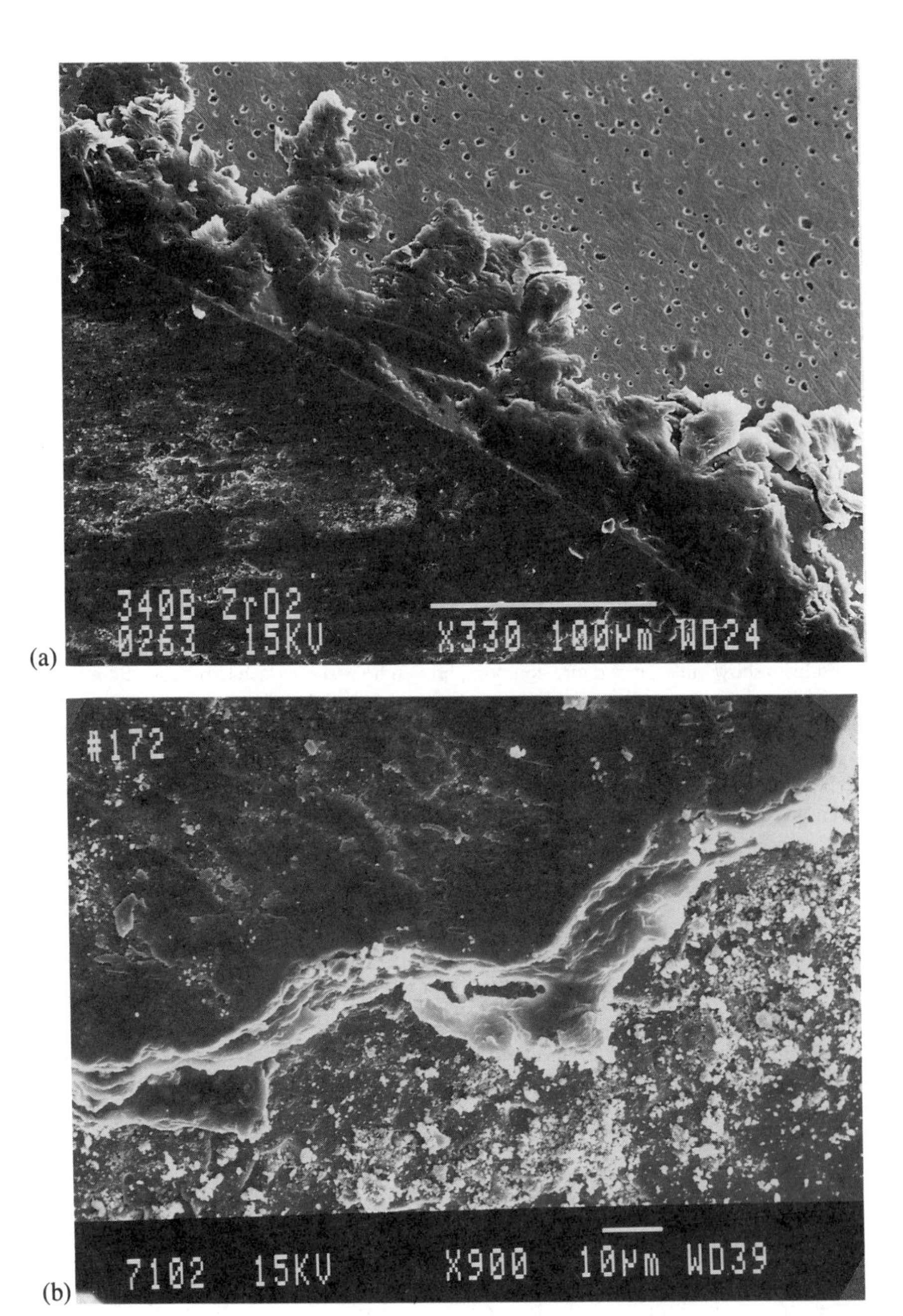

(a)

(b)

Figure 6 Scanning electron micrographs of the trailing edges of (a) ZrO_2 and (b) Al_2O_3 balls shown in Figure 4. Large colonies of flakelike wear fragments were found around the circular edges of balls.

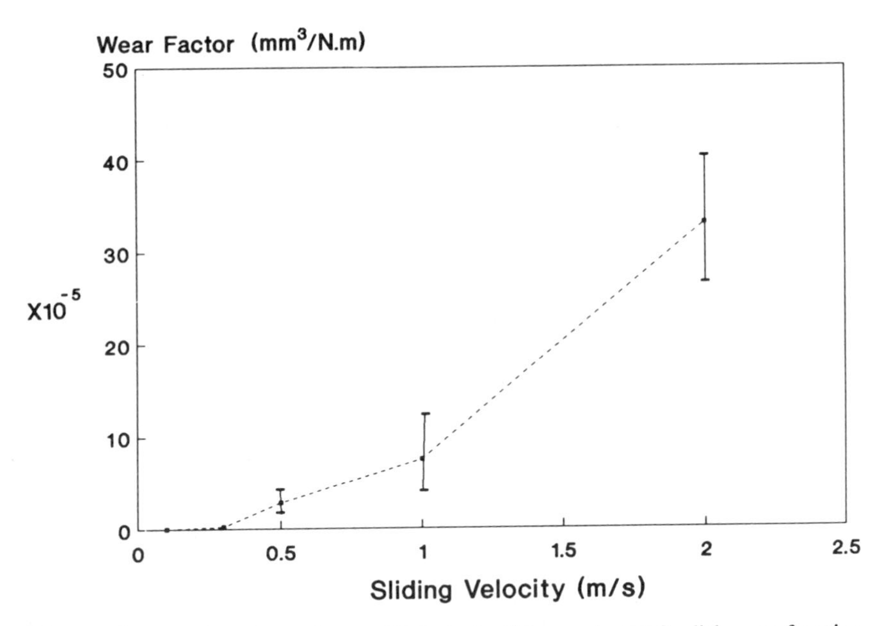

Figure 7 Variation of wear factor of ZrO_2 balls sliding against ZrO_2 disks, as a function of sliding velocity. Test conditions: load, 5 N; relative humidity, > 1%; radius of ball, 4.77 mm; temperature, 23°C.

surface. However, at a higher magnification, it becomes clear that what appeared at lower magnification to be macroscale plastic deformation (Fig. 8a), in fact consists of islands of compacted and/or smeared wear debris particles (Fig. 8b). It is clear that these particles are very small, typically ranging from 10 to 100 nm. The occurrence of minor deformation beneath the sliding surfaces due to dislocation activity and/or twinning cannot be ruled out [41,43]. However, Figure 8 clearly demonstrates the need to be cautious when judging a wear feature indicative of gross plastic flow solely on the basis of microscopic appearance.

In a recent study, Erdemir et al. [48] observed microfeatures indicative of gross plastic flow on rubbing surfaces of Al_2O_3 tested at 4 m/s (see Fig. 9) and attributed them to the thermal softening of the asperities through which frictional heat must be dissipated. As discussed previously, Al_2O_3 lacks the ability to rapidly dissipate the frictional heat from its sliding interfaces. As a result, plastic flow of these areas becomes feasible. According to Terwilliger and Redford [40], some ceramics, including Al_2O_3, can become highly deformable at elevated temperatures.

Transformation-toughened ZrO_2 ceramics were also found to undergo plastic flow during sliding at low velocities [49,57]. The micrograph in Figure 10 presents some features indicative of plastic flow on a transformation-toughened ZrO_2 disk slid at 0.1 m/s. It is believed that owing to its high fracture toughness,

Figure 8 Scanning electron micrographs of a wear track on a flat Al_2O_3 specimen, formed during oscillating against an Al_2O_3 ball at 200°C. (a) Flat plateaus are indicative of gross plastic flow. (b) High magnification scanning electron micrograph reveals that flat plateaus are due to compacted and/or smeared wear debris particles. Test conditions: load, 10 N; relative humidity, 40%; average sliding velocity, 0.05 m/s; radius of ball, 6.35 mm.

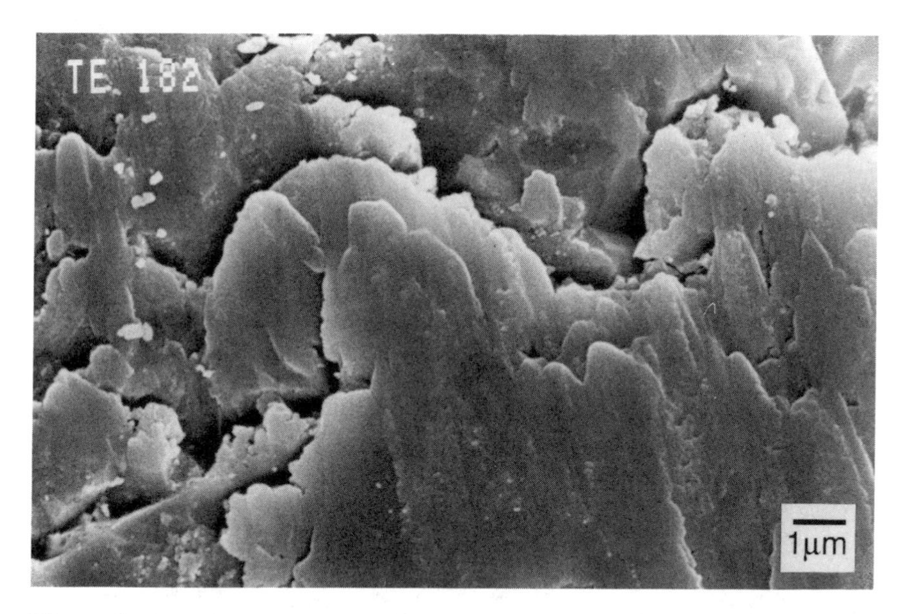

Figure 9 Scanning electron micrograph of a wear scar on an Al_2O_3 ball formed during sliding against an Al_2O_3 disk, presenting physical evidence of plastic flow. Test conditions: load, 5 N, sliding velocity, 4 m/s; relative humidity, > 1%; radius of ball, 4.77 mm; temperature, 23°C.

partially stabilized ZrO_2 can effectively resist initiation of microcracking. Hence, normal and shear stresses can be accommodated by plastic flow, as shown in Figure 10.

III. LUBRICATION OF CERAMICS

Because of their unique combination of mechanical, chemical, and thermal properties, advanced ceramics can be of significant help in solving present and future tribological problems under extreme conditions. However, it is apparent from the foregoing discussion that their wear performance is poor and their friction coefficients generally high. These deficiencies overshadow the favorable properties of ceramics. It appears that for these materials to have significant application possibilities in future tribosystems, new means must be developed to reduce their friction and wear. Specifically, ways must be found to prevent microfracture. The degenerative effects of tribochemical and thermomechanical interactions must be reduced. At present, a series of preventive measures based on vapor and liquid and solid/phase lubrication is being explored.

One potential way to reduce friction and wear in ceramics is to provide a thin

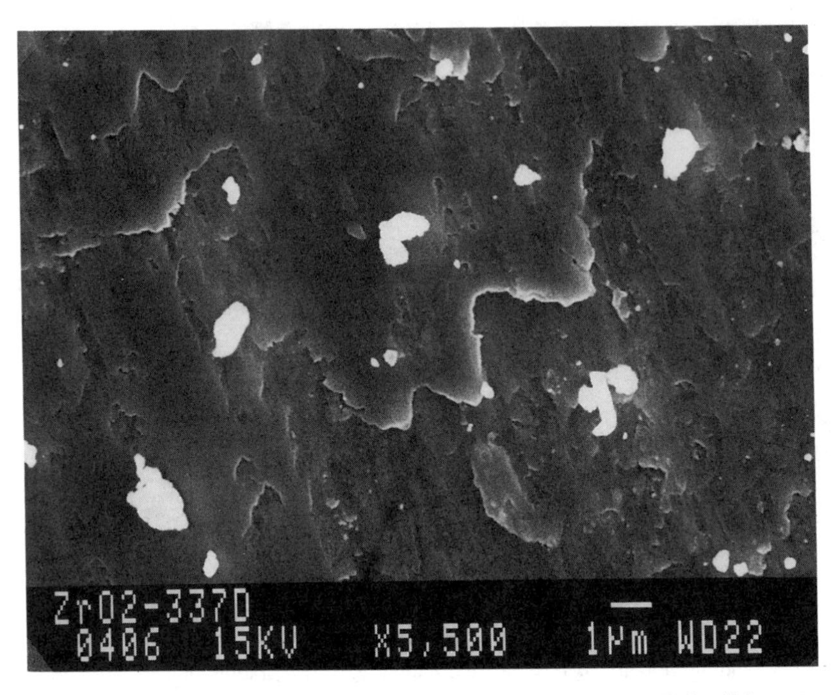

Figure 10 Scanning electron micrograph of a wear track on a ZrO_2 disk, presenting some features indicative of plastic flow. Test conditions: load, 5 N; relative humidity, >1%; sliding velocity, 0.1 m/s; radius of ball, 4.77 mm; temperature, 23°C.

film that can reduce the shear strength of contact interfaces and prevent opposing asperities from coming into direct contact. Sometimes, the rubbing interfaces of ceramics tend to produce such a medium on their own and thus attain low friction and low wear. For example, sliding interfaces of Si_3N_4 were shown to produce a thin silicon oxide film with easy shear capability and to achieve low friction and low wear in a moist environment [38]. Al_2O_3 ceramics were also found to produce soft aluminum hydroxide films on their sliding interfaces and to display low friction when tested in water [52]. Molecular adsorbates and/or thin organic contaminants have also been found to lower the friction and wear of sliding ceramic interfaces [58]. In some ceramics, low friction and wear result from the in situ formation of rolls [59,60] and highly shearable wear debris layers with viscous flow behavior [60]. The rolls may form from soft reaction products resulting from tribochemical interactions in humid air at elevated temperatures.

A. Approaches

It is clear that to achieve low friction and low wear in ceramics, sliding surfaces must be provided with a film that can shear easily and prevent opposing asperities from coming into direct contact. Easy shear at contact interfaces is analogous to low friction, whereas fewer asperity–asperity interactions means low wear. When friction is low, the magnitude of tensile stresses developing behind the moving asperities is also low. As a result, the location of the maximum orthogonal shear stresses is displaced away from the plane of contact and the probability of microcrack initiation on contacting surfaces is reduced [44]. Because of a larger load-bearing area provided by the soft films, the magnitude of contact pressures acting in the normal direction is also reduced. This in turn reduces the probability of radial crack formation away from the contact zone [43].

In the presence of a lubricant film, ceramic interfaces are expected to attain much lower friction coefficients than are metallic interfaces. The reason for this is that the frictional force F is essentially a product of the shear strength s of the lubricant film, multiplied by the real contact area A, resulting from elastic and plastic deformation: $F = s \times A$ [61]. Owing to their much greater mechanical hardnesses and significantly higher elastic moduli, ceramics are expected to establish much smaller real contact areas than are metals under a given normal load. According to the equation above, a smaller contact area translates into a smaller frictional force when a lubricant film is present, as demonstrated several decades ago by Bowden and Tabor [61].

Currently, thin films with low shear strength are produced on ceramic surfaces by at least three methods: vapor, liquid, and solid phase lubrication. The following sections briefly summarize liquid and vapor phase lubrication. Solid lubricant films are then introduced and discussed in greater detail. New deposition technologies available for the application of solid lubricant films also are discussed in detail.

1. Liquid Lubrication

For practical applications, liquid lubrication is the most convenient. However, some problems exist with conventional liquid lubricants when used in ceramic tribosystems. They can function properly only at temperatures well below the projected service temperatures of these systems [22]. When used at elevated temperatures, they tend to break down and leave undesirable deposits on sliding surfaces. A few can function at temperatures up to about 300°C; most cannot. Furthermore, additives used in these lubricants do not seem to respond well to certain ceramic surfaces; hence, the boundary films needed under extreme application conditions are often missing [62]. To solve these problems, advanced liquid lubricants that contain new additives have been developed and are currently being tested on ceramics [63,64].

2. Vapor Phase Lubrication

Another means of providing ceramic surfaces with low shear strength films is vapor phase lubrication. Principally, it involves the production of a thin reaction product film on hot, sliding ceramic surfaces by exposing the surfaces to vaporized lubricants. Although still in the exploratory stage, this method appears to be promising and practical for various tribological applications that involve elevated service temperatures [65]. Thin carbon films resulting from catalytic reactions between carbonaceous gases and ceramic surfaces have also been reported to result in low friction at elevated temperatures [66].

3. Solid Lubrication

For applications involving severe tribological conditions (e.g., high temperature, corrosive media, vacuum environment, high load and speed), solid lubricants may be the only option available for controlling wear and friction in all types of tribosystem. Solid lubricants can be applied on tribological surfaces as thin films or they can be mixed with oils and greases. Solid lubricants of course share with all lubricants the problem of a finite lifetime (e.g., 1–10 million cycles). However, solid lubricants are much more difficult to retain and to replace than are other types of lubricant. A principal means of enhancing retainment lies in improvements in adhesion of solid lubricant films, but even this may not combat long-term chemical degradation. Strong adhesion to metallic surfaces can be achieved with most deposition methods. However, deposition of adherent solid lubricant films on ceramic surfaces has been quite difficult with conventional deposition techniques (e.g., sputtering, ion plating, and chemical vapor deposition).

Sputtering and ion plating of thin solid films rely on a glow-discharge plasma. With electrically conductive substrates (e.g., metals), the discharge plasma is easily sustained and the depositing species are accelerated to substrate surfaces with sufficient energy to establish strong bonding [67]. Briefly, the resultant solid lubricant films are sufficiently adherent to metallic substrates. However, most ceramics are not electrically conductive. As a result, they cannot discharge electrical current and thus cannot sustain a plasma. Instead, they tend to accumulate positive electrical charge and repel depositing species of the same polarity. In extreme cases, the electric field surrounding the substrate is disturbed, and sometimes the whole plasma is extinguished. Consequently, ceramics are not suitable substrates for ion plating and/or sputter deposition of solid lubricant films.

Despite the difficulties described above, a few attempts have been made to apply solid lubricant films on ceramic substrates by conventional deposition methods. Specifically, MoS_2 films have been deposited on SiC by sputtering [68]; amorphous carbon films have been produced on Si_3N_4 by plasma chemical vapor deposition [69]; and boric oxide (B_2O_3) films have been applied on Al_2O_3 by vacuum evaporation [70]. In all these cases, the adhesion of lubricant films to ceramic substrates was dictated by the extent of chemical affinity of film and substrate atoms. In other instances, ceramic surfaces were subjected to ion and/or chemical etching to promote greater chemical interaction, hence stronger bonding. For chemically compatible film–substrate combinations (e.g., B_2O_3 on Al_2O_3), strong bonding was achieved even with the conventional vacuum evaporation technique.

Didziulis et al. [68] used a sputtering method to achieve adherent MoS_2 films on SiC substrates. To improve adhesion of the MoS_2 film, they tried to chemically etch the SiC surfaces before sputter deposition, but found this technique ineffective. MoS_2 films adhered to unetched surfaces better than they did to etched SiC surfaces. Furthermore, compared with films sputtered on steel, MoS_2 films sputtered on SiC surfaces exhibited much shorter lifetimes.

In another study, Miyoshi et al. [69] investigated the friction and wear behavior of amorphous hydrogenated carbon films produced on Si_3N_4 substrates by plasma chemical vapor deposition with methane and butane as the source gases. They reported that the adhesion of amorphous hydrogenated carbon films to Si_3N_4 surfaces was quite strong and that the films were quite capable of affording low friction to sliding interfaces in vacuum and N_2 environments. They noticed that the frictional behavior was dictated by the degree of diamondlike character of the amorphous carbon films. For a more diamondlike carbon film, they measured friction coefficients typical of natural diamond (e.g., 0.08–0.09), whereas for a more graphitelike carbon film, they measured friction coefficients typical of graphite (e.g., 0.1–0.2).

Erdemir et al. [70] investigated the tribological characteristics of vacuum-evaporated boric oxide (B_2O_3) films on Al_2O_3 substrates. They found that boric acid (H_3BO_3), which forms spontaneously on the surfaces of B_2O_3 coatings, is remarkably lubricious. For a sliding pair composed of a sapphire ball and a B_2O_3-coated Al_2O_3 disk, they reported friction coefficients ranging from 0.02 to 0.05 in air with 50% relative humidity. Figure 11, which presents the friction coefficients of various balls sliding against a B_2O_3-coated Al_2O_3 disk under different contact loads, shows that the friction coefficient is dependent on the type of counterface ball. The use of a harder, more rigid ball (e.g., sapphire) results in a lower friction coefficient. This observation can be explained by noting that the area of true contact with a hard, rigid ball will be smaller than that with a soft, less rigid ball. Friction force, which is a product of the true contact area multiplied by the shear strength of the contact interface, will be much smaller when hard, rigid balls are used in sliding contact [61].

Erdemir et al. [70,71] concluded that low friction is a direct consequence of the layered crystal structure of H_3BO_3 films forming on the exposed surfaces of B_2O_3 coating by the spontaneous chemical reaction:

$$1/2\ B_2O_3\ \text{(coating)} + 3/2\ H_2O\ \text{(moisture)} \rightarrow H_3BO_3$$
$$\Delta H_{298} = -45.1\ \text{kJ/mol}$$

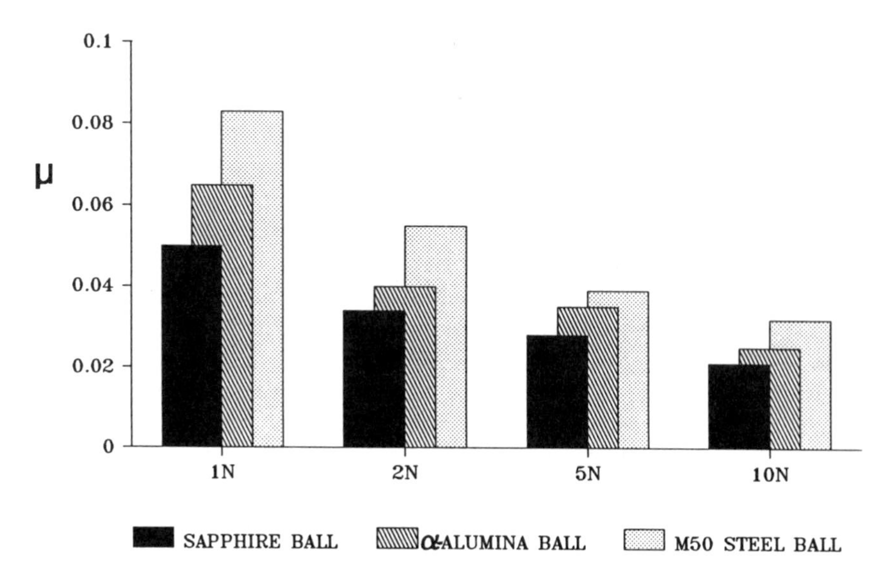

Figure 11 Friction coefficients μ of various balls during sliding against a B_2O_3-coated Al_2O_3 disk under various loads. Test conditions: relative humidity, 50%; sliding velocity, 0.02 m/s; temperature, 23°C; radius of sapphire ball, 3.18 mm; radius of steel and alumina balls, 4.77 mm.

It is important to clarify that the reaction above occurs naturally and that tribochemical acceleration of reaction rates is not required. A thin layer of H_3BO_3 forms everywhere on the exposed surface of B_2O_3. As described in detail in Reference 71, H_3BO_3 crystallizes in a triclinic crystal structure essentially made up of atomic layers parallel to the basal plane. The atoms in each layer are closely packed and strongly bonded together. The bonds between the boron and oxygen atoms are mostly covalent, with some ionic character. Hydrogen bonds strongly hold the planar boron/oxygen groups together. The atomic layers are widely spaced (e.g., 0.318 nm apart) and held together by weak forces (e.g., van der Waals). A depiction of the layered crystal structure of H_3BO_3 is presented in Figure 12. With its layered crystal structure, H_3BO_3 resembles other layered solids well known for their good lubrication capabilities (e.g., MoS_2, and graphite). Therefore, the low friction character of H_3BO_3 is hypothesized to be a direct consequence of its unique crystal structure and bond characteristics [71].

Erdemir [71] proposed that under shear stresses, platelike crystallites can align themselves parallel to the direction of relative motion. Once so aligned, they can slide over one another with relative ease and thus impart the low friction coefficients shown in Figure 11. Figure 13 provides some physical evidence for platelike crystallites with preferred alignment parallel to the sliding surfaces.

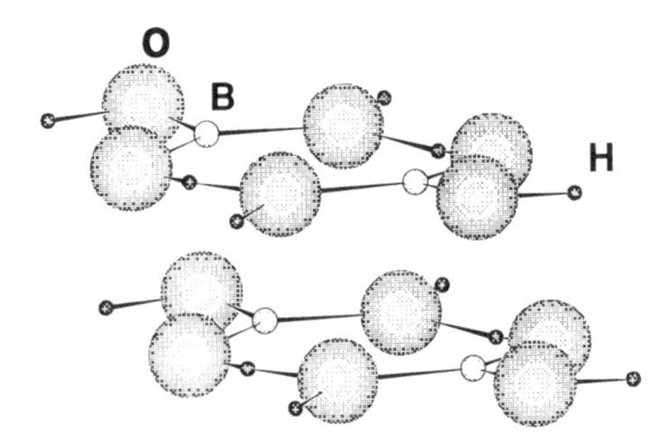

Figure 12 Schematic illustration of the layered crystal structure of boric acid (H_3BO_3).

Because of their excellent chemical compatibility, B_2O_3 films were found to adhere to ion-etched ceramic surfaces quite well [70].

The use of H_3BO_3 for applications involving ceramics at elevated temperatures is not recommended for several reasons. First, beyond about 170°C, H_3BO_3 tends to decompose and eventually turn into B_2O_3, thus losing its layered crystal structure, hence its lubricity. Second, at temperatures beyond about 450°C, B_2O_3

Figure 13 Scanning electron micrograph showing platelike structure of H_3BO_3 crystals formed naturally on the wear track of a B_2O_3-coated Al_2O_3 disk.

becomes liquidlike and tends to react with underlying ceramic substrates. For metals, the chemical reaction is negligible, and low friction can be reinstated by viscous flow lubrication. However, for ceramics, especially for the oxides, the situation is different. Liquid B_2O_3 can react with these ceramics and lead to high corrosive wear. Friction can also be very high because of the high viscosity of the reaction products.

B. Advanced Methods

With the advent of ion beam technology, new deposition methods based on the use of directed high energy ion beams have been developed and introduced to the field of tribology. Ion beam mixing (IBM) and ion beam-assisted deposition (IBAD) are two examples. The primary function of these methods is to impart strong adhesion between the solid lubricant films and the ceramic substrates, thus increasing the lifetime of the lubricant films. In addition, these methods can be effective in controlling film microstructure and morphology. High energy ions used in these processes can result in substantial intermixing of film and substrate atoms, with the result that strong adhesion is achieved. As with ion-plating and sputtering techniques, electrical charging of the substrate is not a problem with ion beam processes. Without altering their kinetic energy, the ions can be neutralized by electrons emitted from a hot filament before striking the substrate surface. One major limitation of these deposition methods is that they are line-of-sight approaches (i.e., only the areas in the direct path of ions will be covered with the adherent coatings). Rotation of specimens during deposition could alleviate this problem to some extent.

1. Ion Beam Mixing

Principles

In ion beam mixing, a flux of high energy ions is used to induce mixing of previously deposited single- or multilayered films with metal or ceramic surfaces. The range of ion energies employed in IBM may vary from a few hundred kiloelectron-volts to several million electron-volts. Ions with such energies can pass through the thin films and penetrate well into the substrates (e.g., about 1 μm at 1 MeV). Mixing is accomplished with the aid of two complementary processes. First, while passing through the surface film, energetic ions collide with the film atoms and drive some of them into the substrates. Second, and most important, because of the large collision cascades associated with high energy ion impact, many vacancies and interstitials are created, thus promoting interdiffusion of film and substrate atoms. Because of the broadening of an otherwise sharp boundary, adhesion of the surface films improves dramatically. Chemical reactions at the film–substrate interface are also feasible and can further increase film adhesion. One of the limitations of this technique is that films thicker than the penetration depths of ions cannot be mixed effectively (≈1 μm at 1 MeV, depending on the ion species and temperature). A schematic depiction of ion beam mixing is shown in Figure 14.

Tribological Experience

Research data pertaining to the tribological properties of ion beam-mixed solid lubricant films on ceramic surfaces are scarce. Wei et al. [74] and Lankford et al. [75] used ion beam mixing to modify the near-surface structure and chemistry, hence the tribological properties, of Si_3N_4 ceramics. According to their data, ion beam mixing of alternate layers of Ti and Ni with Si_3N_4 surfaces produced impressive tribological properties, especially at high temperatures. The results of short-duration friction tests demonstrated that friction coefficients in the range of 0.06–0.09 could be achieved at 800 °C during sliding against TiC pins.

Recently, Bhattacharya et al. [76] accomplished ion beam mixing of sputtered MoS_2 films with sapphire surfaces and reported improvements of two to three orders of magnitude in the functional lifetimes of these films in humid and dry air and in argon environments. Table 2, taken from their work, summarizes the tribological performance of ion beam-mixed and sputtered molybdenum disulfide, MoS_2, films on sapphire substrates.

Kohzaki et al. [77,78] investigated the effect of Ar^+ ion beam mixing on the friction and wear behavior of vacuum-evaporated Nb films on SiC substrates.

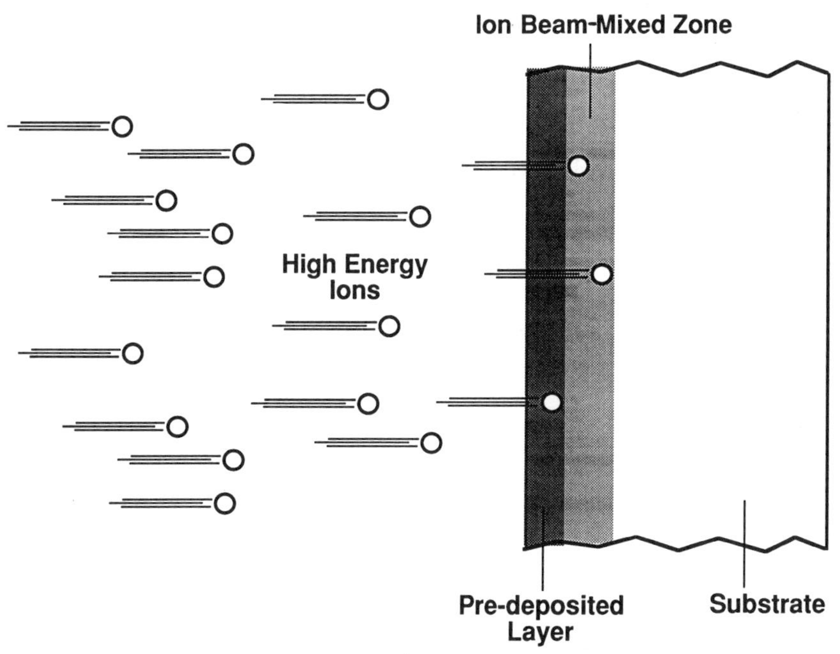

Figure 14 Schematic representation of ion beam mixing (IBM).

Table 2 Comparison of the Tribological Performance of Ag^+ Ion Beam-Mixed MoS_2 with that of Sputtered MoS_2 on Sapphire Substrates[a]

	Steady state friction coefficients			Functional lifetime (number of revolutions)		
MoS_2 film	Dry air	Humid air	Argon	Dry air	Humid air	Argon
Sputtered	0.025	0.18	0.03	13,000	200	15,000
Ion beam-mixed	0.03	0.17	0.02	145,000	5000	345,000

[a]Ion beam mixing conditions: acceleration energy, 2.0 MeV; ion dose, 5×10^{15} Ag^+/cm^2. Tribotest conditions: load, 2 N; sliding velocity, 0.08 m/s; temperature, 20°C; radius of sapphire ball, 3.18 mm.
Source: Reference 76.

They could not measure any wear on SiC substrates mixed with Nb. While sliding against a diamond pin, these ion beam-mixed Nb films exhibited a friction coefficient of $\approx$ 0.04. Without ion mixing, Nb films delaminated rapidly and the wear of underlying SiC substrates was quite substantial.

2. Ion Beam-Assisted Deposition

Principles

As discussed earlier, solid lubricant films can afford low friction and low wear to ceramic surfaces; however, they are effective only as long as they remain intact on these surfaces. Because of high shear forces, films with poor adhesion delaminate quite rapidly and leave the underlying substrate unprotected. The preceding section described IBM as an effective means of achieving strong film adhesion. However, it can be expensive, time-consuming, and sometimes impractical, especially for bulky, odd-shaped substrates. Another ion beam method, IBAD, which eliminates some of the shortcomings of IBM, is an alternative. In IBAD, substrate surfaces and growing films are subjected to concurrent bombardment with energetic ions (e.g., 50 eV to 10 keV) both before and during film formation. The size of the collision cascades is much smaller in IBAD than in IBM, and the penetration depth of ions is on the order of only a few monolayers to several nanometers. However, the striking ions are energetic enough to effectively sputter-clean the surfaces before film deposition and to enhance intermixing with the substrate by generating large numbers of point defects at the onset of film formation. Figure 15 schematically illustrates the IBAD process.

Several mechanisms have been proposed for the excellent adhesion attainable with IBAD-produced thin films on metallic and ceramic substrates [79,80]. Among others, sputter cleaning of the substrate prior to film formation is often postulated to play a major role in the adhesion of films produced by IBAD.

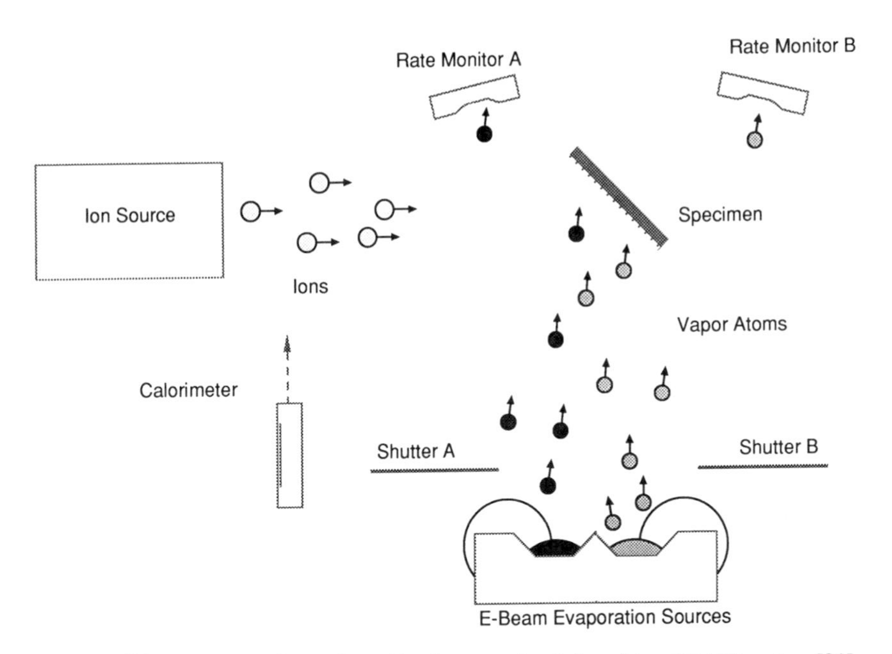

Figure 15 Schematic illustration of ion beam-assisted deposition (IBAD) system [81].

Establishment of a graded shallow interface has also been suggested as an important factor in improved film adhesion. Mechanical interlocking of film and substrate materials at the atomic level, enhanced chemical bonding across the interface, and greater nucleation density for the IBAD-produced films have also been proposed as important factors in improving film adhesion [80].

Concurrent ion bombardment of a growing film was found by Erdemir et al. [81] to alter the film morphology significantly. They demonstrated that the columnar growth morphology of Ag films can be suppressed. As shown in Figure 16a, without the use of concurrent ion bombardment (e.g., vacuum evaporation), Ag films consist of needlelike columnar grains with fine porosity along the boundaries. However, with concurrent ion bombardment (Fig. 16b), a very dense, equiaxial grain morphology is obtained. It is believed that such densification is a direct consequence of the energy imparted to the depositing film atoms by the impinging ions. When ion bombardment is continuous, film atoms can redistribute themselves and fill in vacancies and low atomic density regions (e.g., porous grain boundaries), thereby densifying the microstructure.

In an attempt to clarify the mechanisms involved in the strong adhesion of IBAD-produced films, the chemical and structural characteristics of the interfaces between IBAD-silver films and Al_2O_3 substrates was examined by transmission electron microscopy [81]. Figure 17 shows the details of an interface between

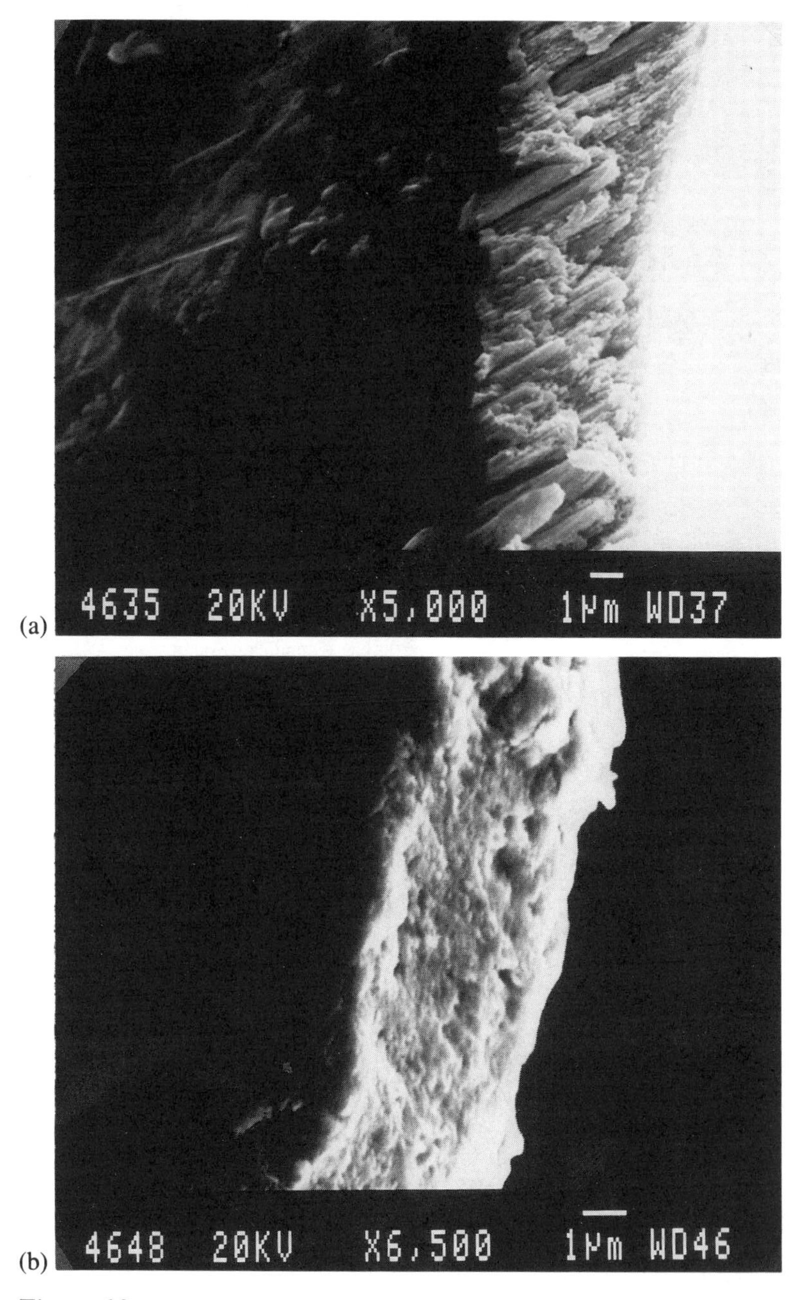

Figure 16 Cross-sectional scanning electron micrographs of Ag films deposited (a) without concurrent ion bombardment (i.e., vacuum evaporation) and (b) with concurrent ion bombardment (i.e., IBAD) [81].

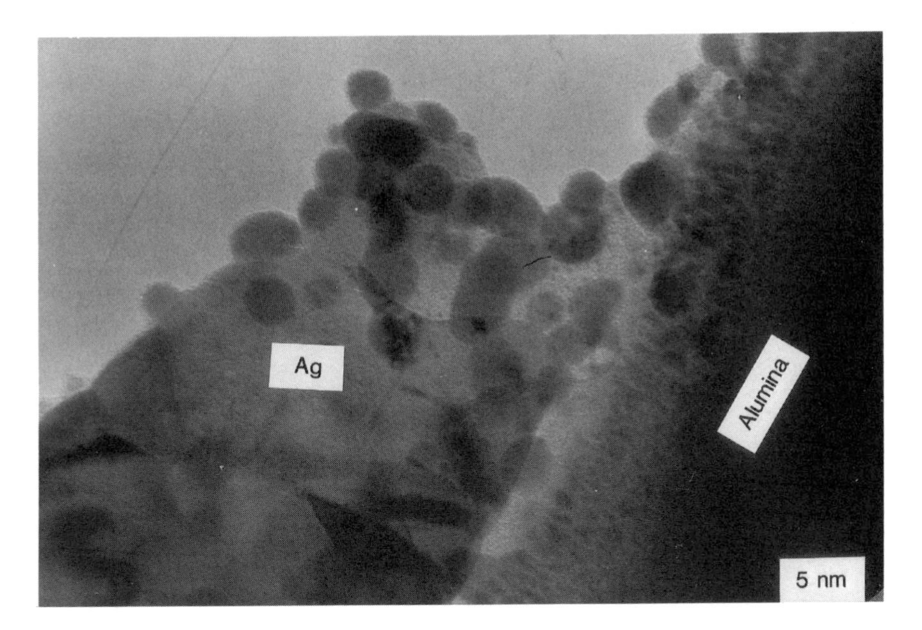

Figure 17 Cross-sectional transmission electron micrograph of the interface between an Ag film produced by IBAD and an Al_2O_3 substrate.

an IBAD-produced Ag film and an Al_2O_3 substrate. The interface is quite sharp, and there is no evidence of substantial intermixing between film and substrate atoms. However, there is no contaminant layer or discontinuity between the Ag film and the Al_2O_3 substrate. These observations suggest that effective sputter cleaning of substrate surfaces and perhaps some chemical bonding across the interface are the major factors in strong film-to-substrate adhesion in this case.

Tribological Experience

To confirm the superior adhesion of IBAD-produced Ag coatings to ceramics, a series of ball-on-disk tests was conducted with a sapphire disk; half of the disk was coated with Ag by the IBAD process; the other half was coated with Ag by a vacuum evaporation technique [81]. The coating produced by vacuum evaporation was rapidly stripped off the sapphire surface, but the coating produced by IBAD remained intact even after 1200 passes, thus verifying the excellent adhesion of IBAD-produced films to ceramic surfaces. Figure 18 shows the surface condition of the sapphire disk before and after wear testing.

Recently, Erck and Fenske [82,83] demonstrated that presputtering of Al_2O_3 and ZrO_2 surfaces with either oxygen ions or a combination of argon and oxygen ions greatly enhances the adhesion of Ag films to these ceramic substrates. Although the precise role of O ions in improved adhesion is not yet known, it

(a)

(b)

Figure 18 (a) Surface condition (a) before and (b) after wear test of a 38 mm diameter sapphire disk. Half of this disk was coated with Ag by vacuum evaporation, the other half was coated with Ag by IBAD. Test conditions: load, 5 N; sliding velocity, 0.57 m/s; temperature, 23°C; sliding distance, 94 m; radius of ball, 3.18 mm [81].

is speculated that, with O ion bombardment, the surface chemical reactivity of Al_2O_3 is increased. As a result, greater chemical bonding is achieved between silver and Al_2O_3 atoms. The results of tribological tests indicated that the wear of Al_2O_3 flats that were Ag-coated by IBAD was unmeasurably low; moreover, the wear of counterface balls sliding against these Ag-coated flats decreased sharply with increasing film adhesion achieved by O + Ar ion bombardment (see Fig. 19) [84]. Figure 19 shows that both the ion type and the duration of sputter cleaning play important roles in the adhesion and wear behavior of Ag films whose nominal thickness ranged from 1 to 2 μm. Exact details of the enhancement in adhesion remain to be determined. The possibility of the formation of AgO is an especially interesting facet to be explored.

As discussed earlier, the poor wear performance of Al_2O_3 and ZrO_2 ceramics during sliding at high velocities is largely associated with the inability of these materials to effectively dissipate frictional heat from their sliding interfaces [48,49]. Essentially all the mechanical work done to overcome friction is converted into heat, and this heat is generated near the areas of real contact (e.g., contacting asperity tips). In some instances, the temperature of these asperities

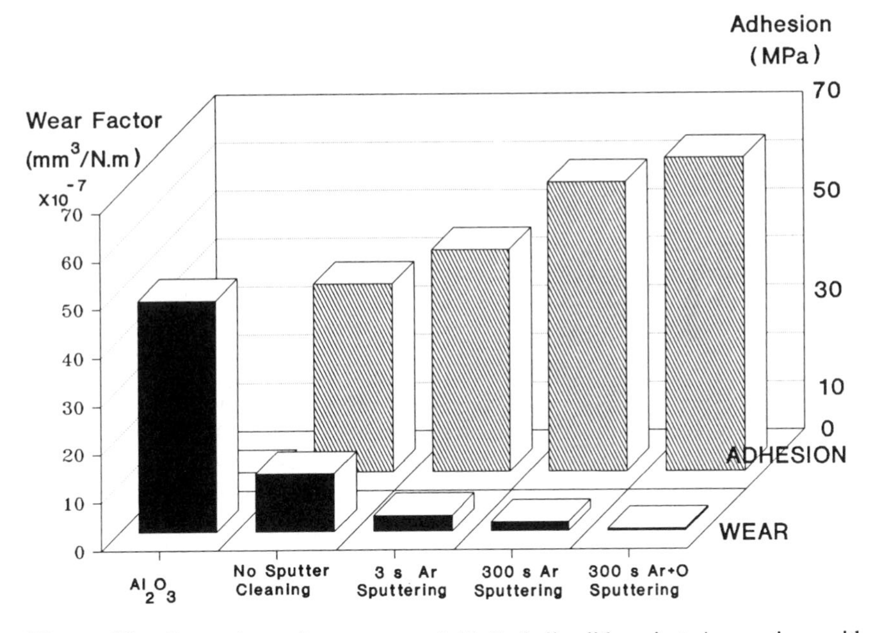

Figure 19 Comparison of wear rates of Al_2O_3 balls slid against Ag coatings with varying degrees of adhesion to flat Al_2O_3 specimens. Test conditions: load, 5 N; average oscillating sliding velocity, 0.05 m/s; relative humidity, 35%; sliding distance, 203 m; radius of ball, 6.35 mm. Adhesion was measured by a pull test machine (Quad Group, Inc., model no. Sebastian 5) [84].

(referred to as flash temperature) may reach a value comparable to the melting point of the base ceramic. As shown in Figure 4, rubbing surfaces of Al_2O_3 and ZrO_2 during sliding tests at 4 and 2 m/s, respectively, contained some evidence of local melting. Large wear fragments (Fig. 6) appear to have been forced out of the sliding interfaces in the molten state, after which they solidified and accumulated around the trailing edges of the balls.

In addition to local melting, large temperature gradients can develop between hot spots and cold surroundings, resulting in the generation of high thermal stresses. Because of the inherent brittleness of most ceramics, cracks can eventually develop and promote wear.

To achieve low friction and wear in ceramics (e.g., ZrO_2, Si_3N_4, Al_2O_3) with poor thermal conductivities (e.g., at 25°C, $k_{Al_2O_3} = 25$ W m^{-1} °C^{-1}, $k_{ZrO_2} = 2$ W m^{-1} °C^{-1}, and $k_{Si_3N_4} = 17$ W m^{-1} °C^{-1}), Erdemir et al. [49,81] proposed the use of thin metallic films that combine high thermal conductivity with low shear strength and high chemical inertness. Specifically, they hypothesized that such thin films can help effectively dissipate the frictional heat laterally from the local contacts of insulating ceramics during sliding at high velocities, thus reducing wear due to thermomechanical stresses and melting. Thin films of Ag and Au appear very promising, because their thermal conductivities k are among the highest (e.g., at 25°C, $k_{Ag} = 420$ W m^{-1} °C^{-1} and $k_{Au} = 317$ W m^{-1} °C^{-1}) of all metals. Because of their low shear strength, these metals have been used to lubricate metallic tribocomponents. Copper, because of its high thermal conductivity (e.g., $k_{Cu} = 401$ W m^{-1} °C^{-1}), may also be effective; however, when compared with Au and Ag, it lacks the high chemical inertness that is necessary for long-term durability, especially at higher temperatures or in aggressive environments.

Recent tribological studies have demonstrated that IBAD-produced Ag films could virtually eliminate the wear of Al_2O_3 and ZrO_2 disks in dry and humid air [48,49]. Uncoated counterface balls sliding against these disks experienced wear that was lower by two to three orders of magnitude than those slid against the uncoated disks, depending on sliding velocity. The rubbing surfaces of uncoated ceramic pairs appeared to have undergone severe microfracture, plastic flow, and local melting. However, no such features were seen on the sliding surfaces of pairs that contained an Ag film. Figure 20, which presents wear test results of ZrO_2 and Al_2O_3 ceramics with and without a Ag film, shows that Ag films can reduce the wear rates of balls quite markedly. When compared with Al_2O_3, ZrO_2 suffers much greater wear losses, but it also benefits most from the Ag film. This finding further supports the hypothesis that high thermal conductivity films can, indeed, be very useful in reducing the wear of low thermal conductivity ceramics, especially when high sliding velocities (≤ 0.5 m/s) are involved.

In its current version (i.e., Version 2.0), the T-MAPS software does not specifically handle coatings. However, assuming that continuum mechanics apply

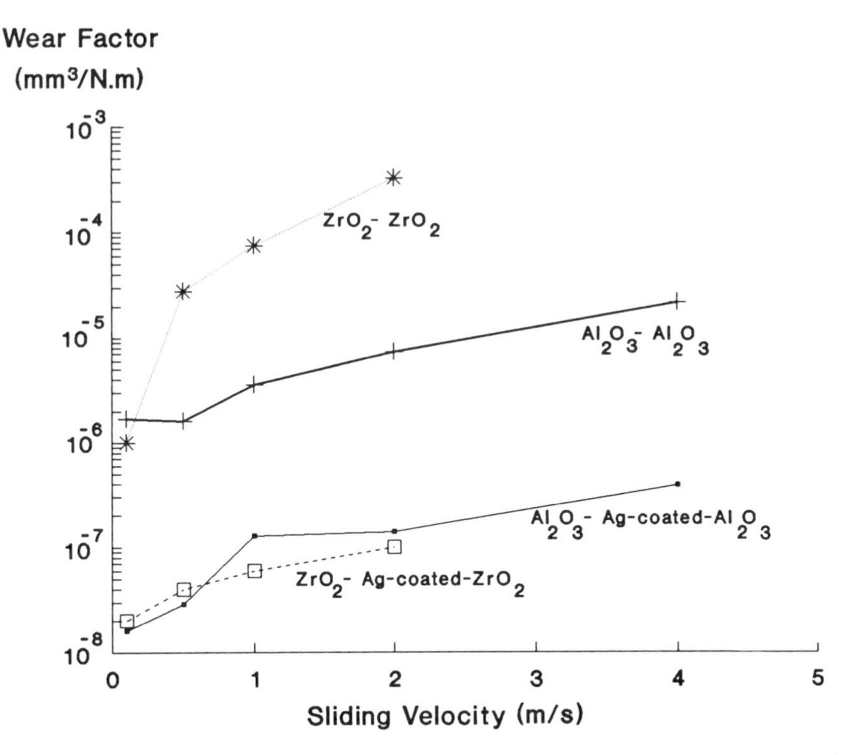

Figure 20 Variation of wear factors of Al_2O_3 and ZrO_2 balls sliding against uncoated and Ag-coated Al_2O_3 and ZrO_2 disks at various velocities. Test conditions: load, 5 N; relative humidity, > 1%; sliding distance, 2000 m; ball radius, 4.77 mm.

and that thermal interactions at the sliding interfaces occur only between the uncoated pins and the Ag applied on a disk, a temperature map (Fig. 21) was created for the sliding pair of ZrO_2/Ag-coated ZrO_2 using the T-MAPS software of Ashby et al. [56]. The flash temperatures predicted for sliding tests at 1 and 2 m/s are essentially the same and very low. When compared with the flash temperatures presented in the ZrO_2/ZrO_2 temperature map (Fig. 3), the flash temperatures shown in Figure 21 are much lower. Material properties used to generate the temperature map in Figure 21 are given in Table 3.

It is important to emphasize that this approach is only a first attempt to account for the heat-dissipating effect of highly conducting thin films. It should be used only to assess the expected flash temperatures, since only short-range, lateral heat flow between asperity contact points is required to reduce asperity heating. For bulk temperature estimates, the assumption of ZrO_2/ZrO_2 contact as in Figure 3 is more appropriate, since the bulk heating is determined by the long-range

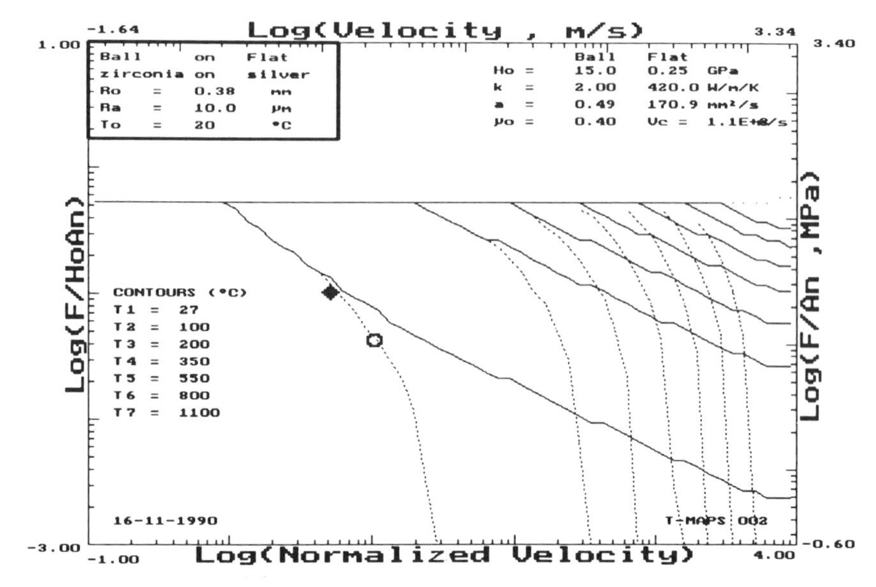

Figure 21 Temperature map of a sliding pair consisting of ZrO_2 ball and Ag-coated ZrO_2 disk: ♦, flash temperature at 1 m/s; ○, flash temperature at 2 m/s. R_0 is nominal contact radius; R_a, radius of single isolated asperity junction; T_0, ambient temperature; H_0, hardness of ZrO_2 at room temperature; k, thermal conductivity; a, thermal diffusivity; μ_0, friction coefficient; v_c, critical velocity (material properties are given in Table 3).

Table 3 Properties of Silver Used in the T-MAPS Software

Property	Units	Ag[a]
Density	kg/m^3	10,490
Young's modulus	GPa	71
Temperature dependence of modulus	—	0.5
Yield strength	MPa	54
Temperature dependence of yield strength	—	0.5
Hardness at room temperature	MPa	250
Melting point	K	1234
Thermal conductivity	$W\ m^{-1}\ K^{-1}$	420
Thermal capacity	$J\ kg^{-1}\ K^{-1}$	234
Latent heat of melting	kJ/kg	72

[a]Data from *Metals Handbook*, 10th ed, Vol. 2, ASM-International; Metals Park, OH, pp. 699–702, 1989, and *Smithells Metals Reference Book*, 6th ed., Butterworths: London, pp. 15.3, 22.189, 1983.

transport of heat through the ceramic substrate to the heat sink where the specimen is secured to the test machine frame.

A comparison of the wear performance of ZrO_2 balls slid against Ag-coated ZrO_2 disks with that of ZrO_2 balls slid against uncoated disks indicated that the wear rate of the balls slid against the Ag-coated ZrO_2 disks was lower by a factor of more than (3000 during sliding at 2 m/s). Based on the temperature maps for uncoated and Ag-coated test pairs and the wear test results, it is reasonable to conclude that the fast lateral dissipation of frictional heat from the local contact areas of ceramics is important for achieving low wear rates. Therefore, the wear-controlling factor at high sliding velocities relates closely to the rapid lateral dissipation of frictional heat from asperity contacts.

Confirming what one would expect from the lower flash temperatures obtained with Ag films, Figures 22 and 23 suggest that these films can be very effective in preventing microfracture and plastic flow of the ceramic materials during sliding contact. These figures show the morphology of a wear track at low and high magnification. It is clear from Figure 22b that the Ag film undergoes extensive plastic deformation to accommodate the surface tensile and normal stresses created during sliding contacts. Because Ag is very soft, it shears easily when subjected to tangential loads. As a result, friction, hence the magnitude of the surface tensile stresses being generated behind the moving asperities, is reduced. As discussed in Section II.A, high tensile stresses can initiate surface cracks, thus promoting wear.

F22,23

Figure 22c shows the condition of a typical wear scar formed on the ball, whereas Figure 22d presents the Ag X-ray map of this wear scar. The X-ray map verifies that some Ag transfers from the disk to the rubbing areas of the pin. This type of transfer of Ag from one sliding face to another can be very beneficial to the functional lifetime of the Ag films. Such a transfer from disk to pin and back to disk will keep Ag within the sliding interface; consequently, Ag will remain at the sliding interface longer and provide prolonged lubrication. However, the transfer may be detrimental to the frictional behavior of the sliding interfaces, because, Ag-to-Ag junctions can easily form between the Ag film on the disk and the transferred Ag on the pin. According to the adhesion theory of friction, the formation and subsequent rupture of these junctions can give rise to high friction [61]. For most ceramic pairs using Ag films, the steady state friction coefficients were in the range of 0.3–0.5 during sliding tests at room temperature.

Figure 23 presents a high magnification scanning electron micrograph of a wear scar formed on an Al_2O_3 ball during sliding against an Al_2O_3 disk coated with an IBAD-Ag film. Figure 23a reveals that transferred Ag was smeared effectively onto the wear scar and that it filled in valleys and internal volume defects (e.g., porosity, grain boundary voids, flaws, etc.). There is no evidence of extensive microfracture or other severe wear activity on this wear scar.

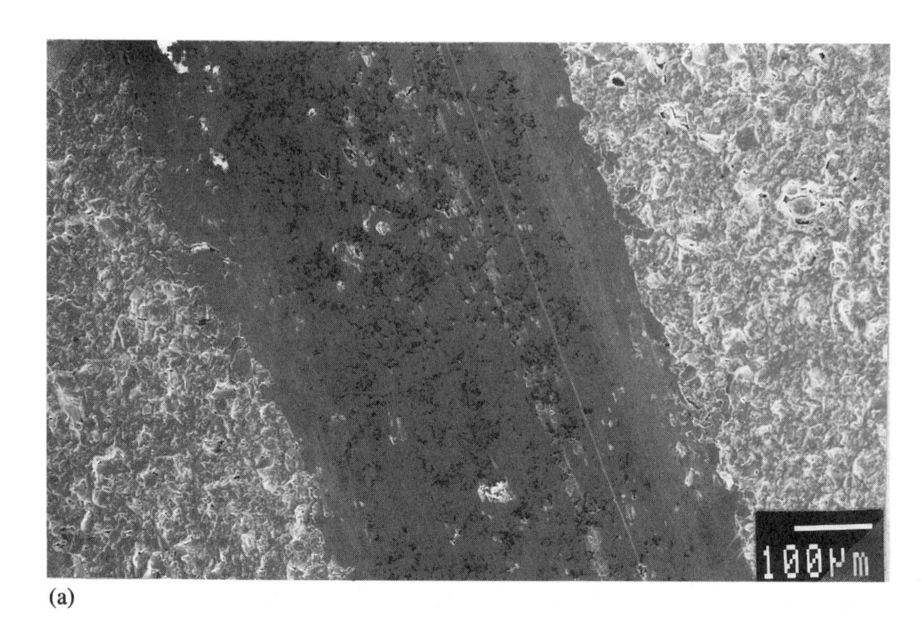

(a)

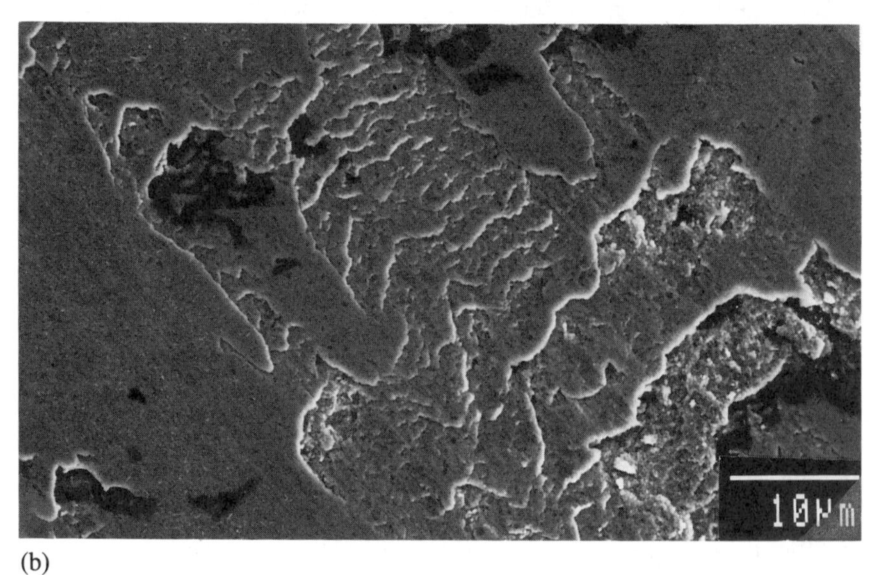

(b)

Figure 22 (a) Low and (b) high magnification scanning electron micrographs of a wear track formed on an Ag-coated Al_2O_3 disk. (c) Low magnification electron micrograph and (d) Ag-X-ray map of a wear scar formed on an Al_2O_3 ball. Test conditions: load, 5 N; sliding velocity, 4 m/s; sliding distance, 2000 m; relative humidity, > 1%; temperature, 23°C.

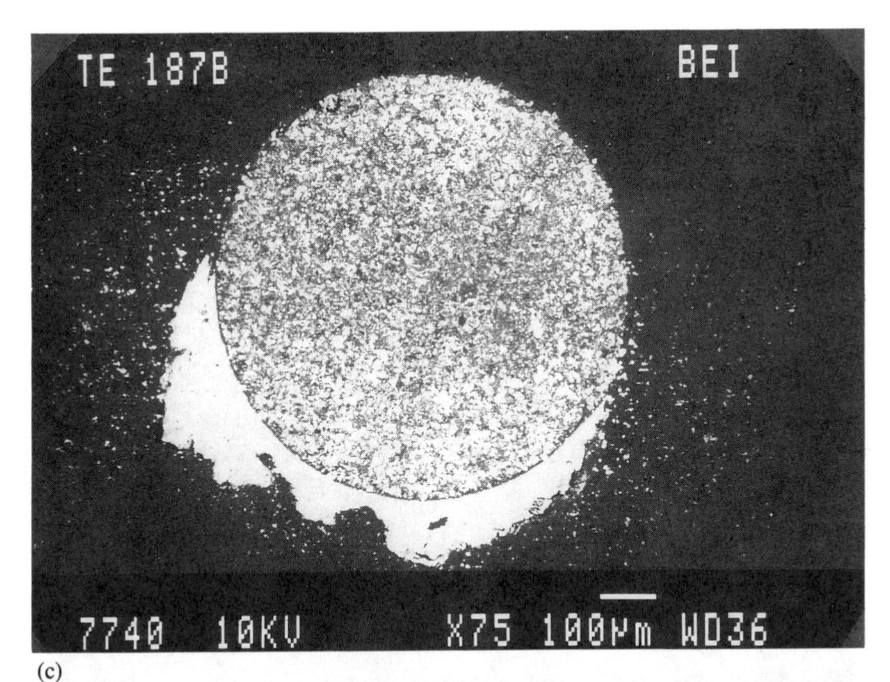
TE 187B
BEI
7740 10KV X75 100µm WD36
(c)

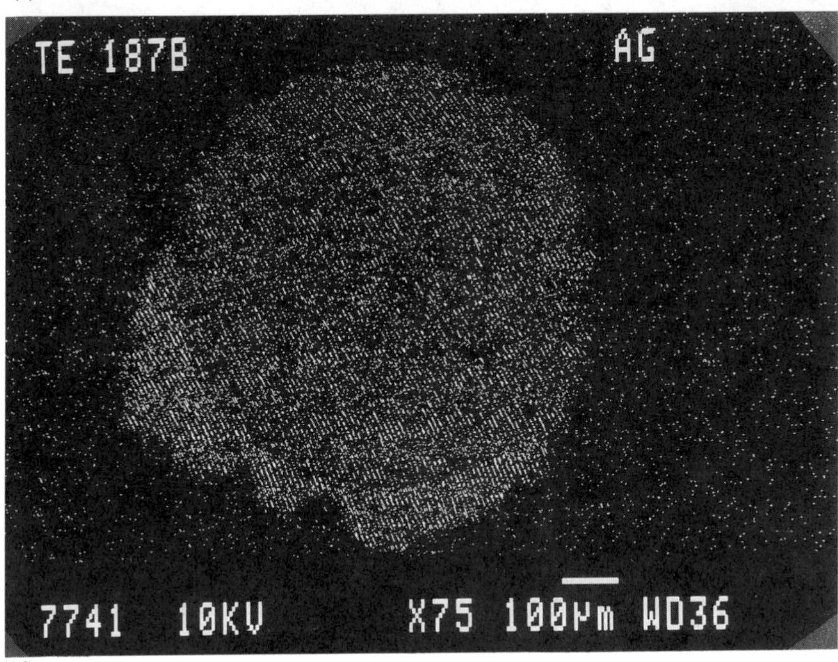
TE 187B
AG
7741 10KV X75 100µm WD36
(d)

(a)

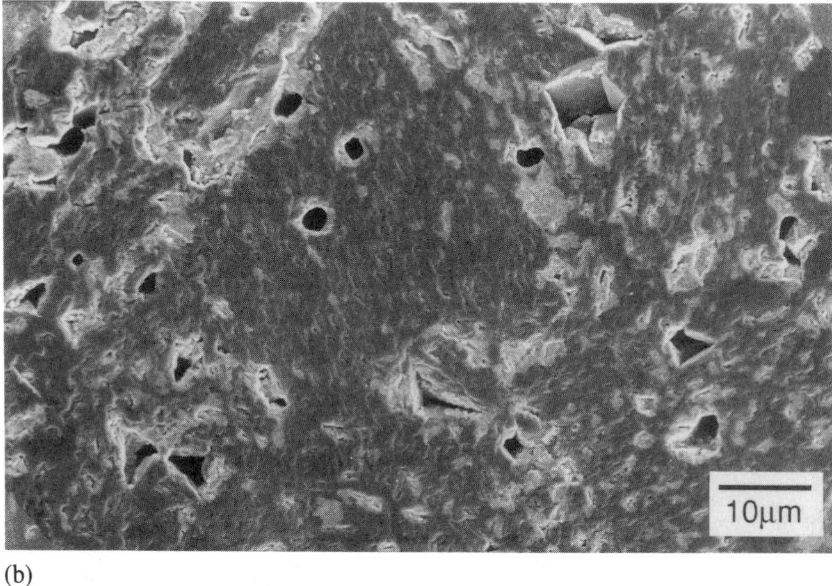

(b)

Figure 23 Surface condition of a wear scar formed on an Al_2O_3 ball (a) before and (b) after acid etching of transferred Ag from the wear scar. Test conditions: load, 5 N; sliding velocity, 4 m/s; sliding distance, 2000 m; relative humidity, $>$ 1%; temperature, 23°C.

However, the micrograph taken from the same wear scar after removal of Ag by acid etching (Fig. 23b) reveals that despite the presence of various internal volume defects, the rubbing surface had not undergone significant microfracture. There is no evidence of crack initiation from the defects. The wear, which can be characterized as a polishing type, was very mild.

In addition to room temperature improvements, Ag coatings imparted significantly improved friction and wear properties to ceramic surfaces at elevated temperatures. Using a reciprocating ball-on-disk apparatus, Erdemir et al. [85,86] observed that adherent silver films were capable of significantly reducing the wear of Al_2O_3, Si_3N_4, and ZrO_2 flats. The wear rates of counterface balls (uncoated) sliding against these Ag-coated flats were lower by factors ranging from one to three orders of magnitude than those of balls slid against uncoated flats. The Ag films used in these experiments were intact on the sliding surfaces even after 110,000 passes under the following test conditions: load, 10 N; average velocity, 0.05 m/s; relative humidity, 20%; tip radius of pin, 6.35 mm. When tested at temperatures higher than 400°C, the Ag films tended to partially lose their adherence and form a network of islands. Deposition of a thin Ti and/or Cr underlayer may have a beneficial effect on the wettability of Ag films at high temperatures, thus extending their useful range.

IV. SUMMARY AND FUTURE DIRECTIONS

This overview demonstrates that ceramics have much to offer for future tribological applications. However, their widespread use in advanced tribosystems must await the development of effective lubricants and lubrication concepts, because both the friction and wear coefficients of these materials are unacceptably high for most commercial applications. The results of earlier investigations suggest that depending on the tribological and environmental constraints, microfracture as well as both tribochemical and thermomechanical interactions may occur at sliding interfaces, controlling the wear behavior of ceramics. The inherent brittleness and usually poor thermal conductivity of these materials are particularly detrimental to their generally poor wear performance. Existing data suggest that with proper lubrication, ceramics may live up to their promise. For applications involving high service temperatures, current liquid lubricants appear to be ineffective. Lubricants and additives of new types are urgently needed. A combination of solid and liquid lubrication may provide short-term solutions to the problems associated with the high friction and wear of ceramics at elevated temperatures. Alternatively, lubrication from vapor phases appears promising and may be practical for ceramic tribocomponents to be used at elevated temperatures.

Strong adhesion is essential for long service life of solid lubricant films. Ion-beam processes, such as ion-beam mixing and ion-beam-assisted deposition, are capable of imparting strong adhesion between solid lubricant films and ceramic substrates. Ion-beam mixing of ceramics with conventional solid lubricants, such as MoS_2, is feasible and appears promising for demanding aerospace applications. For ceramics with poor thermal conductivity, metallic films combining high thermal conductivity with low shear strength and good chemical inertness should be considered. Initial results from diamond and diamondlike carbon-coated ceramics are encouraging. Ways of achieving strong adhesion must be developed for long-term applications. A unique solid lubricant, boric acid, which forms naturally on the surfaces of ceramics containing boric oxide and boron, has recently been discovered. It was shown that this lubricant can impart remarkably low friction coefficients (e.g., 0.02) to sliding ceramic interfaces in humid environments, where MoS_2 is known to be ineffective.

ACKNOWLEDGMENT

This work was supported by the U.S. Department of Energy, Office of Transportation Materials, under contract W-31-109-Eng-38.

REFERENCES

1. *Tribology of Ceramics*, National Materials Advisory Board, National Academy Press: NMAB Publication No. 435, (1988).
2. D. C. Cranmer, *Tribol. Trans. 31*, 164 (1988).
3. S. Jahanmir, in *New Materials Approaches to Tribology, Theory and Applications*, Vol. 140, *MRS Symposium Proceedings*, L. E. Pope, L. L. Fehrenbacker, and W. O. Winer (Eds.), Materials Research Society: Pittsburgh, 1989, p. 285.
4. M. L. Sheppard, *Ceram. Bull. 69*, 1012 (1990).
5. R. N. Katz, Opportunities and prospects for the application of structural ceramics, in *Structural Ceramics*, Vol. 29, *Treatises on Materials Science and Technology*, J. B. Wachtman, Jr. (Ed.), Academic Press: Orlando, FL, 1989, p. 1.
6. S. Jahanmir (Ed.), *Tribology of Ceramics*, STLE Special Publication No. 23, Society of Tribologists and Lubrication Engineers: Park Ridge, IL, 1987.
7. S. Jahanmir (Ed.), *Tribology of Ceramics*, STLE Special Publication No. 24, Society of Tribologists and Lubrication Engineers: Park Ridge, IL, 1987.
8. V. J. Tennery (Ed.), *Ceramic Materials and Components for Engines*, American Ceramic Society: Westerville, OH, 1989.
9. *Proceedings of the 1987 Coatings for Advanced Heat Engines Workshop*, U.S. Department of Energy: Washington, DC, Report Conf. 870762, (1987).
10. C. S. Yust, and R. G. Bayer (Eds.), *Considerations in Ceramic Friction and Wear Measurements: Selection and Use of Wear Tests for Ceramics*, ASTM, STP No. 1010, American Society for Testing and Materials: Philadelphia, 1988.
11. D. C. Larsen, J. W. Adams, L. R. Johnson, A. P. S. Teotia, L. G. Hill (Eds.)

Ceramic Materials for Advanced Heat Engines: Technical and Economic Evaluation, Noyes: Park Ridge, NJ, 1985.

12. L. E. Pope, L. L. Fehrenbacker, and W. O. Winer (Eds.), *New Materials Approaches to Tribology, Theory and Applications*, Vol. 140, *MRS Symposium Proceedings*, Orlando, FL, 1989, Materials Research Society, Pittsburgh, 1989.
13. D. H. Buckley and K. Miyoshi, Tribological properties of structural ceramics, in *Structural Ceramics*, Vol. 29, *Treatises on Materials Science and Technology*, J. B. Wachtman, Jr. (Ed.), Academic Press: p. 293.
14. P. K. Rohatgi, P. J. Blau and C. S. Yust (Eds.) *Tribology of Composite Materials*, ASM International: Metals Park, OH, 1990.
15. R. K. Dwivedi, J. R. Ramberg, A. J. Gesing, and G. D. Webster, in *Ceramic Materials and Components for Engines*, V. J., Tennery (Ed.), American Ceramic Society: Westerville, OH, 1989, pp. 1384–1396.
16. K. A. Blakely, J. R. Schorr, and P. T. B. Shaffer, in *Ceramic Materials and Components for Engines*, V. J., Tennery (Ed.), American Ceramic Society: Westerville, OH, 1989, pp. 1515–1526.
17. P. K. Mehrotra, in *International Conference on Wear of Materials*, K. C. Ludema (Ed.), American Society of Mechanical Engineers: New York, 1987, pp. 301–312.
18. B. M. DeKoven and P. L. Dwyer, in *New Materials Approaches to Tribology, Theory and Applications*, Vol. 140, *MRS Symposium Proceedings*, L. E. Pope, L. L. Fehrenbacker, and W. O. Winer (Eds.), Materials Research Society: Pittsburgh, 1989, pp. 357–362.
19. A. V. Levy and N. Jee, *Wear*, *121*, 363 (1988).
20. K. F. Dufrane and W. A. Glaeser, International Congress and Exposition, Detroit, SAE Preprint No. 870416 (1987).
21. K. F. Dufrane and W. A. Glaeser, in *International Conference on Wear of Materials*, K. C. Ludema (Ed.), American Society of Mechanical Engineers: New York, 1987, pp. 285–291.
22. K. F. Dufrane, *Ceram. Eng. Sci. Proc. 9*, 1409 (1988).
23. R. Kamo, M. Woods, and P. Sutor, *Proceedings of the 1987 Coatings for Advanced Heat Engines Workshop* Castine, ME, Conf. 870762, National Technical Information Service: Springfield, VA, 1987, pp. 73–92.
24. W. J. Lackey, D. P. Stinton, G. A. Cerny, L. L. Fehrenbacher, and A. C. Schaffhauser, *Ceramic Coatings for Heat Engine Materials; Status and Future Needs*, Oak Ridge National Laboratory Report ORNL/TM-8959 (1984).
25. R. N. Katz and J. G. Hannoosh, *Int. J. High Technol. Ceram. 1*, 68 (1985).
26. L. B. Sibley and M. Zlotnick, *Mater. Sci. Eng. 71*, 283 (1985).
27. A. F. McLean and D. L. Hartsock, Design with structural ceramics, in *Structural Ceramics*, Vol. 29, *Treatises on Materials Science and Technology*, J. B. Wachtman, Jr. (Ed.), Academic Press: Orlando, FL, 1989, pp. 27–95.
28. M. L. Torti, The silicon nitride and sialon families of structural ceramics, in *Structural Ceramics*, Vol. 29, *Treatises on Materials Science and Technology*, J. B. Wachtman, Jr. (Ed.), Academic Press: Orlando, FL, 1989, pp. 161–192.
29. C. Suh, Y. Sato, and T. Kojima, in *Ceramic Materials and Components for Engines*, V. J., Tennery (Ed.), American Ceramic Society: Westerville, OH, 1989, pp. 1320–1333.

30. E. Amar, F. Gauthier, and J. Lamon, in *Ceramic Materials and Components for Engines*, V. J., Tennery (Ed.), American Ceramic Society: Westerville, OH, 1989, pp. 1334–1346.
31. B. Bhushan and S. Gray, *ASLE Trans*, *23*, 185 (1980).
32. R. P. Larsen and A. D. Vyas, *International Congress and Exposition*, Detroit, SAE Technical Paper Series, No. 880514 (1988).
33. C. S. Yust and F. J. Cariggnan, *ASLE Trans. 28*, 245 (1984).
34. M. Woydt and K.-H. Habig, *Tribol. Int. 22*, 75 (1989).
35. M. Woydt and K.-H. Habig, *Ceram. Eng. Sci. Proc. 9*, 1419 (1988).
36. W. O. Winer and B.-Y. Ting, *Development of a Theory of Wear of Ceramics*, Oak Ridge National Laboratory Report ORNL/84-7802/1 (1988).
37. T. E. Fischer and H. Tomizawa, *Wear*, *105*, 29 (1985).
38. H. Tomizawa and T. E. Fischer, *ASLE Trans. 29*, 165 (1986).
39. O. O. Ajayi and K. C. Ludema, in *International Conference on Wear of Materials*, K. C. Ludema (Ed.), American Society of Mechanical Engineers: New York, 1987, pp. 349–360.
40. G. R. Terwilliger and R. Redford, *Ceram. Soc. Bull. 53*, 465 (1974).
41. B. J. Hockey, *J. Am. Ceram. Soc. 54*, 223 (1971).
42. R. W. Rice, *Ceram. Eng. Sci. Proc. 6*, 940 (1985).
43. B. R. Lawn, A. G. Evans, and D. B. Marshall, *Am. Cer. Soc. Bull. 63*, 574 (1980).
44. G. M. Hamilton and L. E. Goodman, *J. Appl. Mech. 33*, 371 (1966).
45. S. M. Hsu, D. S. Lim, and R. G. Munro, in *Ceramic Materials and Components for Engines*, V. J., Tennery (Ed.), American Ceramic Society: Westerville, OH, 1989, pp. 1236–1245.
46. A. Skopp, M. Woydt, and K.-H. Habig, *Tribol. Int. 23*, 189 (1990).
47. M. Woydt, D. Klaffke, K.-H. Habig, and H. Czichos, *Wear*, *136*, 373 (1990).
48. A. Erdemir, G. R. Fenske, R. A. Erck, and D. E. Busch, Effect of sliding velocity on the wear Behavior of polycrystalline Al_2O_3 and silver-coated Al_2O_3, presented at the 45th Annual Meeting of the Society of Tribologists and Lubrication Engineers, Denver, May 7–9, 1990.
49. A. Erdemir, D. E. Busch, R. A. Erck, G. R. Fenske, and R. Lee, *Lubr. Eng. 47*, 863 (1991).
50. N. Wallbridge, D. Dowson, and E. W. Roberts, in *International Conference on Wear of Material*, K. C. Ludema (Ed.), American Society of Mechanical Engineers: New York, 1983, pp. 202–211.
51. R. Trabelsi, D. Treheux, G. Orange, G. Fantozzi, P. Homerin, and F. Thevenot, *Tribol. Trans. 32*, 77 (1989).
52. R. S. Gates, S. M. Hsu, and E. E. Klaus, *STLE Trans. 32*, 357 (1989).
53. K. Miyoshi, *Surf. Coat. Technol. 36*, 487 (1988).
54. M. F. Ashby, J. Abulawi, and H. S. Kong, *STLE Trans. 34*, 618 (1991).
55. B.-Y. Ting and W. O. Winer, *J. Tribol. 111*, 315 (1989).
56. M. F. Ashby, H. S. Kong, and J. Abulawi, *T-MAPS: A Program for Constructing Maps for Surface Heating in Unlubricated Sliding*, Version 2.0, Cambridge University Engineering Department, Cambridge, 1990.
57. T. E. Fischer, M. P. Anderson, and S. Jahanmir, *J. Am. Ceram. Soc. 72*, 252 (1989).

58. R. S. Gates, J. P. Yellets, D. E. Deckman, and S. M. Hsu, in *Selection and Use of Wear Tests for Ceramics*, C. S. Yust and R. G. Bayer (Eds.), ASTM STP No. 1010, American Society for Testing and Materials: Philadelphia, 1988, pp. 1–23.
59. D. S. Park, S. Danyluk, and M. McNallan, *Ceram. Trans. 10*, 159 (1990).
60. C. S. Yust and C. E. DeVore, *Tribol. Trans. 33*, 573 (1990).
61. F. P. Bowden and D. Tabor, *The Friction and Lubrication of Solids*, Oxford University Press: Oxford, 1964.
62. P. Studt, *Tribol. Int. 22*, 111 (1989).
63. R. Pool, *Science*, *246*, 444 (1989).
64. R. S. Gates and S. M. Hsu, *STLE Trans. 34*, 398 (1991).
65. E. E. Klaus, J. L. Duda, and S. K. Naidu, in *Formation of Lubricating Films at Elevated Temperatures From the Gas Phase*, National Institute of Standards and Technology, Special Publication No. 744, Washington, DC: Government Printing Office, 1988.
66. J. L. Lauer and B. G. Bunting, *Tribol. Trans. 31*, 339 (1988).
67. J. A. Thornton, Plasmas in deposition processes, in *Deposition Technologies for Films and Coatings*, R. F. Bunshah et al. (Eds.), Noyes: Park Ridge, NJ, 1982, pp. 19–62.
68. S. V. Didziulis, P. D. Fleischauer, B. L. Soriano, and M. N. Gardos, *Chemical and Tribological Studies of* MoS_2 Films on SiC Substrates, Aerospace Corporation: El Segundo, CA, Aerospace Report No. ATR-89 (4752)-4 (1990).
69. K. Miyoshi, J. J. Pouch, and S. A. Alterovitz, *Mater. Sci. Forum*, *52/53*, 645 (1989).
70. A. Erdemir, G. R. Fenske, and R. A. Erck, *Surf. Coat. Technol. 49*, 435 (1991).
71. A. Erdemir, *Lubr. Eng. 47*, 168 (1991).
72. A. Erdemir, G. R. Fenske, R. A. Erck, and F. A. Nichols, *Lubr. Eng. 47*, 179 (1991).
73. A. Erdemir, G. R. Fenske, F. A. Nichols, R. A. Erck, and D. E. Busch, in *Proceedings of the Japan International Tribology Conference*, Nagoya, Japan, 1990, pp. 1797–1802.
74. W. Wei, J. Lankford, and R. Kossowsky, *Mater. Sci. Eng. 90*, 307 (1987).
75. J. Lankford, W. Wei, and R. Kossowsky, *J. Mater. Sci. 22*, 2069 (1987).
76. R. S. Bhattacharya, A. K. Rai, and A. Erdemir, *Nucl. Instrum. Methods*, B59/60, 788 (1991).
77. M. Kohzaki, S. Noda, H. Doi, and O. Kamigaito, *Mater. Lett. 6*, 64 (1987).
78. M. Kohzaki, S. Noda, H. Doi, and O. Kamigaito, *Wear*, *131*, 341 (1989).
79. M. Laugier, *Thin Solid Films*, *81*, 61 (1981).
80. P.-J. Martin, *Gold Bull. 19*, 102 (1986).
81. A. Erdemir, G. R. Fenske, R. A. Erck, and C. C. Cheng, *Lubr. Eng. 46*, 23 (1990).
82. R. A. Erck and G. R. Fenske, *Lubr. Eng. 47*, 640 (1991).
83. R. A. Erck and G. R. Fenske, in *Ion Beam–Solid Interactions: Physical Phenomena*, J. A. Knapp et al. (Eds.), Materials Research Society: Pittsburgh, 1990, pp. 85–90.
84. R. A. Erck, A. Erdemir, and G. R. Fenske, *Surf. Coat. Technol.*, 43/44, 577 (1990).

85. A. Erdemir, G. R. Fenske, F. A. Nichols, and R. A. Erck, *Tribol. Trans. 33*, 511 (1990).
86. A. Erdemir, R. A. Erck, and G. R. Fenske, Ion-beam processing of advanced ceramics for improved tribological behavior, Extended abstract, presented at ESD-Ceramtec 90 Conference, Dearborn, MI, June 1990.

8

Self-Lubricating Ceramic Matrix Composites

Arup Gangopadhyay

Ford Motor Company
Dearborn, Michigan

Said Jahanmir and Marshall B. Peterson

National Institute of Standards and Technology
Gaithersburg, Maryland

ABSTRACT

In certain applications involving high temperatures or vacuum environments, liquid lubricants cannot be used effectively. Self-lubricating materials containing solid lubricants may provide an alternative lubrication method for these applications. This chapter reviews recent results that demonstrate the feasibility of this approach. First, the frictional behavior of alumina and silicon nitride lubricated with nickel chloride intercalated graphite is reviewed. It is shown that the friction coefficients of these ceramics sliding against steel are reduced by graphite lubrication. The reduction in the friction coefficient is shown to be related to the formation of a transfer layer containing a mixture of materials from the surfaces in contact. Considered next are the results on a series of self-lubricating ceramic matrix composites that were prepared by drilling small holes in the ceramics and filling the holes with the intercalated graphite and hexagonal boron nitride solid lubricants. A friction coefficient as low as 0.17 is observed for the silicon nitride–graphite composites. The friction and wear behaviors of self-lubricating composites prepared by casting slurries of zirconia powder mixed with intercalated graphite are also reviewed. The zirconia–graphite composites in sliding contact with steel from room temperature to 400 °C exhibit lower friction coefficients than the cast zirconia sliding against steel. At room temperature, the zirconia–graphite composite is found to have a friction coefficient of 0.11, compared to 0.30 for the cast zirconia materials.

I. INTRODUCTION

High wear resistance combined with high hardness, low density, the ability to retain strength and hardness at elevated temperatures, and corrosion resistance make ceramic materials attractive for applications at high temperatures and in harsh environments. Investigations of the friction and wear behavior of these materials, however, have revealed that they exhibit high friction coefficients, typically 0.5–0.8, under unlubricated sliding conditions [1–8]. The high friction coefficients of these materials limit their use as tribological components. To take advantage of the beneficial properties of advanced ceramic materials, their friction coefficient must be reduced to 0.1 or lower.

It has been observed that such a low friction coefficient (i.e., ≤ 0.1) can be achieved with liquid lubrication [9–14]. However, in certain applications involving high temperatures or vacuum environments, liquid lubricants cannot be used effectively. For example, the bearings and seals in advanced low heat rejection engines and gas turbines are required to operate over a temperature range that exceeds the capabilities of conventional liquid lubricants [15–18]. Also, the moving parts and driving systems in satellites, and the bearings and seals in the turbopumps of launch vehicles and for other space applications require continuous operation under vacuum environments [15,16]. Since liquid lubrication is not feasible for these applications, solid lubricants may be required to meet operational needs.

Solid lubrication of metallic materials has been studied in detail [19–22], but there is only limited information in the literature on the use of solids in the lubrication of ceramics. Recently, Gangopadhyay and Jahanmir [23] studied the friction and wear characteristics of alumina coated with a composite solid lubricant coating containing silver, antimony trioxide, and barium fluoride sliding against uncoated alumina in air. At room temperature, the friction coefficient without the coating was 0.40; the coating reduced the friction coefficient to 0.11. This low value was maintained up to 500 °C; but the solid lubricant coating exhibited a large wear rate. Erdemir et al. [24] observed a considerable improvement in wear resistance of alumina coated with a thin film of silver, but the steady state friction coefficient was only 0.40. Although this is lower than the friction coefficient of uncoated alumina for the test conditions used, it is too high for unlubricated sliding applications. Lankford et al. [25] have reported friction coefficients in the range of 0.06–0.09 for certain material combinations, where the ceramic surfaces were modified by ion implantation with Ti and Ni. Nastasi et al. [26] observed a reduction in the friction coefficient of SiC and TiB_2 by ion implantation of the surfaces with nitrogen.

Although deposited solid lubricant films or thin ion-implanted layers can reduce the friction coefficient and the wear rate, removal of the film by wear can be a problem. This limitation can be avoided by maintaining a constant supply of solid lubricant material at the sliding contact by incorporating the solid lubricant in reservoirs or as a second phase in the ceramic matrix. Van Wyk [27] evaluated this concept for the development of ceramic slider airframe bearings. Lubrication was provided from solid lubricant reservoirs dispersed over the alumina ceramic outer ring. The reservoirs were filled with either graphite or molybdenum disulfide. Solid lubrication has also been considered for silicon nitride bearings at elevated temperatures, where impregnated graphite cages reduced the wear of silicon nitride balls by an order of magnitude [8]. Wedeven et al. [29,30] have reported a low friction coefficient for solid lubricated silicon nitride in a rolling contact at temperatures up to 538°C. The observed low friction coefficient was obtained by burnishing graphite on the contacting surfaces.

This chapter reviews the results of a systematic study conducted to evaluate the feasibility of achieving a low friction coefficient in self-lubricating ceramic matrix composites. First, the feasibility of lubricating alumina and silicon nitride with intercalated graphite was investigated. Second, the friction and wear characteristics of composites, which contained reservoirs filled with intercalated graphite and hexagonal boron nitride solid lubricants, were determined. Finally, self-lubricating ceramic matrix composites were made by casting a slurry of zirconia powder with graphite solid lubricant.

II. LUBRICATION OF ALUMINA AND SILICON NITRIDE

A. Materials and Test Procedure

The ceramic materials selected for this investigation were high purity alumina and hot isostatically pressed silicon nitride. The properties and compositions of these materials are listed in Table 1.

Graphite powder intercalated with nickel chloride ($NiCl_2$), having an average particle size of 100 μm in the planar direction, was chosen as the solid lubricant. It contained 81.5 wt % C, 5.8 wt % Ni, 12.2 wt % Cl, and <0.5 wt % ash. Graphite has a layered lattice structure with weak interplanar van der Waals bonding. This permits easy shear between the layers, resulting in a low friction coefficient. Intercalation of layered lattice materials with additional elements or compounds increases the distance of separation between the layers. This reduces the interplanar bonding energy and allows easy slippage of one plane of atoms over the other, which results in more effective lubrication behavior [31,32].

The friction and wear tests were conducted using a pin-on-ring contact geometry as shown in Figure 1. The pin holder was modified to accommodate two pins [33]—one made of either alumina or silicon nitride and the other made of intercalated graphite. Graphite pins were prepared by pressing the powders in

Table 1 Properties and Compositions of Alumina, Silicon Nitride, and Boron Nitride

	Alumina (Coors AD 998)	Silicon nitride (Norton NBD 100)	Boron nitride (Union Carbide HBN)
Composition, wt %	0.03 Na_2O, 0.01 K_2O 0.02 TiO_2, < 0.1 SiO_2 balance Al_2O_3	0.4–0.6 Mg, 0.18–0.3 Al 0.16–0.35 Fe, 0.05 Mn 0.006–0.03 Ca	43 B, 56 N, 1 (max) O, 0.2 (max) C, 0.01 (max); metallic impurities
Grain size, μm	3	0.1	
Porosity %	0.2	~ 0	
Knoop hardness, GPa	15.2 (at 1000 g)	17.9 (at 500 g)	
Elastic modulus, GPa	345	310	48
Compressive strength, MPa	2071	3000	52
Thermal conductivity, W m^{-1} K^{-1}	29.4	32	17
Fracture toughness, (MPa $m^{1/2}$)	3.8	5.4	
Thermal expansion coefficient, $°C^{-1}$	6.7×10^{-6}	3.5×10^{-6}	8.9×10^{-7}

[a]The information on vendors is included only to identify the materials, no endorsement of the products of any manufacturer is implied.

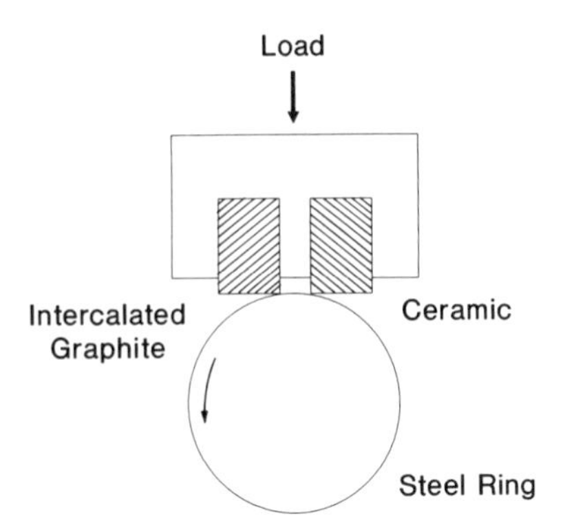

Figure 1 Schematic drawing of the two-pins-on-ring contact geometry

a tool steel die at a pressure of 1.5 GPa. All the pins were 9.5 mm long and 6.25 mm in diameter. AISI 52100 steel rings with a hardness of 58 Rockwell C and a surface roughness of 0.23 μm rms were used as the counterface. The steel rings were 8 mm wide and had a diameter of 35 mm.

Prior to the wear tests, the ceramic pins and the graphite pins were polished with a 1 μm diamond paste and a 600 grit paper, respectively. The ceramic pins and the steel rings were thoroughly cleaned with organic solvents [34].

In addition to alumina–graphite and silicon nitride–graphite combinations, one set of tests were conducted with alumina–alumina, silicon nitride–silicon nitride and graphite–graphite combinations for reference. The tests was conducted at room temperature in the ambient laboratory air, with the relative humidity ranging from 10 to 30%. A normal load of 33 N, a sliding speed of 0.14 m/s, and a sliding distance of 790 m were used in each test. Wear of the steel ring was measured by weight loss measurements and is expressed in terms of the wear coefficient K using the following relationship:

$$K = \frac{WH}{SN\rho}$$

where W is the weight loss, H the hardness, S the sliding distance, N the normal load, and ρ the density.

B. Friction and Wear Data

The friction coefficients of various material pairs sliding against steel are shown in Figure 2a. The reported friction coefficients are the averages of the last 20% of the test cycles. As shown in Figure 2a, the friction coefficient of alumina sliding against steel is 0.48. However, the friction coefficient is reduced to 0.20 for an alumina–graphite combination. The friction coefficient of silicon nitride sliding on steel is 0.45; and by the replacement of one silicon nitride pin with a graphite pin, the friction coefficient is reduced to 0.17. This value is slightly larger than the friction coefficient of intercalated graphite, which is 0.14.

Figure 2b shows the wear coefficients of the steel rings tested against different pairs of materials. It is observed that the wear coefficient of the rings is reduced by a factor of 5 when a graphite pin is used with an alumina pin. However, no appreciable difference in the wear coefficient is observed between silicon nitride and the silicon nitride–graphite combination. The steel ring tested against two intercalated graphite pins showed an increase in weight, indicating the formation of a transfer film on the steel surface. Assuming a uniform distribution, the average thickness of the transfer film calculated from the weight gain was approximately 1 μm.

C. Analysis of Wear Surfaces

The worn surfaces of the ceramic pins, the graphite pins, and the steel rings were examined in a scanning electron microscope (SEM) equipped with an energy-dispersive spectroscope to elucidate the mechanisms responsible for the reduction in the friction coefficient. A micro-Raman spectrometer and a micro-Fourier-transformed infrared (micro-FTIR) spectrometer were also used to analyze the chemical composition of the surfaces. The details of the analysis can be found elsewhere [34].

Figure 3a shows a typical scanning electron micrograph of the worn surface of the steel ring tested with two intercalated graphite pins. A discontinuous transfer film of variable thickness is observed on the surface of the ring. The energy-dispersive spectrum of the area (Fig. 3b) indicates the peaks for Fe and Cr for steel, and a small O peak, possibly for iron oxide. The spectrum also shows the peaks corresponding to Ni and Cl, and a small peak for C. Since energy-dispersive spectroscopy (EDS) is not very sensitive to carbon, which has a low atomic number, the presence of intercalated graphite is primarily indicated by the Ni and Cl peaks.

The worn surfaces of the steel rings tested against alumina and graphite pins were found to contain patches of thin transfer films as shown in Figure 4a. The EDS spectrum of the film (Fig. 4b) is very similar to the spectrum in Figure 3b for the steel ring tested against the graphite pins. These data confirm the presence of intercalated graphite in the transfer film on the steel ring.

Examination of the worn surfaces of the alumina pins tested with the graphite pins also revealed a discontinuous patchy film (Fig. 4c). EDS analysis of the film (Fig. 4d), indicates the presence of intercalated graphite and Fe. (The additional Au peak is due to the sputtered gold coating used to make the sample electrically conductive during SEM examination.) A buildup of reddish-brown wear debris was observed by optical microscopy at the entrance edge of the wear track on the alumina pin. The analysis of the wear debris by EDS indicated that the debris was a mixture of graphite and iron oxide.

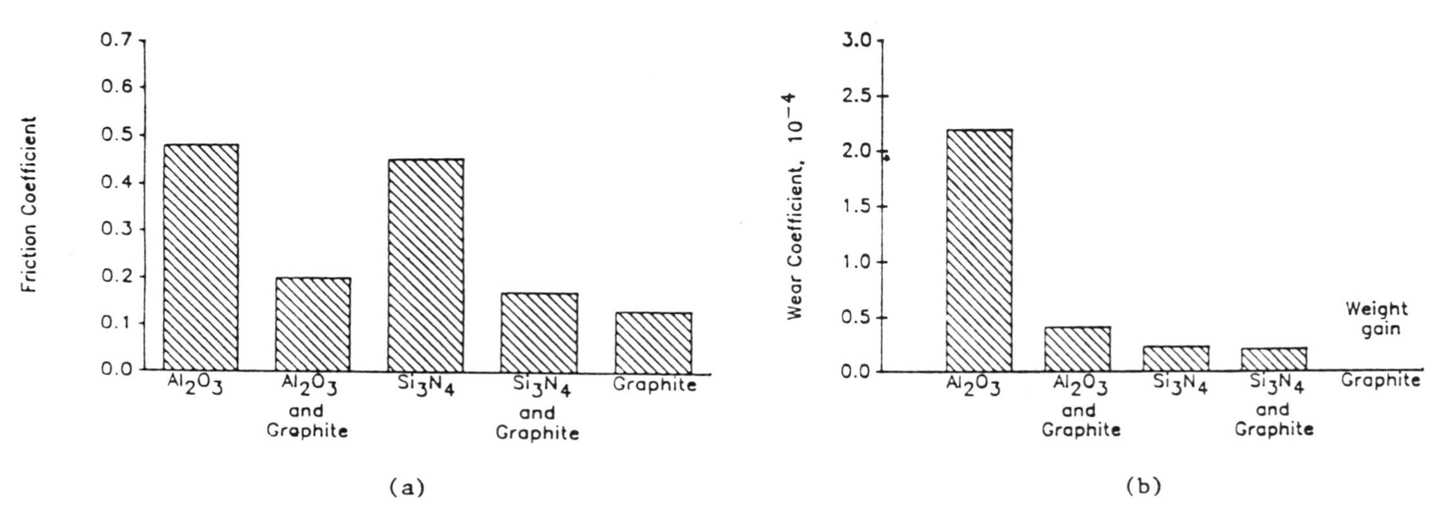

Figure 2 (a) Friction coefficients and (b) wear coefficients of steel ring tested against different material pairs.

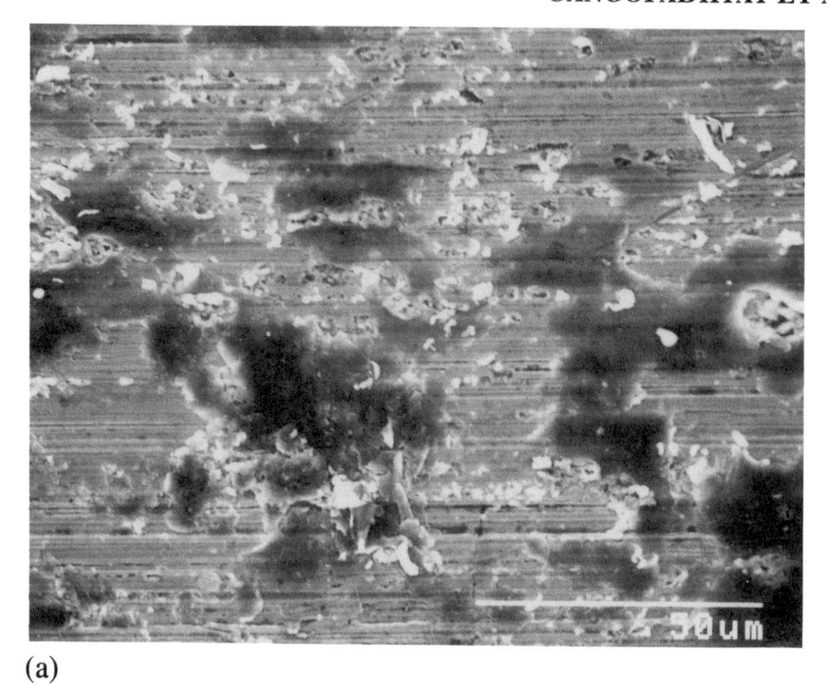

(a)

Intensity (thousand)

Fe

Cl

Fe

C Ni

O

Cr

Fe Ni

Energy, KeV

(b)

Figure 3 (a) Scanning electron micrograph of the worn surface of steel ring tested with graphite pins. (b) Energy-dispersive spectrum of the transfer film.

The worn surface of the steel ring tested with the combination of silicon nitride and graphite pins is shown in Figure 5a. The surface is covered with relatively thick patches of a transfer film. An EDS spectrum of the area in the micrograph (Fig. 5b) is very similar to the spectrum shown in Figure 4b for the steel ring tested against alumina and graphite pins, with the exception of the Si peak. Since no N peak is observed, formation of silicates by tribochemical reactions between silicon nitride and moisture in the air is assumed. The formation of silicates has been detected by FTIR spectroscopy on the worn surface of steel rings tested with silicon nitride pins [34].

Raman spectroscopy was used to gain a better understanding of the nature of graphite in the transfer film. Figure 5c shows a Raman spectrum from the transfer film on the steel ring. A Raman spectrum from an unworn graphite pin (Fig. 5d) is shown for comparison: there is a strong peak at 1575 cm^{-1} and a much weaker peak at 1350 cm^{-1}. These peaks are often referred to as the "graphitic" or "G" band and the "disorder" or "D" band, respectively. In general, the greater the "D" band intensity relative to the "G" band, the smaller the grain size or particle size, whereas the broadening of the peak widths indicates increased interplanar disorder within the graphitic structure [35]. The Raman spectrum from the transfer film on the steel ring (Fig. 5c) indicates that the graphite in the film is different from the unworn graphite. In the transfer film, the intensity of the peak at 1350 cm^{-1} is stronger than that at 1575 cm^{-1}. Based on the intensity ratio of the two peaks, the grain size of graphite in the transfer film can be estimated at 40 Å [36], which is considerably less than the unworn grain size of 100 μm. Also, based on the two peak widths, the graphite structure in the transfer film can be considered to be highly disordered, possibly because after the graphite particles have been transferred onto the steel ring, they are subjected to severe deformation and fracture every time they pass through the contact area.

Examination of the worn surface of silicon nitride tested with one graphite pin against a steel ring also revealed a patchy transfer film (Fig. 6a). The Raman spectrum of the transfer film (Fig. 6b) indicates that the film consists of a mixture of iron oxide and graphite. The shape of the graphite peaks suggests an ordered graphitic structure. However, the wear debris at the entrance side of the wear track on the silicon nitride pin was found to consist of a highly disordered graphite with a grain size of about 40 Å.

Upon examination with Raman spectroscopy, the worn surface of the graphite pin tested with silicon nitride, was also found to be structurally disordered with a grain size of about 40 Å.

In summary, SEM observations and chemical analysis by energy-dispersive and Raman spectroscopy confirmed that sliding of alumina and silicon nitride with intercalated graphite against steel results in the formation of a discontinuous patchy film on the contact surfaces. In the case of alumina, the films contain a mixture of graphite and iron oxide on both the alumina and steel specimens. However, for the silicon nitride sliding system, the transfer films consist of a

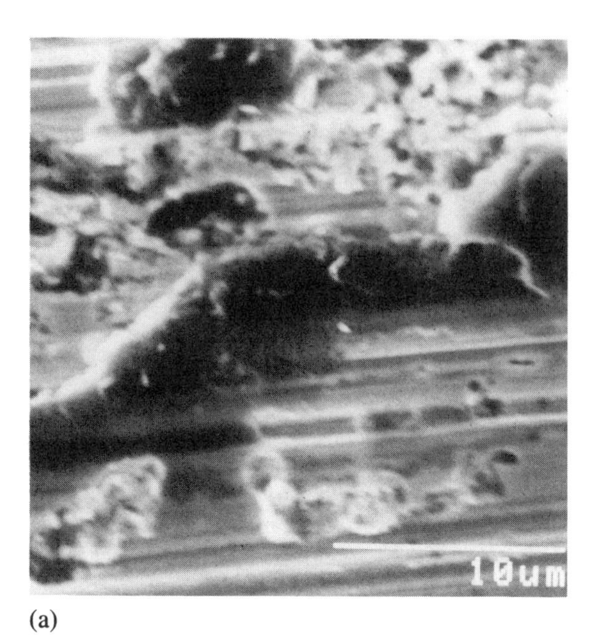

(a)

(b)

Figure 4 (a) Scanning electron micrograph of the transfer film on the worn surface of a steel ring tested with alumina and graphite pins, (b) energy-dispersive spectrum of the film, (c) scanning electron micrograph of the worn surface of the alumina pin, and (d) energy-dispersive spectrum from the worn surface of the alumina pin.

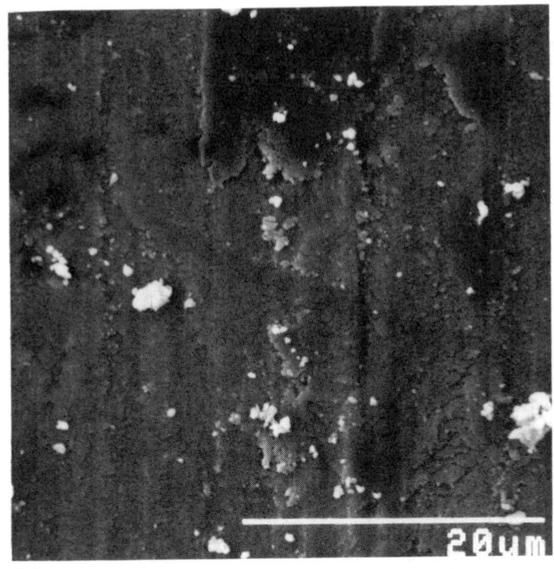
20um

(c)

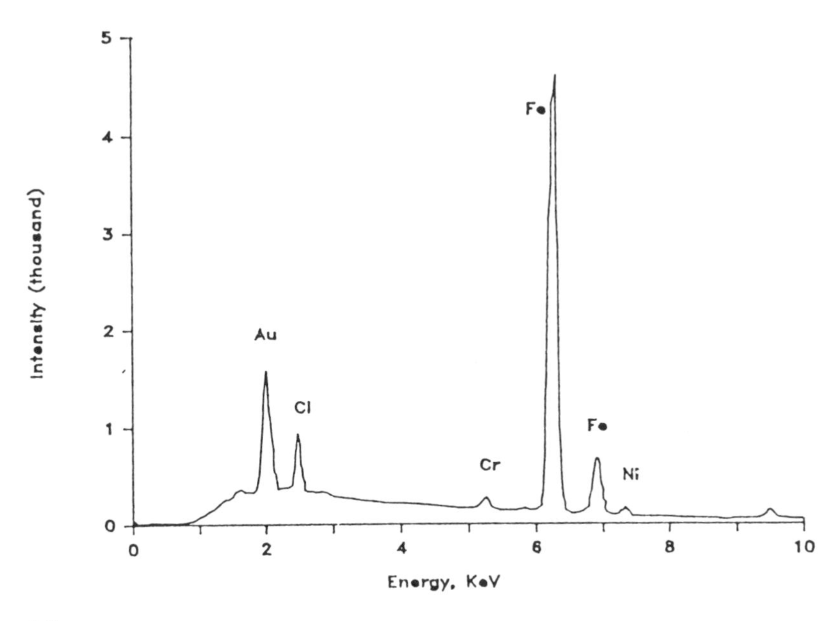
Intensity (thousand)
Energy, KeV
Au
Cl
Cr
Fe
Fe
Ni
0
1
2
3
4
5
0
2
4
6
8
10

(d)

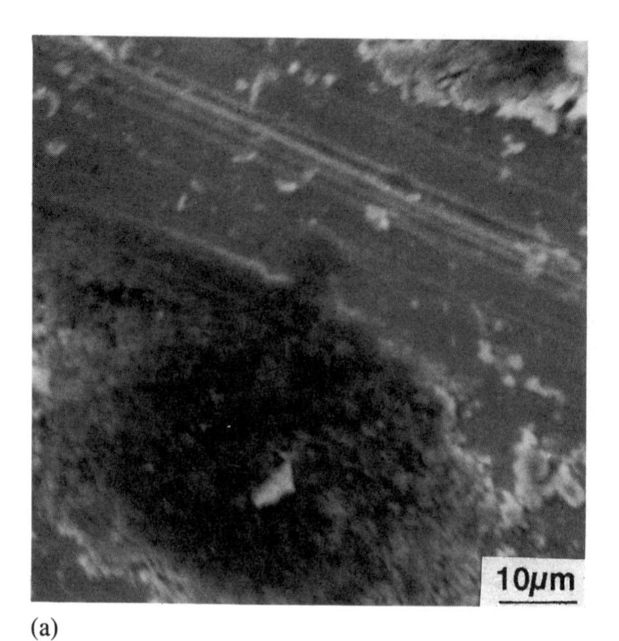

(a)

(b)

Intensity (thousand)

5
4
3
2
1
0

C O Fe Si Cl Cr Fe Fe Ni

0 2 4 6 8 10

Energy, KeV

Figure 5 (a) Scanning electron micrograph of the transfer film on the worn surface of steel ring tested with silicon nitride and graphite pins, (b) energy-dispersive spectrum of the film, (c) Raman spectrum of the film, and (d) Raman spectrum from unworn graphite pin.

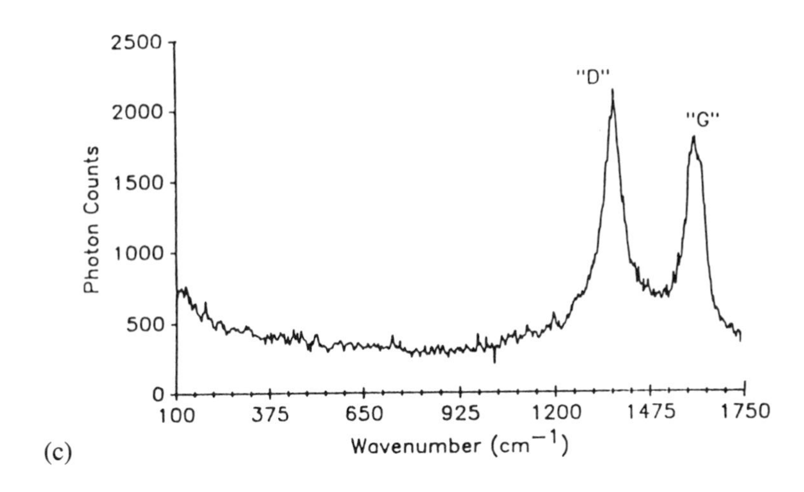

(c)

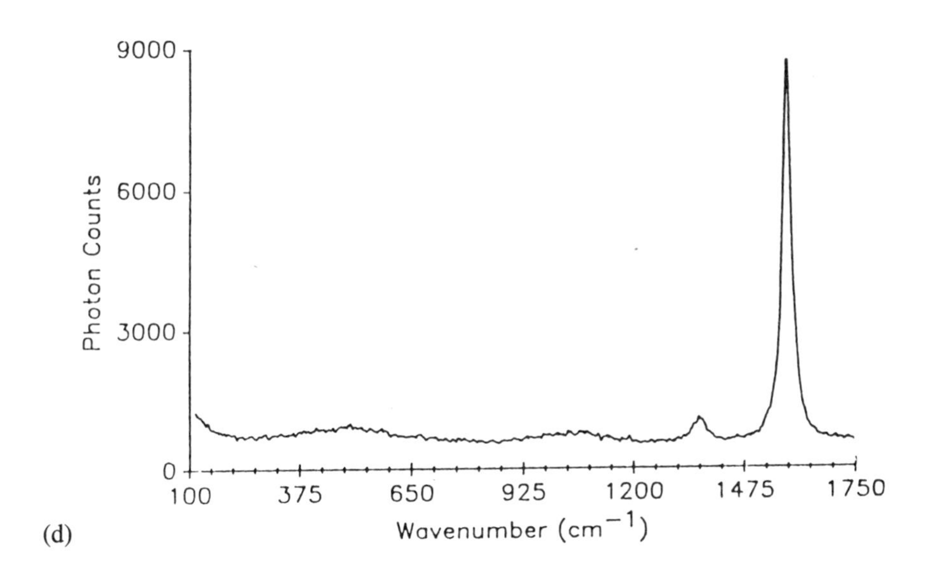

(d)

mixture of iron oxide, very fine particles of graphite, and silicates. The presence of graphite in the transfer film is believed to be responsible for the observed low friction coefficients.

III. SELF-LUBRICATING COMPOSITES

Since the results in Section II demonstrated that alumina and silicon nitride can be lubricated by $NiCl_2$-intercalated graphite, several self-lubricating ceramic

(a)

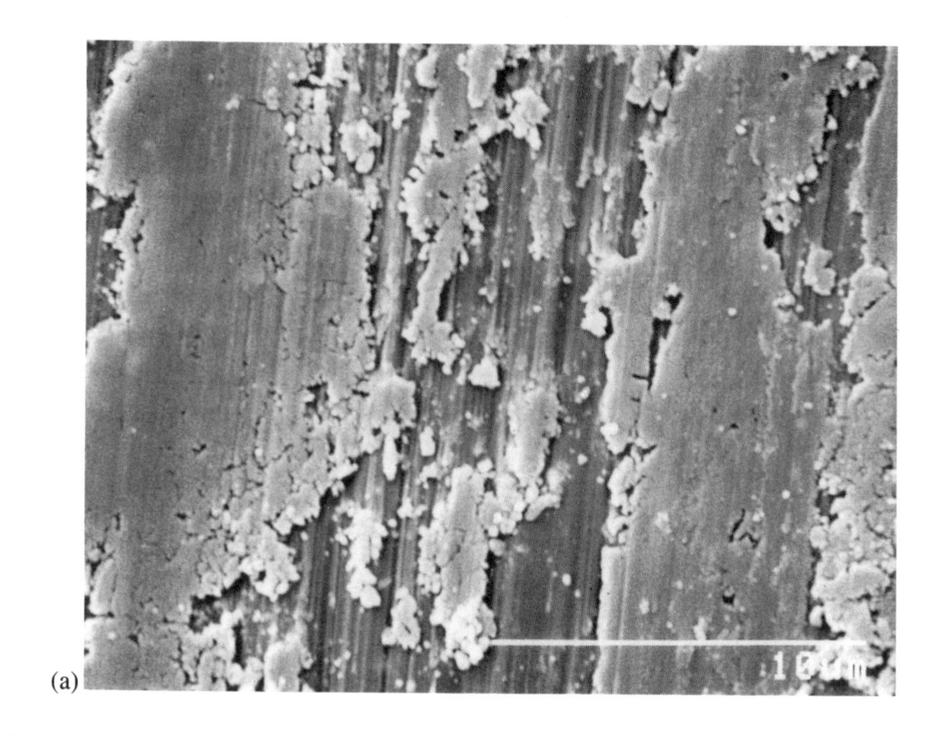

(b)

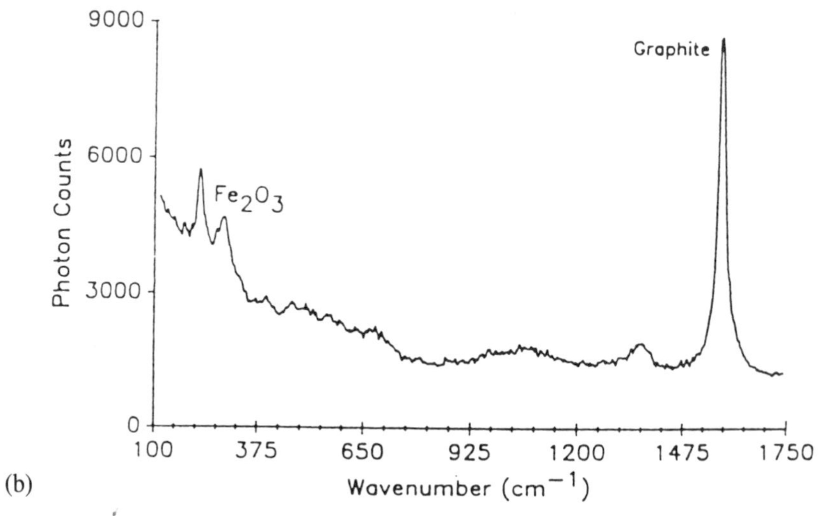

Figure 6 (a) Scanning electron micrograph of the worn surface of silicon nitride, showing transfer film, and (b) Raman spectrum of the transfer film.

matrix composites were prepared and evaluated. Two techniques were used for incorporating solid lubricants in the ceramic matrix: (a) drilling small holes in the alumina and silicon nitride pins, followed by filling the holes with solid lubricants, and (b) casting a slurry containing a mixture of ceramic powders and graphite solid lubricant.

A. Composites Containing Lubricant Reservoirs

1. Sample Preparation and Test Procedure

High purity alumina and hot isostatically pressed silicon nitride ceramics were used for the preparation of the composites. Two types of solid lubricants were used: $NiCl_2$-intercalated graphite and hexagonal boron nitride.

Hexagonal boron nitride has a layered lattice structure, similar to graphite, and therefore, its friction behavior is expected to be similar to that of graphite. Rowe [37] has shown that sintered hexagonal boron nitride exhibits a friction coefficient of 0.25 up to 700 C in air. Miyoshi et al. [38] reported a friction coefficient of 0.10 at 300 C in air during sliding of a diamond hemisphere on ion-beam-deposited boron nitride films on Si, SiO_2, and GaAs substrates.

The hexagonal boron nitride powders used in the present investigation had a particle size of 50 μm and contained 99.4 wt % boron and nitrogen. The impurities consisted of 0.4 wt % oxygen, 0.1 wt % carbon, and 0.05 wt % metallic elements. To compare the frictional behavior of the boron nitride powders with the same material in the solid form, a hot-pressed hexagonal boron nitride was also tested. The chemical composition and the properties of the hot-pressed boron nitride are listed in Table 1.

Composite specimens of alumina–graphite, silicon nitride–graphite, alumina–boron nitride, and silicon nitride–boron nitride were prepared by drilling a series of small holes, 1 mm deep, on the contact surface by ultrasonic machining. Different sizes and spacings of holes were used to investigate the friction and wear behavior of the composites as a function of the amount of solid lubricant in the matrix. The alumina and silicon nitride specimens were 9 mm long and 6.25 mm in diameter. All the holes were drilled in one direction along the diameter of the pins, as shown in Figure 7. The alumina and silicon nitride pins with drilled holes were inserted in a tool steel die and solid lubricant powders were pressed into the holes under a pressure of 0.42 GPa. The excess solid lubricant powder left on the pins after pressing was removed by polishing the samples with a 600 grit paper.

Friction and wear tests were conducted using a pin-on-ring test geometry. The ring material was AISI 52100 steel as described before. The experimental procedure and the test conditions were the same as those described in Section II.A. The wear volume of the pins was calculated from the measured width of the worn area on each pin.

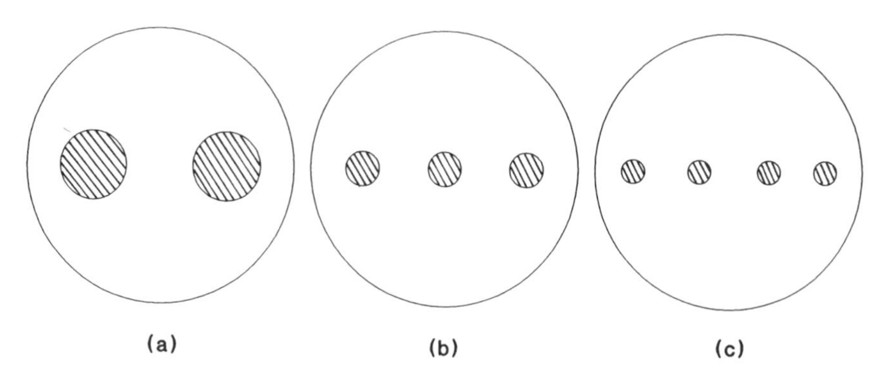

Figure 7 Schematic diagram of the solid lubricant regions in alumina and silicon nitride. Hole diameters (a) 1.65 mm, (b) either 0.76 or 1.14 mm, and (c) 0.52 mm.

2. Composites Containing Intercalated Graphite

The friction coefficients of silicon nitride–graphite and alumina–graphite composites containing various amounts of intercalated graphite are shown in Figure 8a. The area percent of graphite was calculated based on the proportion of graphite regions in contact with the steel counterface in the wear track. It can be observed that the friction coefficient of silicon nitride is reduced from 0.43 to approximately 0.20 with the addition of graphite. Contrary to these results, the friction coefficient of alumina is not affected by the addition of graphite.

The effect of graphite content on the wear rate of the alumina–graphite and silicon nitride–graphite composites is shown in Figure 8b. The wear rates of the silicon nitride composites containing up to 36% graphite are almost the same as those of silicon nitride. However, with further increase in graphite content, the wear rate of the composites tends to increase. In contrast to these results, addition of graphite to alumina has no effect on wear. It can also be seen that the wear rates of the alumina and alumina–graphite composites are lower than those of the silicon nitride and silicon nitride–graphite composites.

The wear rate of the steel rings tested against the ceramics and the composites is shown in Figure 8c. The wear rate of steel sliding against silicon nitride and silicon nitride composites is less than that of steel against alumina and alumina composites. The wear rate of steel decreases slightly with the addition of graphite to silicon nitride.

The worn surfaces of silicon nitride–graphite composites and the steel counterfaces were analyzed by scanning electron microscopy and micro-Raman spectroscopy to elucidate the friction and wear mechanisms. The details of the analysis can be found elsewhere [39]. A discontinuous transfer film is observed on the worn surface of the steel ring tested against the silicon nitride–graphite composites (Fig. 9). Similar to the observations on the steel rings tested with a combination of silicon nitride and graphite pins, EDS analysis indicated that the

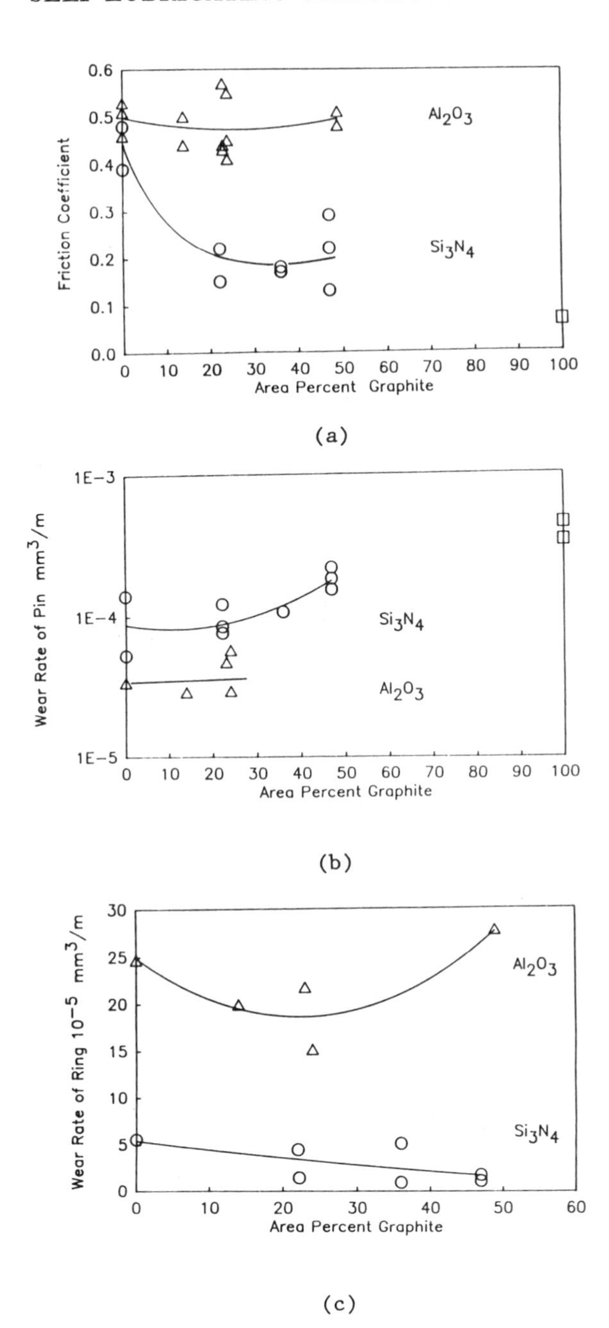

Figure 8 (a) Friction coefficients, (b) wear rates of alumina–graphite and silicon nitride–graphite composites, and (c) wear rates of steel rings, as a function of the amount of graphite at the contact area.

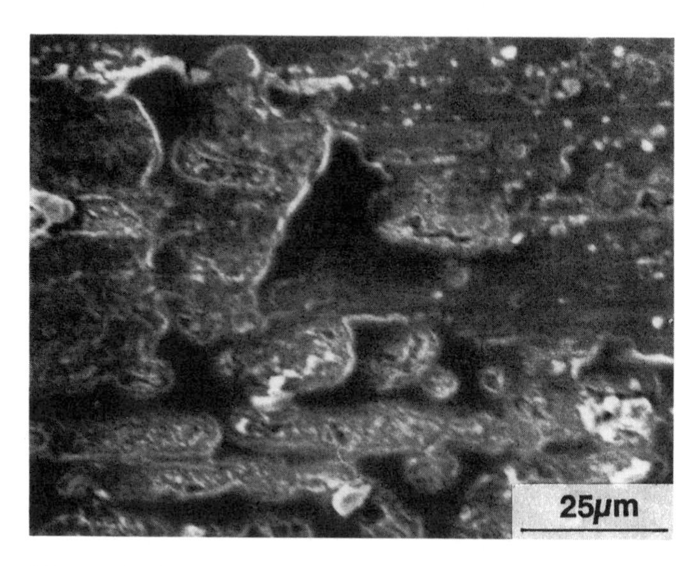

Figure 9 Scanning electron micrograph of the worn surface of steel tested with silicon nitride–graphite composite.

(a)

Figure 10 (a) Scanning electron micrograph of the worn surface of silicon nitride–graphite composite, (b) energy-dispersive spectrum of the transfer film on silicon nitride regions, and (c) Raman spectrum of the transfer film.

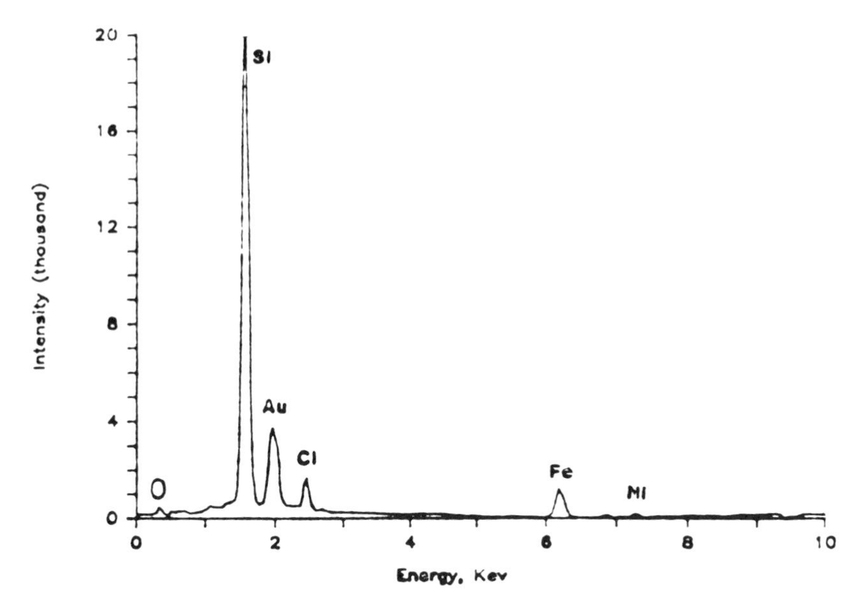
Intensity (thousand)
20
16
12
8
4
0
Si
Au
Cl
O
Fe
Ni
0
2
4
6
8
10
Energy, Kev

(b)

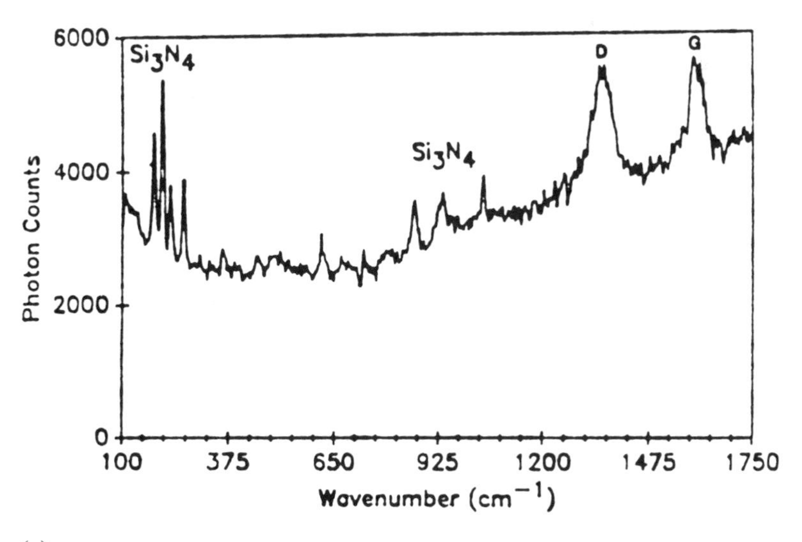
Photon Counts
6000
4000
2000
0
Si3N4
Si3N4
D
G
100
375
650
925
1200
1475
1750
Wavenumber (cm−1)

(c)

film contains graphite, iron oxide, and silicates. Analysis of the film on the steel ring by Raman spectroscopy confirmed that the transfer film contains Fe_2O_3 and disordered graphite with a particle size of about 30 Å.

Figure 10a shows a typical SEM micrograph of the worn surface of a silicon nitride–graphite composite containing a graphite-filled hole, part of which is covered with a transfer film. The silicon nitride regions on the wear scar are also partly covered by a discontinuous film. The energy-dispersive spectrum of the transfer film on the silicon nitride regions within the wear scar (Fig. 10b) indicates that the film contains intercalated graphite and iron oxide. The Raman spectrum (Fig. 10c) confirms that graphite in the transfer film is structurally disordered and has a grain size of 30 Å. The Raman spectrum also shows the silicon nitride peaks corresponding to the substrate.

The transfer film that partially covers the graphite regions was also analyzed by EDS. The film was found to contain Si in addition to C, Ni, and Cl. In some cases the transfer film contained Fe and O, suggesting the presence of iron oxide. A possible explanation for the formation on the pins of films with mixed compositions is that the silicates were formed on the silicon nitride surface, transferred to the steel counterface, then transferred back to the composite pin. In this process, iron oxide can also transfer to the silicon nitride surface, making a complex film containing silicates, iron oxide, and graphite.

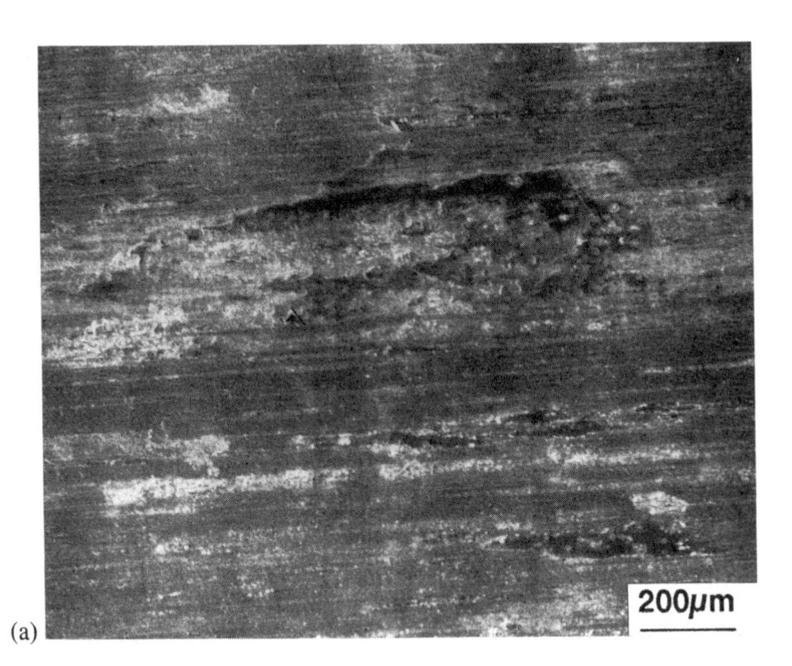

Figure 11 Scanning electron micrographs of the worn surfaces of (a) steel ring and (b) alumina–graphite composite. (c) Raman spectrum of the compacted layer in (b).

The worn surface of the steel ring tested against the alumina–graphite composite appeared quite smooth and polished, although there was a pile-up of reddish wear debris on the sides of the wear track. The worn surface contained a small amount of transfer film as shown in Figure 11a. EDS analysis of the wear surface indicated the presence of only a small amount of graphite.

In a typical SEM micrograph of the wear track on an alumina–graphite composite containing 22% graphite (Fig. 11b), a graphite region almost fully covered by wear

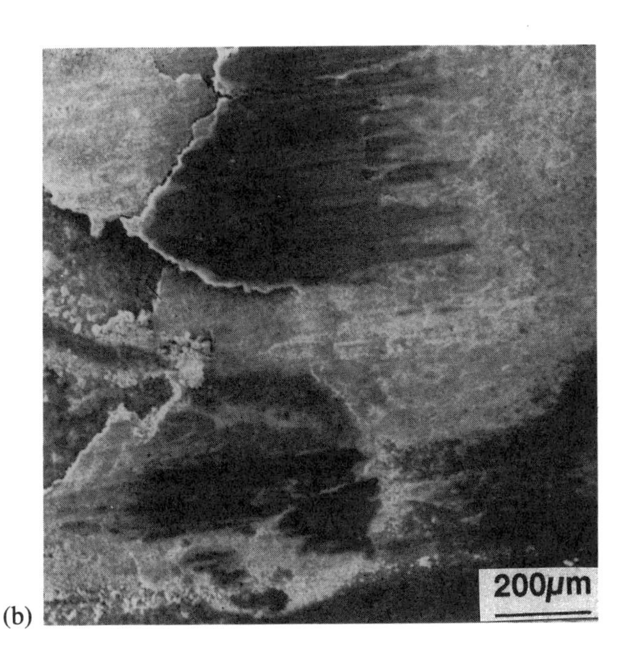

(b)

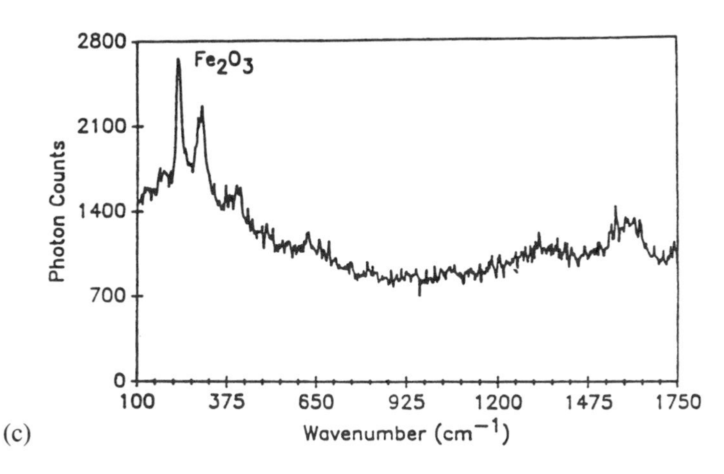

(c)

debris can be seen. The alumina surface is also seen to be covered in some areas by a thick layer of compacted debris. EDS analysis of the compacted layer showed only peaks of Fe and Cr, which correspond to the material transferred from the steel counterface. The Raman spectrum of the compacted layer (Fig. 11c) shows the presence of Fe_2O_3 and a weak signal from graphite, indicating that only a small amount of graphite may be mixed with iron oxide.

In summary, this investigation has demonstrated that the friction coefficient of silicon nitride sliding against steel can be reduced from 0.43 to 0.20 by making a composite with intercalated graphite. The reduction in the friction coefficient is due to the formation of a transfer film containing graphite, iron oxide, and silicates. However, the friction coefficient of alumina sliding against steel is not reduced by using the same approach; although in the contact geometry featuring two pins on a ring, in which one pin was made of alumina and the other of graphite, a friction coefficient as low as 0.20 was obtained. The difference in behavior between the alumina and silicon nitride appears to be related to the restricted supply of graphite at the contact area.

3. Composites Containing Boron Nitride

The friction coefficient of hot-pressed hexagonal boron nitride measured in two-pins-on-ring tests was 0.07. It was therefore expected that addition of boron nitride to the ceramics would reduce the friction coefficient. The friction coefficients of alumina and silicon nitride containing 45 and 55% hexagonal boron nitride at the contact area are shown in Figure 12a: clearly, addition of boron nitride has no effect on the friction coefficient, although a slight increase in the friction coefficient is observed for the silicon nitride composites.

The wear rates of the composites containing boron nitride are shown in Figure 12b. The wear rate of the alumina–boron nitride composite containing 55% boron nitride could not be measured reliably because of the presence of a large amount of wear debris at the contact area, which hindered the precise measurement of the size of the wear scar. It can be seen from Figure 12b that the composites have a lower wear rate than alumina and silicon nitride.

The worn surfaces of silicon nitride–boron nitride composites and the steel counterfaces were examined in the SEM. The worn surface of the composite (Fig. 13a) shows that the boron nitride regions are partially covered with debris, which appeared reddish-brown in an optical microscope. Raman spectroscopy confirmed that such debris consists primarily of Fe_2O_3. The silicon nitride regions on the worn surface appeared shiny, although in some regions brown wear debris was observed. Raman spectroscopy of the silicon nitride regions in the wear track (Fig. 13b) indicates the presence of iron oxide and only a weak peak due to boron nitride. The worn surface of the steel ring tested against the silicon nitride–boron nitride composite appeared smooth, and iron oxide and boron nitride were detected by Raman spectroscopy in only a few regions.

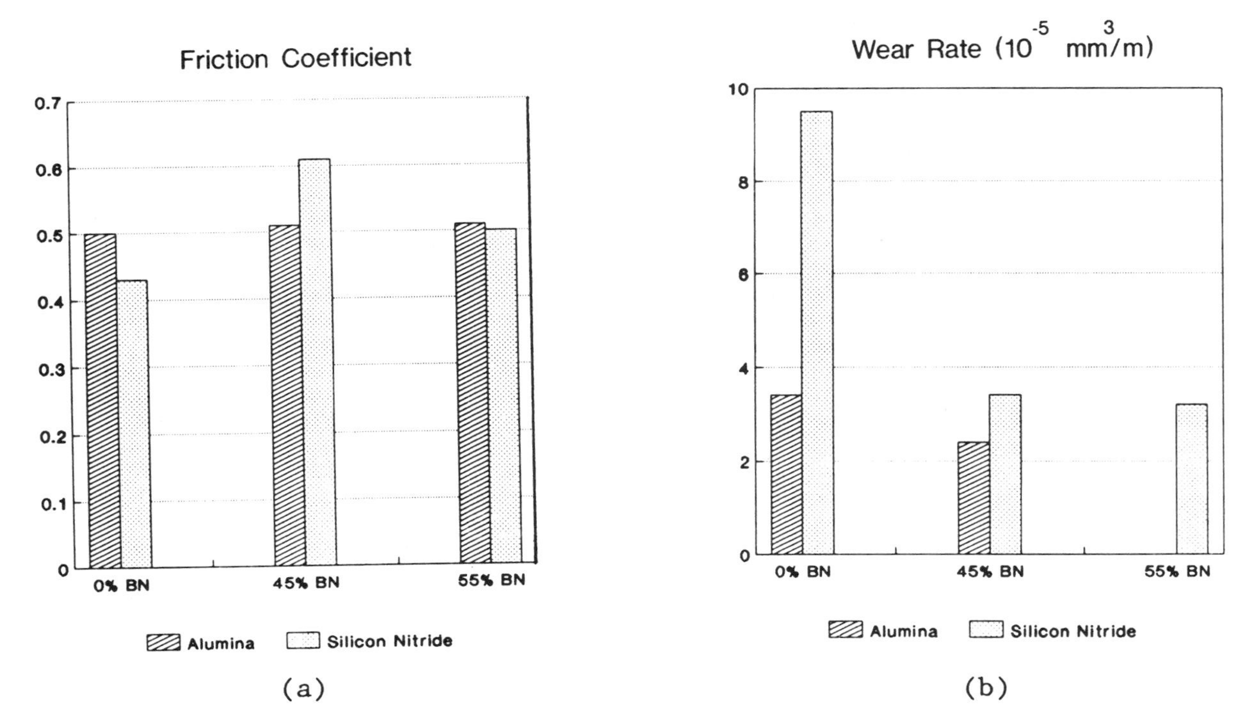

Figure 12 (a) Friction coefficients and (b) wear rates of alumina–boron nitride and silicon nitride–boron nitride composites.

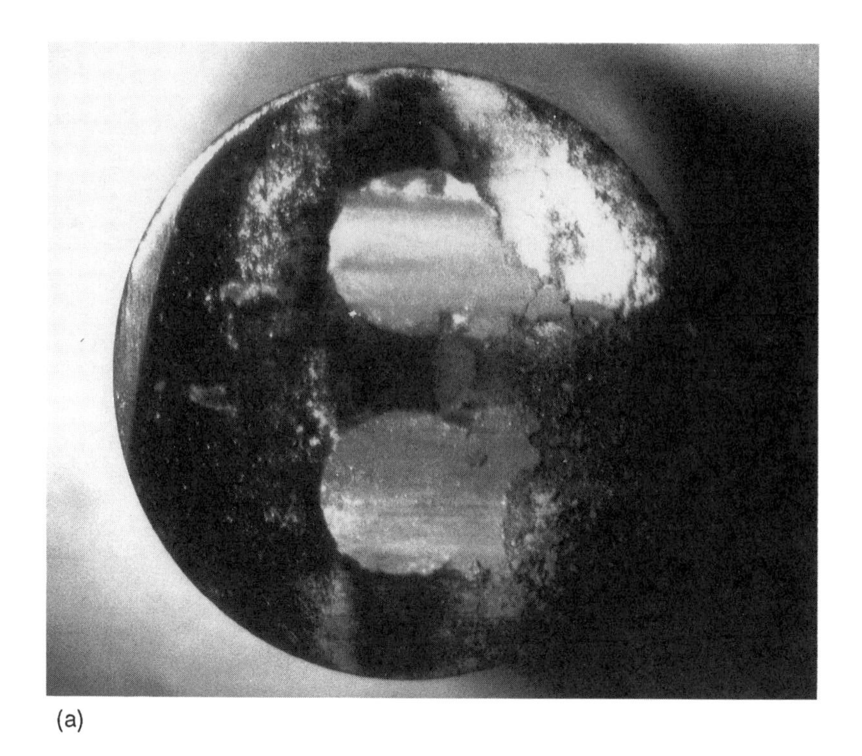

(a)

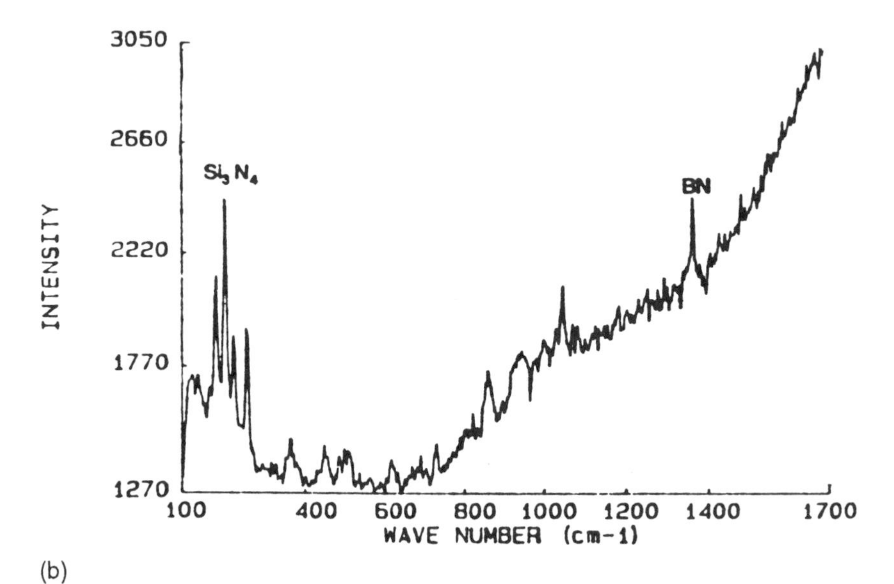

(b)

Figure 13 (a) Scanning electron micrograph of the worn surface of silicon nitride–boron nitride composite and (b) Raman spectrum of the wear debris on the boron nitride.

The worn surface of an alumina–boron nitride composite was also found to be covered with brown wear debris. The wear surface of the steel ring appeared polished and Raman spectroscopy could not detect any boron nitride.

The ineffectiveness of boron nitride in lubricating either alumina or silicon nitride can be attributed to the formation of an insufficient amount of lubricating transfer films on the contacting surfaces. The limited formation of the transfer film may be due to (a) restricted supply of boron nitride at the contact area because the wear debris partially covers the solid lubricant regions, and (b) poor adhesion of boron nitride to the steel and the ceramic surfaces.

B. Composites Prepared by Casting

1. Composite Preparation and Test Procedure

The technique of composite preparation represented a very simple design: the supply of solid lubricant at the contact areas was controlled by the selection of different sizes of solid lubricant filled regions. It was, however, of interest to prepare composites with uniformly distributed solid lubricant in the ceramic matrix. Preparation of such a composite was done by mixing a castable zirconia powder with intercalated graphite. Castable zirconia was chosen as the matrix material because silicon nitride was not available in the castable form. The castable zirconia powder, supplied by Aremco Corporation,* consisted of zirconia particles as the major constituent, although it contained alumina particles as impurities and silicates as the binder.

Zirconia–graphite composite disks with a diameter of 55 mm and a thickness of 6.25 mm were made by thoroughly mixing castable zirconia powder with 20 vol % intercalated graphite. This composition was selected because in the earlier study a maximum friction reduction was observed at this concentration of intercalated graphite. The powder mixture was then blended with approximately 8 wt % distilled water to make a slurry, which was packed in polyethylene bags and allowed to settle for 24 hours. The slurry was then cured, first at 93 °C for 3 hours, followed by another cure at 123 °C for 3 hours. Zirconia reference disks were also prepared by the same technique. Test coupons 12 mm × 12 mm × 2 mm were cut from the zirconia and the zirconia composite disks for friction and wear testing.

Optical micrographs of polished sections of the zirconia and zirconia–graphite composite are shown in Figure 14. The microstructure of the cast zirconia consists of globular alumina particles and a few large zirconia particles in a matrix of zirconia. The microstructure of the zirconia–graphite composites confirmed that

*Information on vendors is included only to identify the materials; no endorsement of the product of any manufacturer is implied.

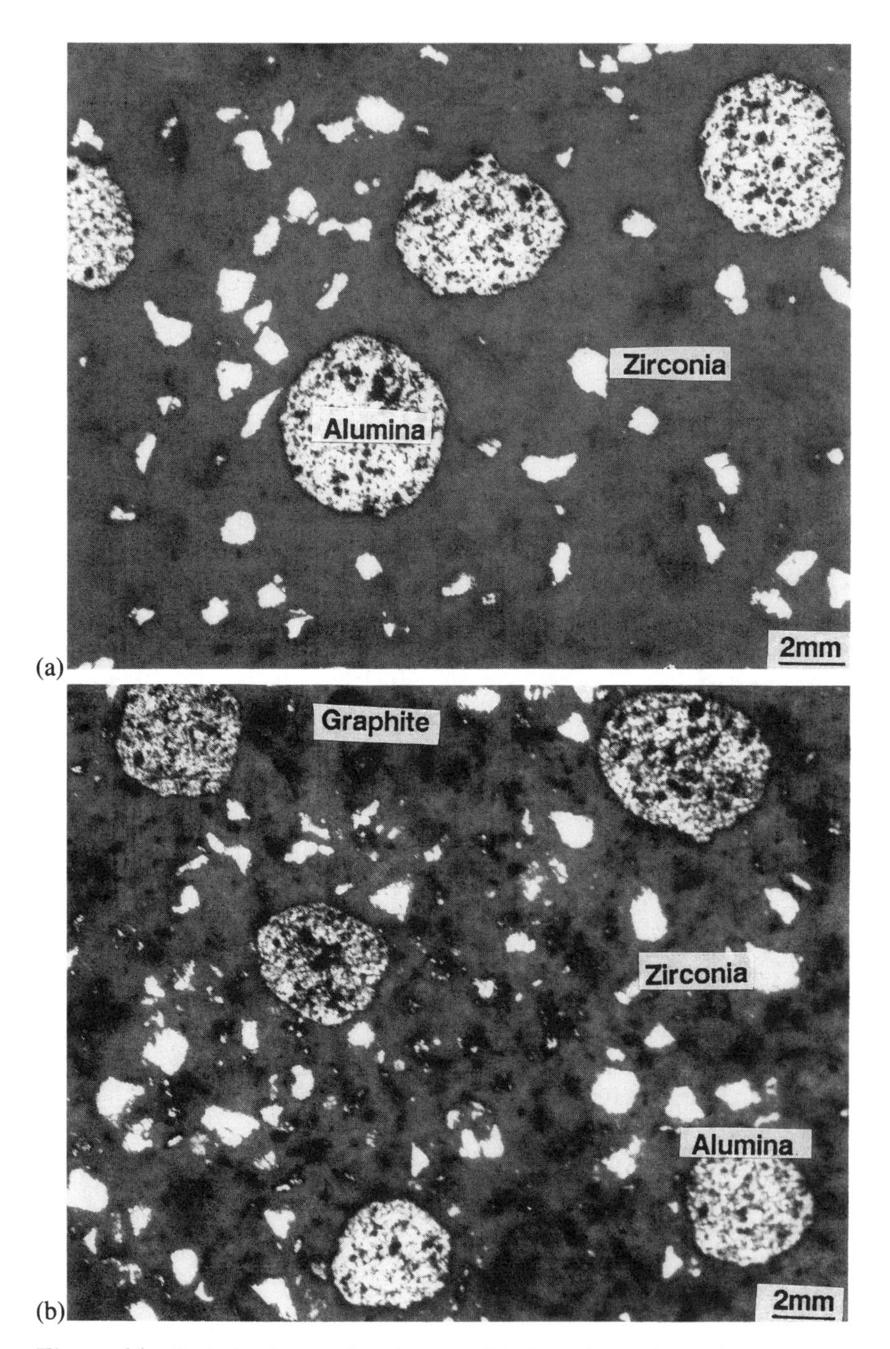

Figure 14 Optical micrographs of the polished sections of (a) zirconia and (b) zirconia–graphite composite.

the graphite particles were uniformly distributed in the matrix. Hardness measurements conducted with a Rockwell hardness tester on the cast zirconia and zirconia–graphite composites did not yield any reliable data because the samples fractured under the indenter. No other mechanical properties were measured on the cast samples.

Friction and wear tests were conducted on zirconia and zirconia–graphite composites using a reciprocating ball-on-flat test apparatus. The ball was held fixed and the flat was made to reciprocate. The details of the apparatus can be found elsewhere [23]. The balls used for the investigation had a diameter of 12.5 mm and were made of AISI M50 steel. The experiments were conducted at 23 C (room temperature), 200 C, and 400 C, at an applied load of 29 N, a sliding speed of 1.3 mm/s, and in the ambient laboratory air with a relative humidity of 65%. The duration of each test was one hour. After the tests, the wear volume of the flats was calculated from three to five surface profilometer traces across the wear track and perpendicular to the sliding direction. The wear volume of the balls was calculated from the measured diameter of the wear scar using the procedure recommended by Whittenton and Blau [40].

2. Friction and Wear Data

The friction coefficients of zirconia and zirconia–graphite composites sliding against steel, as a function of test temperature, are shown in Figure 15. The friction coefficients remained fairly constant during each test. The reported values are the averages of the last 10 passes. It can be seen that the variation in the

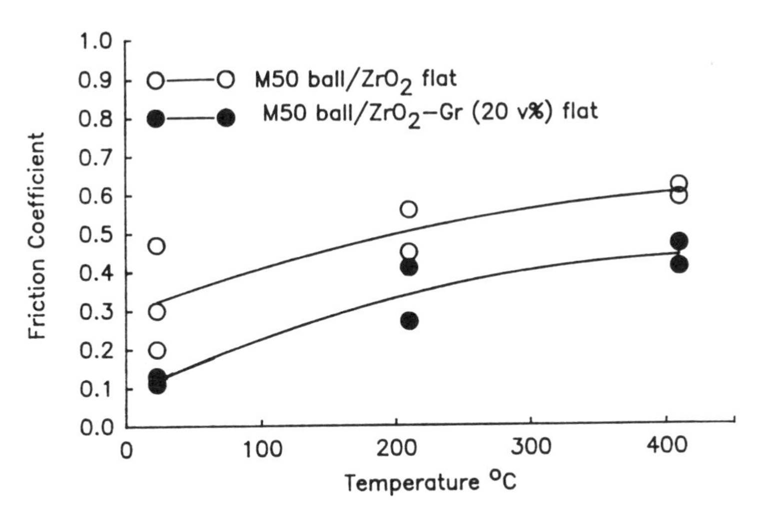

Figure 15 Friction coefficients of zirconia and zirconia–graphite composites as a function of test temperature.

friction coefficients of zirconia is reduced as the test temperature is increased. At room temperature, the average friction coefficient of zirconia is 0.35; addition of graphite reduces the friction coefficient to 0.11. The friction coefficient of both zirconia and the zirconia–graphite composites increases with increasing temperature. However, in the temperature range used, the friction coefficients of zirconia–graphite composites remain less than those of zirconia.

Addition of graphite to the castable zirconia did not have much effect on the wear rate. The wear rates of both zirconia and zirconia–graphite composites were on the order of 0.1 mm^3/m, which is a high value. This can be attributed to the low strength of the castable zirconia, as demonstrated by the hardness measurements. In addition, the wear rates of the steel balls were generally the same when tested against the zirconia and the zirconia–graphite composite. An average wear rate of 10^{-3} mm^3/m was obtained for the wear of steel against either material. This value was almost independent of temperature. However, the steel balls tested against zirconia exhibited a larger variability in the wear rate compared to those against the composites.

3. Analysis of the Wear Surfaces

The worn surfaces of zirconia tested at room temperature appeared shiny and polished. However, examination under a scanning electron microscope revealed fracture in several areas (Fig. 16a). Microcracks were also observed at the interface between the alumina particles and the matrix. An EDS spectrum from the worn area is shown in Figure 16b. The Al, Zr, Si, and Ca peaks are due to castable zirconia and the binder phase. The Fe peak indicates transfer of the steel counterface material to the zirconia wear track.

SEM examination of the worn surface of the zirconia–graphite composite tested at room temperature also revealed a shiny wear track containing several microcracks and occasional fracture. An example is shown in Figure 17a. The Ni, Cl, and Fe peaks in the EDS spectrum of the wear track (Fig. 17b) confirm the presence of intercalated graphite and transfer of steel. Graphite was observed all over the wear track, whereas in the unworn regions, graphite could be observed only where the graphite particles were located. This suggests that the graphite particles smear over the contact area as a result of sliding, thus forming a thin lubricating film.

Examination of the worn surface of the steel balls exhibited a patchy transfer film. An example is shown in Figure 17c for the steel ball that was tested against the zirconia–graphite composite at 400 °C. An EDS spectrum from the transfer film (Fig. 17d) shows peaks of Cl and Ni in addition to peaks of Al, Zr, Si, and Ca, indicating transfer of the composite to the steel surface. The Fe, Cr, and V peaks are from the M50 steel.

In summary, the analysis of the worn surfaces of the zirconia–graphite composites and the steel counterfaces confirmed that a transfer film containing

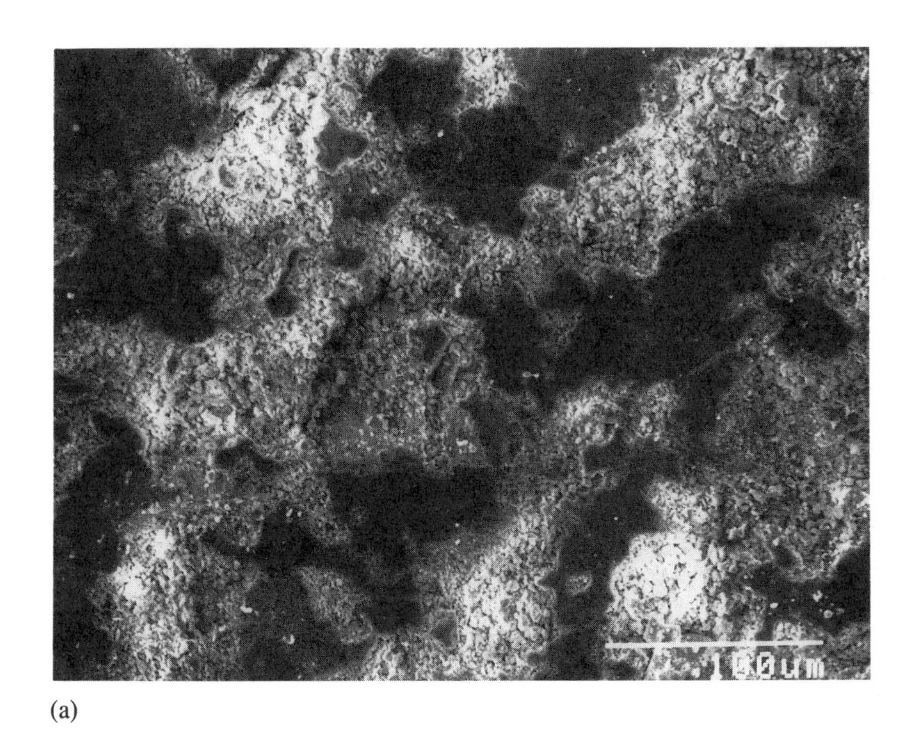

(a)

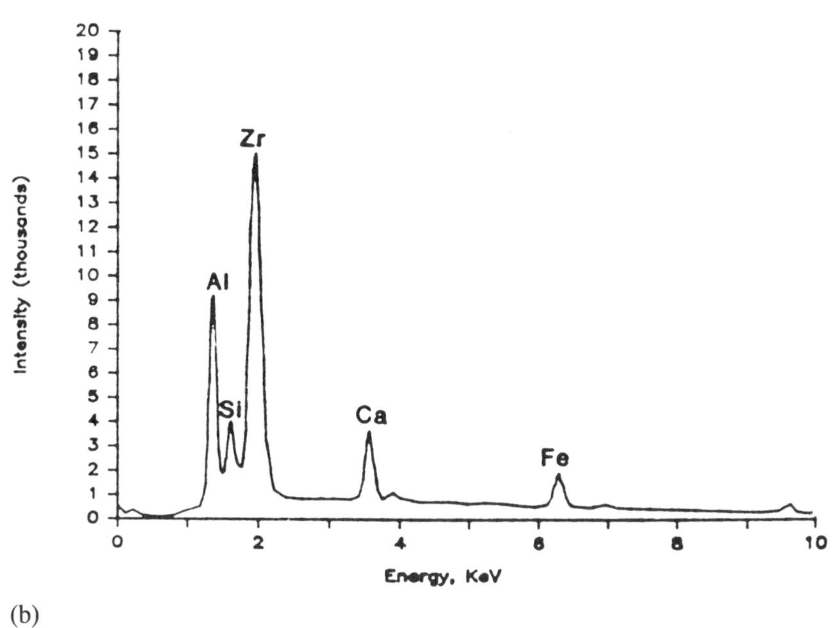

(b)

Figure 16 (a) Scanning electron micrograph of the worn surface of zirconia and (b) energy-dispersive spectrum of the worn surface.

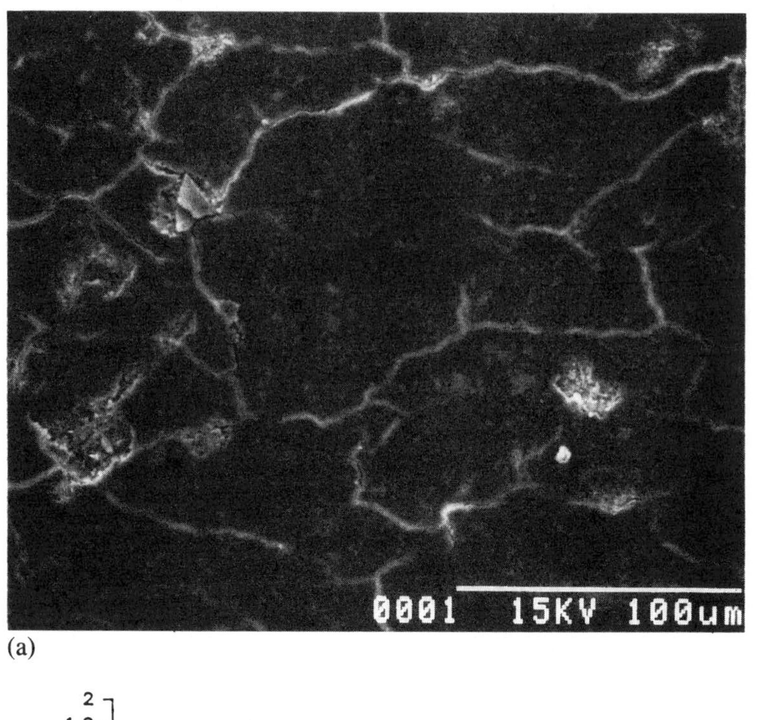

(a)

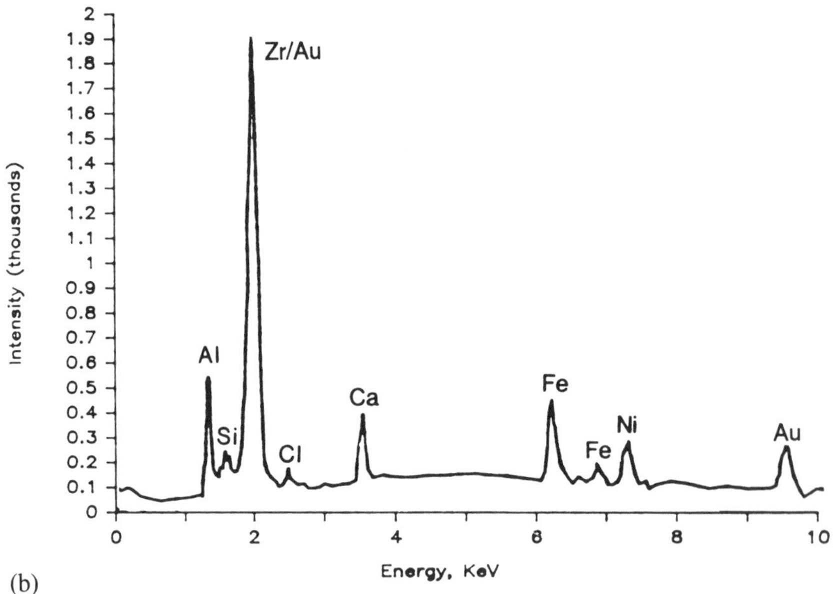

(b)

Figure 17 (a) Scanning electron micrograph of the worn surface of zirconia–graphite composite, (b) energy-dispersive spectrum of the worn surface, (c) scanning electron micrograph of the worn surface of steel showing transfer film, and (d) energy-dispersive spectrum of the transfer film.

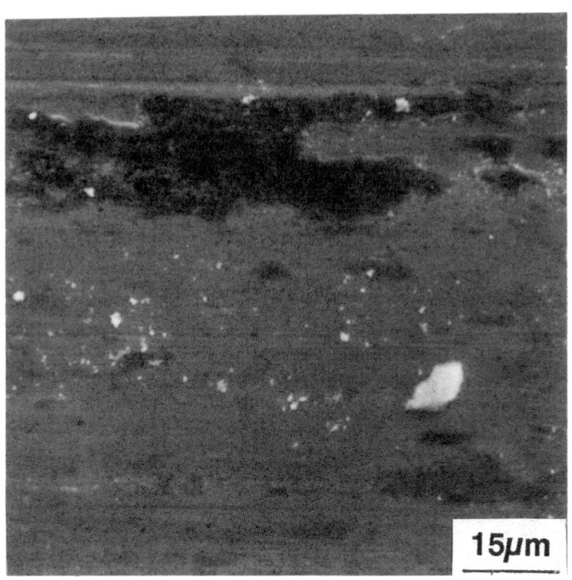
15μm

(c)

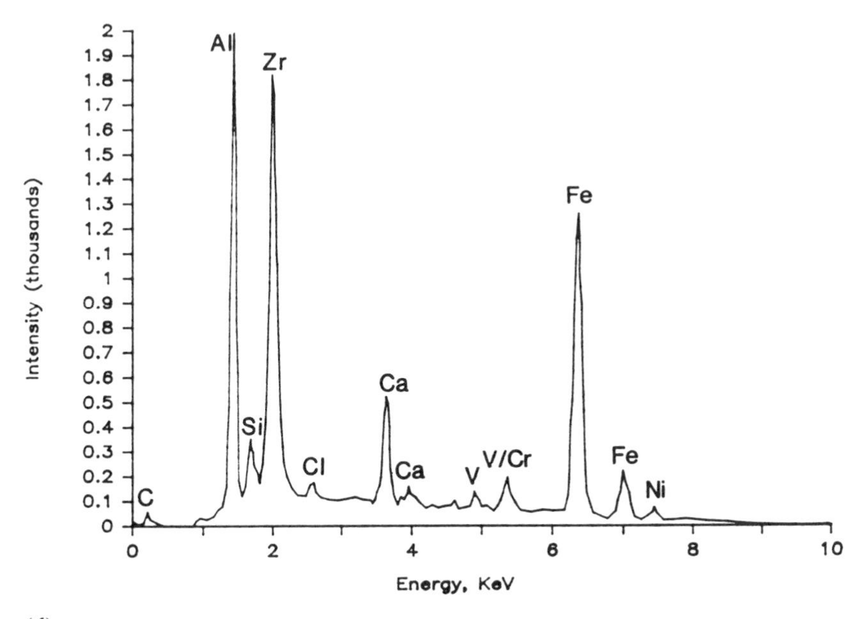
Al
Zr
Fe
Ca
Si
Cl
Ca
V
V/Cr
Fe
Ni
C
Intensity (thousands)
Energy, KeV
0
0.1
0.2
0.3
0.4
0.5
0.6
0.7
0.8
0.9
1
1.1
1.2
1.3
1.4
1.5
1.6
1.7
1.8
1.9
2
0
2
4
6
8
10

(d)

graphite forms on the contact surfaces. This film is believed to be responsible for the lower friction coefficients of the zirconia–graphite composites compared to those of the zirconia.

IV. SUMMARY AND DISCUSSION

The results presented in this chapter clearly establish the feasibility of achieving low friction coefficients when using self-lubricating ceramic matrix composites with $NiCl_2$ intercalated graphite as the solid lubricant. It is, however, important to recognize that the proper combination of the ceramic matrix, solid lubricant, and counterface must be selected for particular operating conditions. The primary factor in achieving self-lubrication is the formation of an adherent transfer film with a low shear strength.

The friction measurements from a two-pins-on-ring contact geometry demonstrated a reduction in friction coefficient for silicon nitride sliding against steel from 0.45 to 0.17, and for alumina against steel from 0.48 to 0.20, when intercalated graphite was used as the solid lubricant. The reduction in friction coefficient was found to be due to the formation of a transfer film on the ceramic and the steel contact surfaces. The transfer film consisted of a mixture of graphite and iron oxide as well as a possible contribution from the ceramic materials. In the case of silicon nitride, tribochemical reaction products such as silicates were found in the transfer film. It was also observed that the graphite in the transfer film differed significantly from the original graphite material. The graphite in the transfer film had a more disordered structure, and its grains were much finer (40 Å) than the grains in the original graphite (100 μm).

These results led us to prepare composites by drilling holes of various sizes in alumina and silicon nitride, which were subsequently filled with either intercalated graphite or hexagonal boron nitride. A friction coefficient as low as 0.20 was observed for silicon nitride–graphite composites sliding against steel. The reduction in the friction coefficient was due to the formation of a transfer film consisting of graphite, iron oxide, and silicates. In the case of alumina–graphite composites, no reduction in friction coefficient was observed. This is contrary to the results of the tests obtained from the two-pins-on-ring contact geometry, where the supply of the solid lubricant was governed by the wear of the graphite pin. The larger wear rate of the graphite pin compared to the wear rate of the ceramic pin resulted in a sufficient amount of graphite lubricant to form a transfer film. In the case of the composites, the supply of solid lubricant was controlled by the wear rate of the ceramic matrix. The low wear rate of alumina restricted the supply of graphite at the contact area. Also, the large wear rate of the steel counterface produced iron oxide, which covered the graphite regions and further restricted the graphite supply.

Replacement of graphite by hexagonal boron nitride in alumina or silicon nitride did not produce a reduction in the friction coefficient. The wear debris generated during the test partially covered the boron nitride filled regions, thus restricting the quantity of the solid lubricant at the contact area. Boron nitride did not appear to adhere well to either the steel or the ceramic surfaces. Since adhesion of the solid lubricant at the contacting surfaces is important for the formation of a transfer film [41], poor adhesion of boron nitride may be a factor in preventing a reduction in the friction coefficient.

Composites were also prepared by mixing intercalated graphite powder with zirconia powder into a slurry followed by casting. This technique provided a uniform distribution of the solid lubricant in the matrix. At room temperature, the addition of graphite reduced the friction coefficient of zirconia from 0.35 to 0.11 in sliding contact with steel. At elevated temperatures, the zirconia–graphite composites also exhibited lower friction coefficients than zirconia. Based on the examination of the wear surfaces, the reduction in friction coefficient was found to be related to the smearing of graphite over the contact area and the formation of graphite-containing transfer film on the steel counterface. The friction coefficient of both zirconia and zirconia–graphite composite increased with increasing temperature. Because the friction behavior of graphite is sensitive to the presence of moisture and other contaminants [42], the increase in the friction coefficient of zirconia–graphite composite at 200 C is believed to be due to the loss of these contaminants from the surface. Wedeven et al. [30] have observed a similar effect of graphite lubrication of silicon nitride on the friction coefficient. The increase in friction coefficient of zirconia–graphite composites from 200 to 400 C may be due to the decomposition of the intercalation compound (i.e. $NiCl_2$), which decomposes at 325 C.

One of the limitations of castable composites is their low strength, which results in a high wear rate. To overcome this limitation, composites should be prepared by commonly used ceramic processing techniques utilizing high temperatures and pressures. Also, ceramic fibers or whiskers can be added to increase the strength. Recently, Kustas et al. [43] attempted to fabricate self-lubricating ceramic materials by mixing hollow carbon spheres with silicon nitride and silicon carbide powders followed by hot-pressing. Reduction of friction coefficient by over more than 50% was observed for the silicon nitride composites.

In conclusion, our research indicates that the tribological properties of ceramic matrix composites containing a solid lubricant can be improved through proper selection of the solid lubricant, the ceramic matrix, and the counterface material. The selection of the materials should be such that a lubricíous transfer film forms on the contacting surfaces. The major advantage of these composites over the deposition of thin solid lubricating films on ceramic surfaces is that a constant supply of solid lubricant can be maintained, which can prevent catastrophic failure caused by film breakdown.

ACKNOWLEDGMENTS

The authors acknowledge the assistance of B. E. Hegemann and Z. Hu for obtaining the micro-Raman and the micro-FTIR spectra. The authors also acknowledge the Ford Motor Company for providing support to AKG in the preparation of the manuscript.

REFERENCES

1. K. Miyoshi and D. H. Buckley, *ASLE Trans.* *22*(1), 79 (1979).
2. T. E. Fischer, M. P. Anderson, S. Jahanmir, and R. Salher, *Wear*, *124*, 133 (1988).
3. S. Jahanmir and T. E. Fischer, *Tribol. Trans.* *31*(1), 32 (1988).
4. D. H. Buckley and K. Miyoshi, *Wear*, *100*, 333 (1984).
5. M. Bohmer and E. A. Almond, *Mater. Sci. Eng.*, *A105–A106*, 105 (1988).
6. O. O. Adewoye and T. F. Page, *Wear*, *70*, 37 (1981).
7. J. C. Sikra, J. E. Krysiak, P. R. Eklund, and R. Ruh, *Ceram. Bull* *53*(8) (1974).
8. J. Breznek, E. Breval, and N. H. McMillan, *J. Mater. Sci.* *20*, 4657 (1985).
9. W. Holzhauer, R. L. Johnson, and S. F. Murray, in *Wear of Materials*, K. C. Ludema (Ed.), American Society of Mechanical Engineers: New York, 1981, p. 676.
10. R. S. Gates, S. M. Hsu, and E. E. Klaus, *Tribol. Trans.* *32*(3), 357 (1989).
11. A. K. Gangopadhyay, F. B. Janon, M. E. Fine, and H. S. Cheng, *Tribol. Trans.* *33*(1), 96 (1990).
12. H. Tomizawa and T. E. Fischer, *ASLE Trans.* *30*(1), 41 (1987).
13. P. A. Willermet, *ASLE Trans.* *30*(1), 128 (1987).
14. J. F. Braza, H. S. Cheng, and M. E. Fine, *Tribol. Trans.* *32*(4) (1989).
15. W. Wei, K. Beaty, S. Vinyard, and J. Lankford, Sliding seal materials, presented at the 25th Automotive Technology Development Contractors' Coordination Meeting, Dearborn, MI, 1987.
16. K. M. Taylor, L. B. Sibley, and J. C. Lawrence, *Wear*, *6*, 266 (1963).
17. M. Nishimura, *JSME Int. J.* *31*(4), 661 (1988).
18. P. D. Fleischauer and M. R. Hilton, in *New Materials Approach to Tribology: Theory and Applications*, Vol. 140, *MRS Symposium Proceedings*, L. E. Pope, L. L. Fehrenbacher, and W. O. Winer (Eds.), Materials Research Society: Pittsburgh, 1989, p. 9.
19. M. R. Hilton and P. D. Fleischauer, in *New Materials Approach to Tribology: Theory and Applications*, Vol. 140, *MRS Symposium Proceedings*, L. E. Pope, L. L. Fehrenbacher, and W. O. Winer (Eds.), Materials Research Society: Pittsburgh, 1989, p. 227.
20. E. W. Roberts and W. B. Price, in *New Materials Approach to Tribology: Theory and Applications*, Vol. 140, *MRS Symposium Proceedings*, L. E. Pope, L. L. Fehrenbacher, and W. O. Winer (Eds.), Materials Research Society: Pittsburgh, 1989, p. 251.
21. H. E. Sliney, *Thin Solid Films*, *64*, 211 (1979).

22. H. E. Sliney, T. P. Jacobson, D. Deadmore, and K. Miyoshi, *Ceram. Eng. Sci. 7*, 1039 (1986).
23. A. K. Gangopadhyay and S. Jahanmir, *Tribology of Composite Materials*, ASM International: Metals Park, OH, 1990, p. 337.
24. A. Erdemir, G. R. Fenske, R. A. Erck, and C. C. Cheng, *Lubr. Eng. 46*(1), 23 (1990).
25. J. Lankford, W. Wei, and R. Kossowsky, *J. Mater. Sci. 22*, 2069 (1987).
26. M. Nastasi, R. Kossowsky, J. P. Hirvonen, and N. Elliot, *J. Mater. Res. 3*(6), 1127 (1988).
27. J. W. Van Wyk, Ceramic airframe bearing, presented at the 30th Annual Meeting of the American Society of Lubrication Engineers, Atlanta, 1975.
28. B. Bhusan and L. B. Sibley, *ASLE Trans. 25*(4), 417 (1982).
29. R. A. Pallini and L. D. Wedeven, *Tribol. Trans. 31*(2), 289 (1988).
30. L. D. Wedeven, R. A. Pallini, and N. C. Miller, *Wear, 122*, 183 (1988).
31. A. A. Conte, Jr., *ASLE Trans. 26*, 200 (1983).
32. W. E. Jamison, *ASLE Trans. 14*, 62 (1971).
33. A. W. Ruff, M. B. Peterson, A. K. Gangopadhyay, and E. Whittenton, Wear and friction characteristics of self-lubricating copper-intercalated graphite composites, in *Proceedings of the Fifth European Tribology Conference*, Helsinki, Finland, 1889.
34. A. K. Gangopadhyay, S. Jahanmir, and B. E. Hegemann, in *Mechanics of Coatings*, D. Dowson, C. M. Yaylor, and M. Godet (Eds.), Elsevier Science Publishers: London, 1990, p. 63.
35. F. Tuinstra and J. L. Koenig, *J. Chem. Phys. 53*, 1026 (1970).
36. R. Al-Jishi Lespade, and M. S. Dresselhaus, *Carbon, 20*, 427 (1982).
37. G. W. Rowe, *Wear, 3*, 274 (1960).
38. K. Miyoshi, D. H. Buckley, J. J. Pouch, S. A. Alterovitz, and H. E. Sliney, *Surf. Coat. Technol. 33*, 221 (1987).
39. A. K. Gangopadhyay and S. Jahanmir, *Tribol. Trans. 34*(2), 257 (1991).
40. E. Whittenton and P. J. Blau, *Wear, 124*, 291 (1988).
41. J. K. Lancaster, *J. Tribol. 107*, 437 (1985).
42. R. H. Savage, *J. Appl. Phys. 19*, 1 (1948).
43. F. M. Kustas, S. P. Rawal, and M. S. Misra, presented at the Annual Meeting of the Society of Tribologists and Lubrication Engineers, Denver, 1990.

9

Tribological Behavior of Whisker-Reinforced Ceramic Composite Materials

Charles S. Yust

Oak Ridge National Laboratory
Oak Ridge, Tennessee

ABSTRACT

The utility of ceramics as wear-resistant materials may be limited by their susceptibility to fracture. In recent years the fracture toughness of ceramics has been improved through the introduction of secondary phases into the microstructure. Zirconia-toughened alumina and whisker-reinforced alumina are two examples of this emerging class of materials. Since the onset of severe wear in ceramics is preceded by fracture, improvements in fracture toughness also imply a potential for improved wear resistance. The studies reported in this chapter indicate that the increased complexity of the toughened ceramic microstructures can result in more complex wear responses. Although the composite body is toughened and fracture suppressed, the wear debris of the secondary phase may abrade the matrix, and the toughening mechanism may depend on a microstructural condition that is detrimental, rather than beneficial, with regard to wear. In addition, the strengthening phase(s) may interact chemically with the environment, producing either desirable or undesirable results. Results obtained on whisker-reinforced alumina and silicon based ceramics are presented and discussed, with reference to the complex microstructural relationship to wear.

I. INTRODUCTION

Improvements in the mechanical properties of ceramics, in particular the fracture toughness, achieved by the incorporation of ceramic whiskers in the microstructure, have made the application of composite ceramics as structural elements and machine components more feasible [1–3]. Toughening by alteration of the phase content of the microstructure is another means of mechanical property improvement [4,5]. Whiskers may also be used in conjunction with matrix transformation toughening processes generated by the use of metastable particulate phases. The tetragonal form of zirconia in alumina is one example. Other investigations of toughening have examined the incorporation of both particulate phases and other reinforcing elements in one composite body [6]. Phases added for mechanical property improvement may also yield tribological benefits by virtue of chemical interaction with the environment, as in the case of a silicon nitride composite containing both silicon carbide whiskers and titanium carbide particles [7,8]. The combination of ceramic whiskers with other phases in a ceramic matrix, therefore, may lead to microstructures of greatly increased complexity, albeit improved tribological properties. The tribological response of such complex ceramic bodies is only now being explored in detail. In general, approaches based on microstructural considerations have led to improvements in the flexural strength and fracture toughness of monolithic ceramics.

Studies of the wear of ceramics have demonstrated that fracture plays a significant role in that process [9–11]. Stresses induced by both thermal and mechanical effects can introduce microcracks or initiate cracks from existing flaws. Cyclically varying stresses, which may be imposed by both rolling and sliding mechanical systems, present the possibility of fatigue crack formation and a consequent rapid transition in wear mode. In combined rolling and sliding wear, an initial period of mild wear followed by a transition to severe wear has resulted from the formation of surface fractures [12]. The recent developments in toughened ceramics, therefore, suggest a potential for improved wear response through inhibition of crack initiation and propagation.

This chapter reviews the current technical literature on wear of whisker-reinforced ceramic composites. The available results are presently limited, but there is information on whisker-reinforced alumina (WRA), zirconia-toughened alumina (ZTA), whisker-reinforced ZTA (WRZTA), and whisker-reinforced silicon nitride (WRSN).

II. WEAR OF WHISKER-REINFORCED ALUMINA

The enhancement of mechanical properties, especially fracture toughness, accompanying the inclusion of silicon carbide whiskers in an alumina matrix has prompted the study of the wear resistance of this composite. The improved properties of WRA and some of the other composites considered in this chapter are summarized in Table 1. The composite is typically formed by mixing an alumina powder with silicon carbide whiskers, then hot-pressing the mixture in a graphite die. Representative flexural strength and toughness values for a body

Table 1 Composition and Properties of Whisker Composites

	Al_2O_3	WRA [4,13]	ZTA [28]	WRZTA [28]	Si_3N_4	WRSN
Composition, wt%						
Al_2O_3	99.9	80	93.0	74.4		
ZrO_2			7.0	5.6		
SiC_w		20		20.0		20
Si_3N_4					100	80
Zirconia, %						
as monoclinic			6	35		
as tetragonal			94	65		
Four-point bend strength, MPa	350	650–800	500	650	900	770
Fracture toughness (K_{IC}), MPa · $m^{1/2}$	4.6	8.0–8.5	6.1	7.4	6	6.9
Thermal conductivity, W m^{-1} K^{-1}	39	35	23	30	15	
Hardness, GPa	15.5	20.5	17.0	16.5	15.0	
Density, $kg/m^3 \times 10^{-3}$	3.96	3.80	4.00	3.75	3.31	3.22

containing 20 vol % whiskers are, respectively, 650–800 MPa and 8–8.5 $MPa{\cdot}m^{1/2}$ [13]. The wear of a material of this composition has been investigated by Yust et al. [14] in unlubricated sliding using a pin-on-disk test apparatus. Tests were performed in nitrogen at room temperature and at 673 K, at a sliding velocity of 0.3 m/s, at normal loads of 2.2 to 8.9 N, and test durations of 1–4 hours. Both the pin and the disk members of the sliding couple were prepared from the composite. In each of the tests at room temperature, mild wear was observed (wear corresponding to a wear factor $< 10^{-6}$ $mm^3/N{\cdot}m$), while at 673 K only severe wear (wear factor $> 10^{-6}$ $mm^3/N{\cdot}m$) occurred. The room temperature mild wear surfaces were partially covered by a layer of adhered wear debris, but the undamaged surface of the disk was visible through openings in the debris layer (Fig. 1). In contrast, the high temperature, severe wear surfaces exhibited a substantial loss of material from both the disk and pin, and surface fracture was evident within the disk wear groove (Fig. 2). Values for the logarithm of the wear factor for both the pins and the disks are plotted in Figure 3. The wear factor values reflect the surface conditions observed; that is, the pins and disks tested at room temperature have wear factors in the range of 10^{-8} to 10^{-9} $mm^3/N{\cdot}m$, while the wear factors for the tests at 673 K are in the range of 10^{-4} to 10^{-5} $mm^3/N{\cdot}m$.

Friction and wear results for unlubricated pin-on-disk sliding of the alumina composite containing 20 vol % silicon carbide whiskers in air at temperatures to

Figure 1 Scanning electron micrograph showing wear debris accumulated on the disk wear track of a whisker-reinforced alumina composite specimen. Openings in the debris layer reveal the undamaged disk surface beneath the debris. Sliding direction from bottom to top.

1073 K also have been reported by Yust and Allard [15]. The results of the tests again demonstrated the almost immediate onset of severe wear at 673 K. The mild wear response at room temperature was confirmed in these tests, and it was shown that mild wear was also the result at 1073 K. The wear factors for the pins and disks used in these latter tests are also presented in Figure 3. At room temperature, the disk surfaces were only lightly marked and a significant amount of wear debris was not produced. The corresponding pin-on-disk wear factors are approximately 10^{-9} $mm^3/N \cdot m$. At 673 K, severe wear, typically starting within the first kilometer of sliding, was observed, and the wear factors had the appropriate values. The mechanism of mild wear was observed to be complex at 1073 K, involving oxidation of the exposed surfaces of the silicon carbide whiskers and the formation of a protective wear debris layer on the disc surface. The composition of the debris layer was identified by Auger analysis as a mixture of aluminum, silicon, and oxygen, and the layer had the appearance of a tightly adherent protective film. Electron-beam sputter etching also was used to remove

Figure 2 Scanning electron micrograph showing severe wear surface of a whisker-reinforced alumina composite disk tested at 673 K. Debris agglomerates and fracture surface are evident. Sliding direction is from bottom to top.

part of the debris layer to reveal an undamaged disk surface [15]. The data in Figure 3 show a consistent response for unlubricated sliding of the composite in both air and nitrogen and indicate that wear of the whisker-reinforced composite is greatest at 673 K. The same response has been observed in wear tests on other alumina-based microstructures.

The friction coefficient values observed in unlubricated sliding of the alumina–silicon carbide whisker composite show trends similar to those noted for the wear factors (Fig. 4). At room temperature the values were in the range of 0.4–0.6, although some extreme values lie outside this range, while the range of values at 673 K extends from 0.6 to 1.28. At 1073 K, the friction coefficient returns to 0.5 to 0.6. The observed difference in friction coefficient values may be due in part to the additional work input required in the severe wear interface to propagate cracks, to compact interfacial debris, and to overcome the increased sliding resistance of an irregular surface.

Fracture during unlubricated sliding of the alumina–silicon carbide whisker composite surfaces is apparently more readily initiated and/or propagated at 673 K. One cause of this response may be desorption of surface contaminants and/or

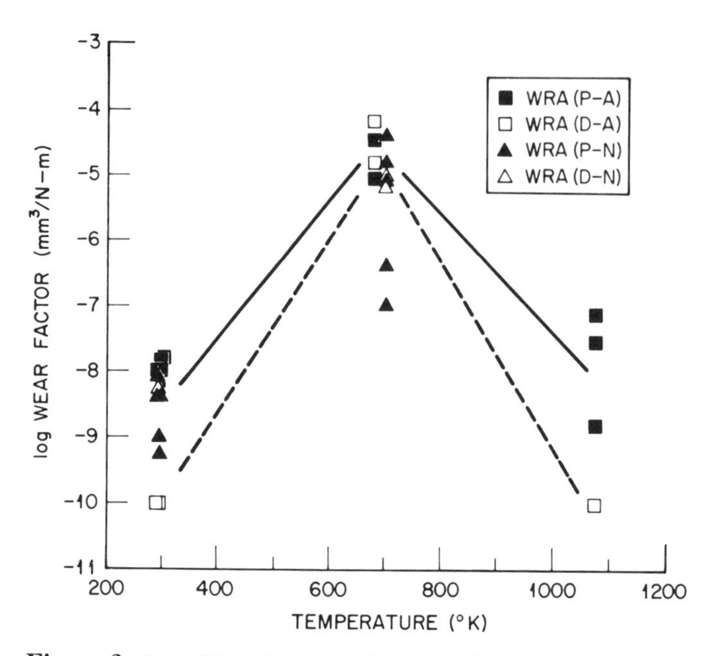

Figure 3 Logarithm of the wear factor as a function of temperature for whisker-reinforced alumina (WRA) tested in nitrogen and air atmospheres: P, pins; D, disks; A, argon; N, nitrogen. The values for the whisker-reinforced compositions at 673 K reflect the transition to severe wear at this temperature.

formation of reaction products at elevated temperatures, resulting in increased sliding friction. Consequently, the magnitude of the surface tensile stress behind the moving slider increases, resulting in a greater tendency for fracture. Gates et al. [16] have reported the formation of an aluminum hydroxide reaction product in the presence of water in a sliding alumina interface, which serves as a solid lubricant. However, at 673 K the reversion of the lubricating hydroxide to α-alumina is favored and the beneficial effect of the reaction is lost. Initiation of surface fractures can be observed at whisker–matrix boundaries in some of the room temperature mild wear surfaces, and the formation and propagation of similar fractures in the increased tensile stress field at 673 K may facilitate the transition to severe wear at this temperature. The apparent dilemma for wear application of WRA is that although the matrix–whisker interface may be a potential fracture nucleation site, a certain amount of debonding at this interface is required for effective fracture toughening of the composite by the whiskers [17]. Composites in which the whisker–matrix interface is securely bound and the whiskers are firmly fixed to the matrix do not exhibit any significant enhancement of mechanical properties.

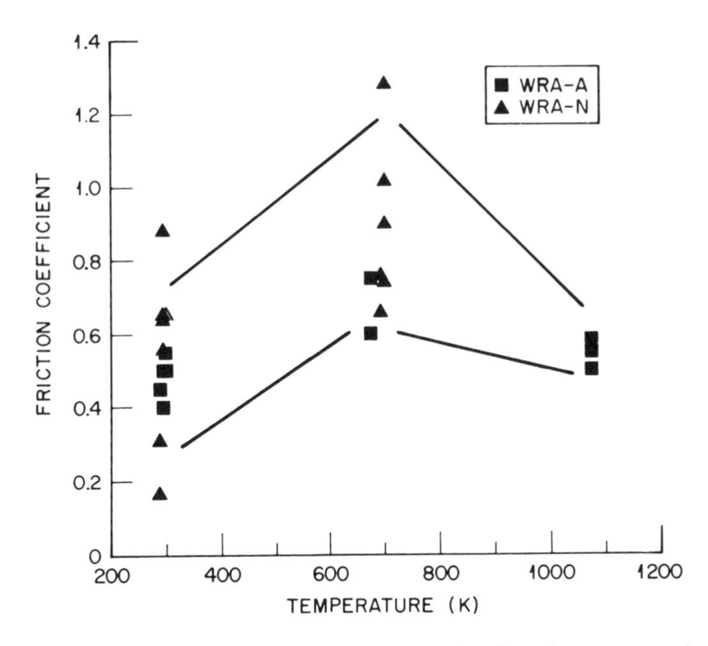

Figure 4 Friction coefficient values as a function of temperature for WRA. Although the data for the nitrogen tests are dispersed, there is an indication of greater friction during severe wear at 673 K.

Wear debris gathered from the surface of WRA specimens severely worn at 673 K has been examined to reveal the particle morphology [18]. Optical inspection of the debris particles at low magnification showed them to be agglomerates of smaller particles. Viewed in the scanning electron microscope at increasingly greater magnifications, they continued to have the form of an agglomerate (Fig. 5). Viewed at even very large magnifications with a transmission electron microscope, the agglomerated condition persisted. Dark field observation of the edges of such agglomerates clearly showed that the ultimate particle size of the debris is approximately 50–100 Å. Chemical electron-dispersive X-ray analysis (EDAX) revealed that the particles were either aluminum oxide or silicon carbide and that silicon oxide particles were not present to an important degree. Single-crystal particles of aluminum oxide and silicon carbide greater in size than 100 nm were also observed in the debris. Debris particles exhibiting the whisker morphology were not observed, indicating that significant comminution had occurred. These larger particles are either shards from the original surface fractures or the residue of larger shards that have not been completely comminuted in the wear track. The maximum size of a shard produced by fracture at the wear surface may be limited by the size of the mean separation

Figure 5 Typical debris agglomerate from a severe wear surface of a whisker-reinforced alumina specimen. It is evident that the agglomerates are comprised of individual particles much smaller than 0.1 μm.

between whiskers in the composite. During the period of retention of the debris in the moving interface, the as-formed particles are comminuted. The result is the observed condition of a few larger particles mixed into the fine-particle debris.

DellaCorte et al. [19] have reported on the behavior of a WRA composite at temperatures to 1473 K, and the results of their study, which are consistent with those discussed in this chapter, are presented in detail in the next chapter of this volume.

One of the important elements to be considered in any tribosystem is the state of stress at the contact region. Pin-on-disk tests utilizing a spherically tipped pin pressed against a flat surface form a Hertzian contact. At this contact, stress fields are formed in the near-contact regions of both the pin and the disk, and rotation of the disk beneath the pin results in the application of a cyclic stress in the wear path region. The stress fields may include tensile as well as compressive stress elements, and subsurface damage effects may occur, even in hard materials, if appropriate conditions are used. In titanium carbide, for example, the motion of dislocations near stressed surfaces at elevated temperatures has been demonstrated by Brookes and Parry [20], as has the work hardening of a hard body by repeated traversals of a cone of softer material. Our experiments at 1073 K on the WRA composite under a contact stress of 150 MPa for 4 km of sliding have shown similar results [18]. The surface appearance of the specimen was that of mild wear and minimal material loss. The subsurface region of the wear track, however, was found by transmission electron microscopy to contain significant accumulations of dislocations in grains to a depth of 1000 nm (Fig. 6). The total number of stress cycles imposed on this sample was 6.7×10^4. The accumulation of dislocations can ultimately lead to fracture initiation and surface fragmentation. Observations such as these emphasize that subsurface damage processes of a cumulative nature (i.e., fatigue processes including both deformation and/or fracture) may result from cyclic stresses applied to a wear interface. The potential for fatigue in ceramics is receiving increased attention, as evidenced by recent results [21], and the possibility for delayed transitions in wear mode, due at least in part to fatigue, should not be ignored.

Sliney and Deadmore [22] have studied the sliding wear of WRA composites having whisker contents of 8, 15, and 25 vol % against a nickel-based alloy, Inconel 718 (IN-718). They utilized a tribometer in which the flat surfaces of two diametrically opposed blocks are pressed against the edge of a rotating cylinder. Sliding wear and friction results from room temperature to 1073 K were reported. The wear factor for the composite sliding on IN-718 was found to decrease from 10^{-5}–10^{-6} $mm^3/N{\cdot}m$ at room temperature to 10^{-7}–10^{-8} $mm^3/N{\cdot}m$ at 1073 K. At each temperature level studied the wear factor decreased with increasing whisker content. In contrast to the whisker composite results obtained by Yust et al. [14] for sliding against alumina, a prompt transition from mild to severe wear was not observed at 673 K by Sliney and Deadmore. The friction coefficient likewise diminished steadily from about 0.45 at room temperature to 0.25–0.30 at 1073 K. The decrease in friction coefficient was attributed by the authors to the oxidation of the metal alloy and transfer of the oxide to the ceramic surface, forming a tenacious lubricating film. The presence of a lubricious layer is most likely responsible for the suppression of the wear mode transition observed in composite–ceramic sliding.

The wear behavior of whisker-reinforced ceramics in unlubricated sliding against a bearing steel has been reported by Fukuda et al. [23]. A pin-on-disk configuration was used to test hemispherically tipped pins sliding on JIS-SUJ2 bearing steel hardened to a Vickers hardness of 8.5 GPa. The pins tested included alumina, alumina with 30% silicon carbide whiskers, silicon nitride, silicon nitride with 10, 20, and 30% silicon carbide whiskers, and silicon carbide. The sliding velocity was 0.1 m/s; an applied load of 28 N and a total sliding distance

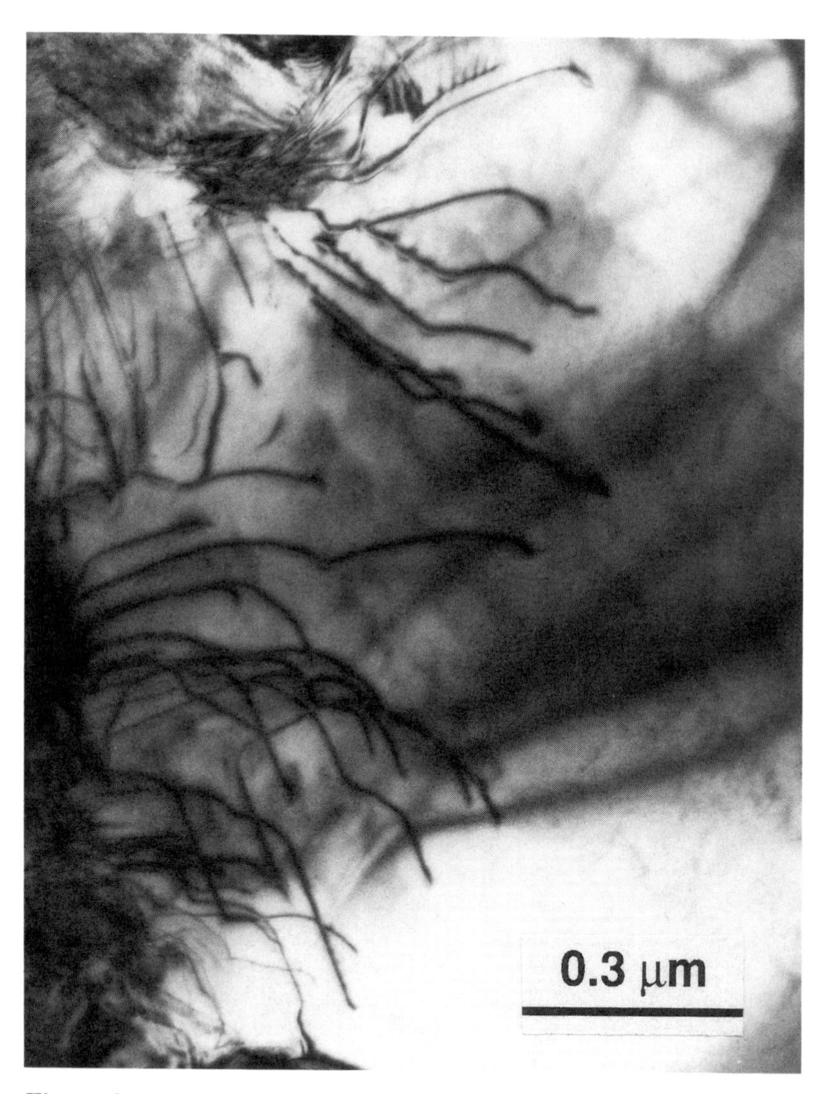

Figure 6 Transmission electron photomicrograph of the dislocation content of a grain 100 μm below the wear surface of a whisker-reinforced alumina composite worn at 1073 K at a contact pressure of 150 MPa.

of 1000 m were used. The tests were performed in dry air at room temperature. The ceramic pins showed a small decrease in wear with increased whisker content. At whisker contents greater than 20%, the wear of the composites ($\approx 1 \times 10^{-6}$ mm^3/N·m) was lower than that of any of the separate constituents. The wear of the steel disk was about an order of magnitude greater than that of the pins ($\approx 1 \times 10^{-5}$ mm^3/N·m) and was greatest for the case of alumina or the alumina composite. Investigation of the wear interfaces indicated the transfer and retention of iron on the alumina surfaces to a greater extent than on silicon carbide surfaces. The authors attribute the difference in iron transfer to varied compatibility between the iron and the ceramic phases and the wear mechanism to adhesion processes.

The technical literature also includes references to the use of WRA as a cutting tool. Billman et al. [24], for example, compared the wear behavior of a cemented carbide, sialon, and an alumina–titanium carbide composite with that of WRA in machining nickel-based and ferrous alloys, indicating the conditions for which WRA cutting tools provided superior wear resistance. Their results also showed an increased tool–workpiece reactivity for WRA as compared to monolithic alumina, an observation of relevance to all tribological applications of WRA. The effect of thermal shock on the wear of WRA used as a cutting tool has been investigated by Wayne and Buljan [25]. These investigators found that the susceptibility of WRA to thermal shock was not significantly better than that of monolithic alumina and that the SiC whiskers reacted chemically with ferrous alloys. This latter result is amplified by the work of Barrett and Page [26], who investigated the reactions of WRA with a range of liquid iron–nickel alloys. The result of 2-minute immersion in the molten alloys was a severe preferential attack of the whiskers, apparently by the decomposition of the silicon carbide to form carbides with the molten metals. Infiltration of the reaction zones in the composite specimens to depths of 60 μm suggested a highly interconnected whisker structure. In spite of the limitations suggested by the foregoing results, WRA is being used with great success as a cutting tool and has potential for many other applications.

III. ZIRCONIA-TOUGHENED ALUMINA: WITH AND WITHOUT WHISKERS

The persistent fracture and rapid, detrimental wear mode transition of WRA at 673 K suggested the substitution of a composite containing an additionally toughened alumina matrix. The addition of zirconia to WRA produces a duplex composite with greater fracture toughness than either of the simple composites [6,27]. Accordingly, specimens were prepared from a zirconia-toughened alumina (ZTA), both with and without silicon carbide whisker reinforcement. The specimens were prepared from powders supplied by a commercial source, and were hot-pressed in graphite dies [28]. The powder used to prepare ZTA contained 7.0 wt % unstabilized zirconia; the powder used to prepare the whisker-reinforced ZTA (WRZTA) contained 5.6 wt % unstabilized zirconia and 20 wt % silicon carbide whiskers. These specimens were tested without lubrication against alumina in a high temperature pin-on-disk apparatus. The test atmosphere was air at temperatures of 293, 673, and 1073 K.

The wear behavior of WRZTA as a function of temperature is similar to that of WRA in that the wear factor is greater at 673 K than at room temperature or 1073 K (Fig. 7). At room temperature, the WRZTA wear rate was in the mild regime, but material loss from the disk surface was detectable by contact profilometry and by microscopic inspection (Fig. 8). At 1073 K, measurable amounts of wear are introduced, but the values remain in the mild wear regime. The wear response at 673 K is characterized by a consistent transition to severe wear within the first kilometer of sliding. The wear debris was composed of agglomerates of very fine particles, and debris particles in the shape of whiskers were not observed. In contrast, the ZTA specimens showed only very mild wear to be operative at all three temperature levels studied (Fig. 7). Wear marks on

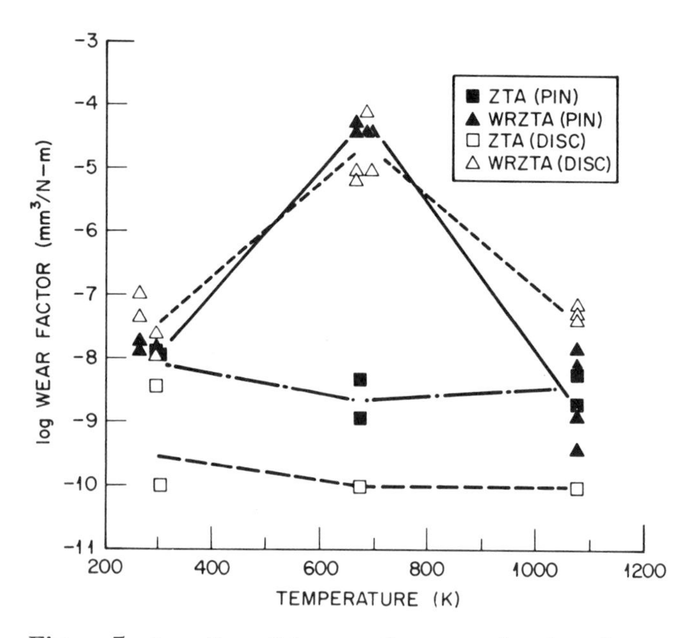

Figure 7 Logarithm of the wear factor as a function of temperature for ZTA and WRZTA. Severe wear of the WRZTA occurs in the first kilometer of sliding at 673 K. ZTA exhibits mild wear at each test temperature. The values of –10 for the ZTA disks indicate no measurable loss of material from the disk surface.

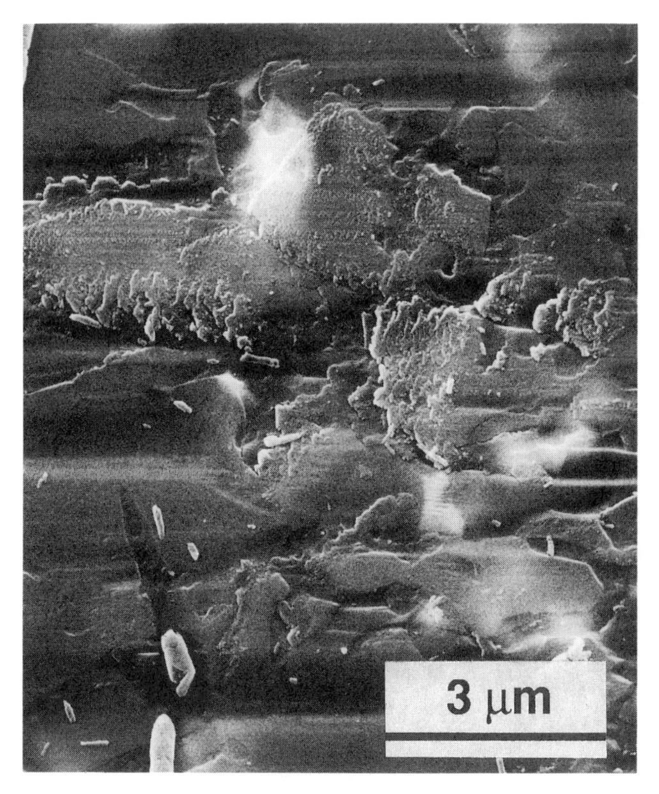

Figure 8 Wear track of a WRZTA disk worn at 293 K. Some material loss has occurred, although the wear rate is within the mild wear regime.

the ZTA disks were scarcely visible, and contact profilometry did not reveal any measurable loss of material. Correspondingly small wear scars were observed on the alumina counterfaces. The surface condition of ZTA after sliding 4 km at 293 K is shown in Figure 9. Very light abrasion marks on the surface in the direction of sliding and small rolls of wear debris are the only evidence of sliding on this surface.

The friction coefficient values for ZTA and WRZTA are shown in Figure 10. As was observed for WRA, friction values are greatest for WRZTA at 673 K, where severe wear is operative. The friction values for ZTA, however, which experienced only mild wear at each temperature, are also greatest at 673 K.

The increased friction for ZTA at 673 K may be due to desorption of adsorbed surface species or decomposition of reaction products as discussed. The decrease in friction at 1073 K in WRZTA is most likely related to the formation of surface oxides on the exposed silicon carbide whiskers. Earlier studies of debris on silicon

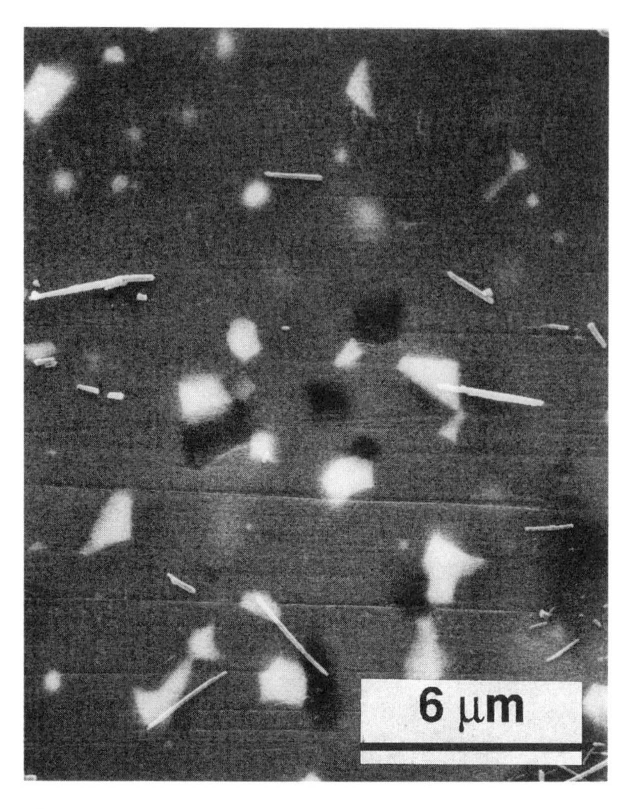

Figure 9 Wear track of a ZTA disk tested in air at 293 K. The wear track is delineated by very light abrasion marks. The white grains are zirconia distributed throughout the alumina matrix. The needlelike forms are rolled debris.

carbide whisker-reinforced material suggests that the debris layer consists of a mixture of oxides [18]. In specimens tested at 1073 K, the disk wear path on ZTA contains only small, isolated debris patches, while the WRZTA wear path is covered by a continuous layer of debris. This difference in interfacial debris content may contribute to the difference in observed friction coefficient at 1073 K, although the specific reason for the larger friction decrease on ZTA is not clear.

Although whisker reinforcement provides enhanced fracture toughness and zirconia additions to the alumina whisker composites further increase toughening, the results of these tests show that the WRZTA is not more wear resistant than ZTA. The difference in response is particularly evident at 673 K, where WRZTA makes a transition to the severe mode almost immediately, while ZTA is essentially unworn. The results appear to place the benefit of whisker additions

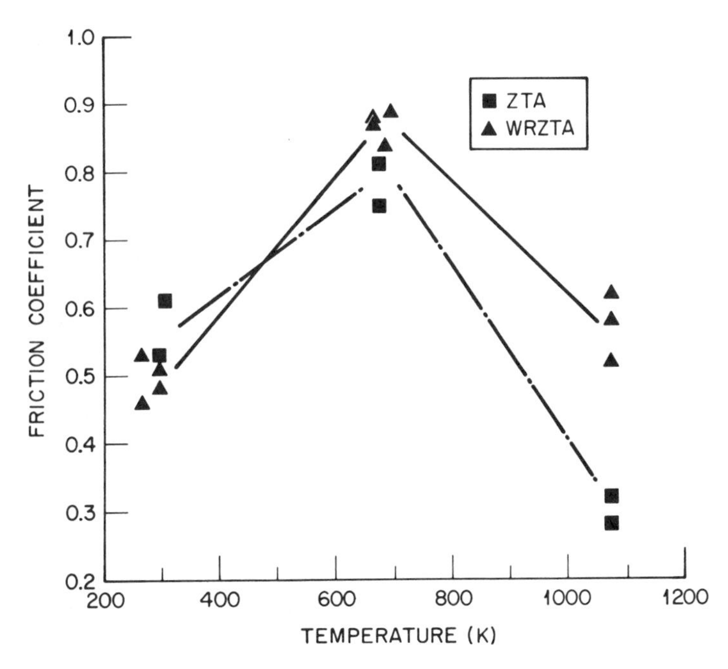

Figure 10 Friction coefficient values as a function of temperature for ZTA and WRZTA. Friction is greatest for each composition during severe wear at 673 K.

for wear resistance greatly in doubt for this ceramic. X-Ray investigation of the phase content of the zirconia-containing material sheds some light on the cause of the observed response.

The zirconia in the ZTA was found by X-ray analysis to consist of 94% tetragonal phase zirconia and 6% monoclinic phase zirconia [28]. In the WRZTA composite the relative proportions of tetragonal and monoclinic phases are 65 and 35%, respectively. During the period of cooling from the hot-pressing temperature, the stresses generated in the three-phase microstructure apparently promote a greater transformation of tetragonal phase to monoclinic phase than is the case in the whisker-free ZTA composite. The phase transformation is accompanied by the formation of microcracks around the transformed regions [5,29]. The resultant microcracks in and near the wear surface are exposed to the surface tensile stress formed behind the sliding contact. Since the friction coefficient has been determined to be greatest at 673 K the largest value of tensile stress is imposed on the surface at this temperature. The surface tensile stress at 673 K is sufficient in magnitude to promote the extension of the microcracks, thereby promoting the transition to severe wear. The apparently confusing wear results can thus be rationalized, and a possible approach to improvement suggested. If suppression of the extent of the transformation of the tetragonal phase can be achieved by compositional or processing variations, the decreased extent of microcrack formation should improve the wear response of WRZTA.

Studies of abrasive wear, erosion, and sliding wear are among the papers appearing in the recent literature on the wear of ZTA. Trabelsi et al. [30] describe the sliding wear behavior of alumina–zirconia composites having a range of zirconia contents. The study included alumina specimens containing unstabilized zirconia contents of 5–20 vol % and yttria-stabilized zirconia contents of 20 and 45 vol %. Sliding of ceramic blocks against an AISI 52100 steel ring was investigated in air and in water. The results are not directly comparable with those reported here because test conditions differed, but they do have some related points. Trabelsi et al., for example, reported that for zirconia contents greater than 10 vol %, the tetragonal-to-monoclinic phase transformation induces the formation of microcracks in the material, leading to a decrease in wear resistance [30]. The ZTA samples produced for the study by Yust and DeVore [28] contained less than 10 vol % zirconia, principally as the tetragonal phase (94%), as noted above. The wear resistance of this material was excellent, in general agreement with the expected absence of microcracking at this zirconia content. The WRZTA produced by Yust and DeVore contained only 65% zirconia as the tetragonal phase [28]. This result also is consistent with the view of Trabelsi et al. that the deterioration of wear resistance is a consequence of the formation of microcracks.

Krell and Blank [31] have examined the wear of zirconia-toughened aluminas of two slightly varied compositions. Based on their observation of the doping effects on the grain boundary fracture toughness and the consequent stimulation of subcritical crack growth, the authors anticipate increased wear in ZTA containing glassy grain boundary phases [32–34]. The compositions tested were intended to provide amorphous grain boundary phases in ZTA. Wear was measured by determining the mass loss of isostatically pressed and conventionally sintered grinding balls rotated in cylindrical alumina jars. Twenty balls 10 mm in diameter were rotated in water for 100 hours in a horizontal orientation and 5 hours in a vertical position on a planetary mill. The results showed that the ZTA microstructures formed had significantly reduced, rather than increased, wear rates as compared to Al_2O_3. The authors attribute the observed wear response to some synergetic combination of microcracking due to ZrO_2 phase transformation and that due to alteration of the grain boundary toughness. They recommend high resolution transmission electron microscopy to clarify phase transformations during wear as one approach to resolution of this unexpected result.

Studies by Breder et al. [35] have shown that the erosive wear rate of ZTA is not significantly different from that found for commercial aluminas. The impact of eroding particles triggers the tetragonal–monoclinic transformation in the surface material, and the associated microcracking is assumed to result in monoclinic grains that are only weakly bonded to the matrix [35].

The abrasive wear of a whisker-reinforced ZTA has been evaluated by Bohmer and Almond [36]. Fine-grained ZTA and ZTA containing 17 wt % SiC whiskers were obtained from a commercial source for use in the study. Abrasive wear resistance was measured by sliding in abrasive slurries using alumina, silicon carbide, and boron carbide particles, and by self-mated unlubricated sliding at room temperature in a pin-on-disk machine. High purity alumina was used as a basis for comparison. It was found in the abrasion tests that the whisker-reinforced material was superior in wear resistance to the ZTA without whiskers. The authors attributed this result to the increased hardness and toughness of the whisker-reinforced material. The dry sliding wear tests showed that there was no significant difference in the friction coefficient (≈ 0.4) obtained in the self-mated sliding of ZTA and that of self-mated WRZTA. The friction coefficient for sliding of alumina on both ZTA and WRZTA determined by Yust and DeVore [28] was about 0.5 at room temperature. Bohmer and Almond reported a better wear resistance for WRZTA, apparently based on a limited number of tests. Based on a pin displacement measurement, they concluded that greater wear at room temperature occurs on the specimens without whiskers, although in one case the increased wear was due to the formation of a chip in the sliding interface and subsequent abrasion of the disk surface. In contrast, the work of Yust and DeVore [28] showed that whisker-reinforced material had the greatest wear, particularly at 673 K.

IV. SILICON NITRIDE REINFORCED WITH SILICON CARBIDE WHISKERS

The friction and wear behavior of WRSN has been studied at room temperature [37]. The tests were conducted in reciprocating motion and included both unlubricated and lubricated conditions. Both silicon carbide and silicon nitride spheres have been used as counterface materials to assess the role of each composite constituent in opposition to the composite. In unlubricated sliding at normal loads of 1 and 10 N, the silicon nitride sphere caused larger wear rates than the silicon carbide spheres (Fig. 11). The friction coefficient was approximately the same in both cases, about 0.5. The wear rate was not affected by the load for silicon nitride as the counterface but did vary with silicon carbide. The profiles of both the pin and the disk wear surfaces were found to conform closely, and the surface of the wear tracks was found to consist of alternating bands of smooth surface and fracture surface (Fig. 12). The retention of smooth surface regions after the removal of material to a depth of 2 μm suggests that the unlubricated sliding wear process involves sequential fracture and polishing mechanisms.

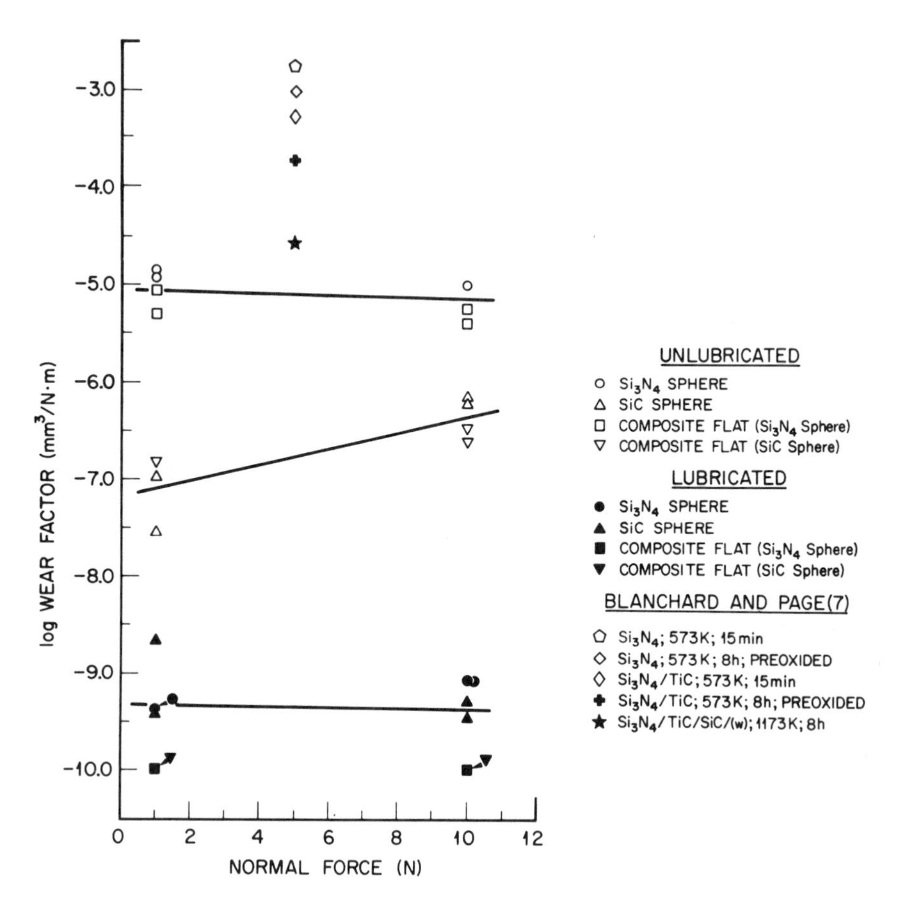

Figure 11 Wear factors for sliding of silicon carbide and silicon nitride spheres on the whisker-reinforced silicon nitride composite. The silicon carbide sphere produces less wear than silicon nitride in unlubricated sliding but is more sensitive to the applied force.

In lubricated reciprocating sliding, using a fully formulated, synthetic engine oil (SAE 5W-30), the wear rate of the composite against both silicon carbide and silicon nitride spheres was significantly lower than that of the unlubricated couples and about the same with both counterface materials (Fig. 11). Small wear scars formed on the ends of the spheres, and the wear tracks on the disks were visible only by virtue of dark deposits formed along the edges of the wear path. Surface profilometry did not reveal the removal of any material from the disk surface. Unidirectional lubricated sliding, using a pin-on-disk apparatus, yielded wear factor values similar to those obtained in reciprocating motion even though significantly greater applied force values were used. Friction coefficient

Figure 12 Wear surface of a whisker-reinforced silicon nitride flat after removal of material to a depth of 2 μm. The wear surface contains both regions of fracture surface (left part of field) and smooth areas. Intersections of whiskers with the smoothed surface are evident.

values in unidirectional sliding were in the range of 0.02 at all conditions examined.

Ishigaki et al. [38] have also studied the unlubricated sliding of a silicon carbide whisker-reinforced silicon nitride (WRSN) composite. A flat-on-ring test geometry was used, in which composite specimens were worn against rotating monolithic silicon carbide and silicon nitride rings. At a sliding velocity of 0.31 m/s and an applied force of 5 N, the logarithm of the wear factor for the composite containing 20 wt % whiskers sliding against silicon carbide was observed to be –5.5, while that for the same composite sliding against silicon nitride was about –5.2. These values were determined on the sample surface oriented perpendicular to the hot pressing direction. Comparison of these data with the results of Yust and DeVore [28], quoted above for unlubricated reciprocal sliding on WRSN

(Fig. 11), shows good agreement for the case of silicon nitride sliding on the composite. The logarithm of the wear factor obtained by Yust and DeVore was –5.0, which compares well with the value of Ishigaki et al., which was –5.2. For silicon carbide sliding on the composite there is a divergence of results. Yust and DeVore reporting –6.5 while Ishigaki et al. report –5.5, an order of magnitude greater. Comparable values of friction coefficient, 0.5–0.6, were observed in these studies.

V. SILICON NITRIDE MATRIX WITH MULTIPLE SECONDARY PHASES

The combination of silicon nitride as a matrix with multiple secondary phases has yielded particularly interesting wear results. Blanchard-Ardid and Page, for example, have achieved improvement of both friction and wear response by incorporating silicon carbide whiskers and particulate titanium carbide in a silicon nitride matrix [7,8]. A recent study by these investigators compared the mechanical properties and the friction and wear response of the unreinforced matrix, the matrix plus each reinforcing phase separately, and the matrix plus both reinforcing phases [7]. The compositions containing titanium carbide had the largest fracture toughness value (7.2–7.9 $MPa \cdot m^{1/2}$) with or without whiskers, while the greatest flexural strength (578 MPa) was attained when only particulate titanium carbide was present. Whisker additions alone did not significantly enhance either fracture toughness or flexural strength, and in fact, appeared to degrade the strength as compared to the whisker-free matrix.

Pin-on-flat tests in the reciprocating mode indicated that the wear and friction behavior were substantially improved by the TiC additions in tests at 1173 K. The cause of this improvement is the formation of titanium oxide on the surface by oxidation of the carbide phase, and the subsequent shearing of this layer to form a lubricating film [7]. In tests of 15-minute duration at 573 and 773 K, the specimens containing TiC were much less worn than either silicon nitride or silicon nitride with whiskers, and at 973 and 1173 K they were essentially unworn. The friction coefficient correspondingly decreased at 1173 K from about 0.7 to 0.5. Preoxidized surfaces also showed diminished wear, as did the TiC-containing compositions in 8-hour tests at 1173 K. These results demonstrate how incorporated phases may improve both mechanical and tribological properties.

Although not reported, the wear factor values for the experiments of Blanchard and Page can be estimated from the wear track cross sections and test parameters published by these authors. The resultant values at 573 K are compared with those measured by Yust and DeVore in Figure 11 [37]. Silicon carbide was used as a counterface material in both the Blanchard–Page and Yust–DeVore experiments, and the test conditions were sufficiently similar to yield a meaningful comparison. The velocity was lower in the Blanchard–Page experiments (0.08 m/s vs. 0.3 m/s in the Yust–DeVore work), but neither velocity is great enough to invalidate the comparison. The wear factor values reported by Yust and DeVore for unlubricated sliding of the silicon nitride–silicon carbide whiskers composite are approximately two orders of magnitude smaller than that of Blanchard and Page. This result is consistent with the difference in mechanical properties cited in each of the studies (771 MPa vs. 436 MPa for flexural strength and 6.9 $MPa{\cdot}m^{1/2}$ vs. 5.7 $MPa{\cdot}m^{1/2}$ for fracture toughness), the greater values being those of the Yust–DeVore material. The wear and friction response at elevated temperatures, however, is clearly improved by the inclusion of TiC.

The effect of longer test duration was evaluated by Blanchard and Page at 1173 K [7]. The extent of wear after 8 hours of testing was compared in silicon nitride containing only TiC in one case, and both TiC and SiC whiskers in the other. The whisker-containing specimen formed a measurable wear groove (and yielded a wear factor of 2.7×10^{-5} $mm^3/N{\cdot}m$), whereas the specimen containing only TiC did not show any measurable evidence of wear. The result suggests that the presence of the whiskers, which diminished the strength and did not increase the fracture toughness in these specimens, induces a less favorable wear behavior. The behavior of a composite retaining good strength and fracture toughness, such as that used in the Yust–DeVore experiments, when combined with TiC particles, is worthy of further investigation.

VI. SUMMARY

Wear mechanisms in whisker-reinforced ceramics can be complex, and they depend on the mechanical and chemical behavior of each of the microstructural constituents, both singly and in combination, in a particular environment. Results presented in this survey indicate that the increased strength and fracture toughness produced by the inclusion of whiskers into ceramic matrices may be nullified by other microstructural considerations. For example, the introduction of the whiskers creates interfaces with specific properties that affect the behavior of the composite. Some of the wear studies have indicated a potential for fracture initiation at whisker–matrix interfaces, although paradoxically some bonding loss at this interface is necessary to achieve improved toughness in the composite. Similarly, the prospect for enhanced matrix toughening by inclusion of a secondary toughening phase, as in the case of zirconia in alumina, may lead to undesired consequences such as microcracking accompanying the tetragonal-to-monoclinic transformation. In addition, the role of interface chemistry has been illustrated by the oxidation of the silicon carbide whiskers in elevated temperature tests, as has the beneficial oxidation of a secondary phase to form a solid lubricant. These examples also suggest that the evaluation of the wear mechanisms of ceramics will become increasingly difficult as more sophisticated, multiphase ceramics are produced.

Examples have been presented of the transition from mild to severe wear, which is a characteristic not only of the wear of ceramics but of other materials, as well. For most applications of ceramics (i.e., seals, bearings, engine components, etc.), this transition must not occur during the anticipated lifetime of the ceramic component. A major objective of increased toughening of ceramics through whisker incorporation is the suppression of crack formation and crack growth, the processes that may lead to surface fragmentation and the wear mode transition. In addition to preserving the original surface finish and dimensions, the maintenance of a mild wear status minimizes debris formation and the subsequent degradation of other system components.

Most ceramic wear testing to date indicates that unlubricated use of ceramics as machine components is not likely to be successful, except under conditions of very limited contact. Even under lubricated sliding conditions, stresses are applied to wear surfaces that can ultimately lead to surface degradation through generation of preliminary subsurface damage involving either deformation or fracture. Whisker incorporation or other toughening mechanisms, however, may ultimately yield sufficiently durable ceramic microstructures to make the use of ceramic machine components generally practicable. The evidence presented here emphasizes the need for continued attention to the relationship among microstructure, mechanical properties, and wear behavior.

There are many conditions of use in which ceramics can be and are reliably applied today. Additional applications are being recognized, many of which involve high rates of motion, for example, sliding velocities in excess of 2 m/s. Most wear studies, however, are performed at much lower velocities, and the need for high velocity information remains unsatisfied. Future technologies will turn more frequently to the use of ceramics for rigorous service, especially as the quality of these materials is improved. The demand for wear-resistant ceramics will require a clearer understanding of the wear processes in increasingly complex microstructures.

ACKNOWLEDGMENTS

Research sponsored by the Office of Advanced Transportation Materials, Tribology Program, U.S. Department of Energy, under contract DE-AC05-84OR21400 with Martin Marietta Energy Systems, Inc.

REFERENCES

1. P. F. Becher and G. C. Wei, Toughening behavior in SiC-whisker-reinforced alumina, *Commun. Am. Ceram. Soc. 67*, 259–260 (1984).

2. G. C. Wei and P. F. Becher, Development of SiC-whisker-reinforced ceramics, *Am. Ceram. Soc. Bull. 64*, 298–304 (1985).
3. R. C. Garvie, R. H. Hannink, and R. T. Pascoe, Ceramic steel? *Nature (London) 258*, 703–704 (1975).
4. D. L. Porter and A. H. Heuer, Mechanisms of toughening in partially stabilized zirconia (PSZ), *J. Am. Ceram. Soc. 60*, 183–184 (1977).
5. A. H. Heuer, Transformation toughening in ZrO_2-containing ceramics, *J. Am. Ceram. Soc. 70*, 689–698 (1987).
6. P. F. Becher and T. N. Tiegs, Toughening behavior involving multiple mechanisms: Whisker reinforcement and zirconia toughening, *J. Am. Ceram. Soc. 70*, 651–654 (1987).
7. C. R. Blanchard and R. A. Page, Effect of silicon carbide whisker and titanium carbide particulate additions on the friction and wear behavior of silicon nitride, *J. Am. Ceram. Soc. 73*, 3442–3452 (1990).
8. C. R. Blanchard-Ardid and R. A. Page, Improved contact damage resistance of a $Si_3N_4/TiC/SiC_{wh}$ composite, *Ceram. Eng. Sci. Proc. 9*, 1443–1452 (1989).
9. M. V. Swain, Microscopic observations of abrasive wear of polycrystalline alumina, *Wear, 35*, 185–189 (1975).
10. N. S. Eiss and R. C. Fabiniak, Chemical and mechanical mechanisms in wear of sapphire on steel, *J. Am. Ceram. Soc. 49*, 221–226 (1966).
11. C. S. Yust and F. J. Carignan, Observations on the sliding wear of ceramics, *ASLE Trans. 28*, 245–252 (1985).
12. L. D. Wedeven, R. A. Pallini, and N. C. Miller, Tribological examination of unlubricated and graphite lubricated silicon nitride under traction stress, in *Wear of Materials—1987*, K. C. Ludema (Ed.), American Society of Mechanical Engineers: New York, 1987, p. 333.
13. P. F. Becher and T. N. Tiegs, Temperature dependence of strengthening by whisker reinforcement: SiC whisker-reinforced alumina in air, *Adv. Ceram. Mater. 3*, 148–153 (1988).
14. C. S. Yust, J. M. Leitnaker, and C. E. DeVore, Wear of an alumina–silicon carbide whisker composite, *Wear, 122*, 151–164 (1988).
15. C. S. Yust and L. F. Allard, Wear characteristics of an alumina–silicon carbide whisker composite at temperatures to 800 C in air, *Tribol. Trans. 32*, 331–338 (1989).
16. R. S. Gates, S. M. Hsu, and E. E. Klaus, Tribochemical mechanism of alumina with water, *Tribol. Trans. 32*, 357–363 (1989).
17. P. F. Becher, C. H. Hsueh, P. Angelini, and T. N. Tiegs, Toughening behavior in whisker-reinforced ceramic matrix composites, *J. Am. Ceram. Soc. 71*, 1050–1061 (1988).
18. C. S. Yust and L. F. Allard, Microanalytical characterization of wear damage in an alumina–silicon carbide whisker composite, in *Ceramic Materials and Components for Engines*, V. J. Tennery, (Ed.), American Ceramic Society: Westerville, OH: 1989, pp. 1212–1224.
19. C. Della Corte, S. C. Farmer, and P. O. Book, Tribological characteristics of silicon carbide whisker-reinforced alumina at elevated temperatures, Chapter 10, this volume.

20. C. A. Brookes and A. R. Parry, Some fundamental aspects of the mechanical wear of hard ceramic crystals due to sliding, *Mater. Sci. Eng. A105–A106*, 143–150 (1988).
21. A.-P. Nikkilä, and T. A. Mäntylä, Cyclic fatigue of silicon nitrides, *Ceram. Eng. Sci. Proc. 10*, 646–656 (1989).
22. H. E. Sliney and D. L. Deadmore, Friction and wear of oxide–ceramic sliding against IN-718 nickel base alloy at 25 to 800°C in atmospheric air, NASA Technical Memorandum No. 102291 (August 1989), 26 pp.
23. K. Fukuda, Y. Sato, T. Sato, and M. Ueki, Wear properties of SiC-whisker-reinforced ceramics against bearing steel, in *Tribology of Composite Materials*, P. K. Rohatgi, P. J. Blau, and C. S. Yust (Eds.), ASM International: Metals Park, OH: 1990, pp. 323–328.
24. E. R. Billman, P. K. Mehrotra, A. F. Shuster, and C. W. Beeghly, Machining with Al_2O_3-SiC-Whisker Cutting Tools, *Ceram. Bull. 67*, 1016–1019 (1988).
25. S. F. Wayne and S. T. Buljan, The role of thermal shock on wear resistance of selected ceramic cutting tool materials, *Ceram. Eng. Sci. Proc. 9*, 1395–1408 (1989).
26. R. Barrett and T. F. Page, The interactions of an Al_2O_3-SiC whisker-reinforced composite ceramic with liquid metals, *Wear, 138*, 225–237 (1990).
27. T. N. Tiegs and P. F. Becher, Whisker-reinforced ceramic composites, in *Ceramic Materials and Components for Engines, Proceedings of the Second International Symposium*, Lübeck-Travemunde, April 14–17, 1986, Deutsche Keramische Gesellshaft: Bonn, Germany: 1986, pp. 193–200.
28. C. S. Yust, and C. E. DeVore, Wear of zirconia-toughened alumina and whisker-reinforced zirconia-toughened alumina, *Tribol. Trans. 33*, 573–580 (1990).
29. A. G. Evans, Toughening mechanisms in zirconia alloys, in *Advances in Ceramics*, Vol. 12, N. Claussen, M. Ruehle, and A. H. Heuer (Eds.), American Ceramics Society: Westerville, OH, 1981, pp. 193–212.
30. R. Travelsi, D. Treheux, G. Orange, G. Fantozzi, P. Homerin, and F. Thevenot, Relationship between mechanical properties and wear resistance of alumina–zirconia ceramic composites, *Tribol. Trans. 32*, 77–84 (1989).
31. A. Krell and P. Blank, On abrasive wear of zirconia-toughened aluminas, *Wear, 124*, 327–330 (1988).
32. A. Krell, J. Woltersdorf, E. Pippel, and D. Schulze, On crack-propagation-related phenomena in Al_2O_3 + ZrO_2, *Phil. Mag. A, 51*, 765–776 (1985).
33. A. Krell, E. Pippel, and J. Woltersdorf, On grain boundary strength in sintered Al_2O_3 sintered in air and hydrogen, *Phil. Mag. A, 53*, L11 (1986).
34. A. Krell and W. Pompe, The influence of subcritical crack growth on the strength of ceramics, *Mater. Sci. Eng. 89*, 161 (1987).
35. K. Breder, G. DePortu, J. E. Ritter, and D. Dalle Fabbriche, Erosion damage and strength degradation of zirconia-toughened alumina, *J. Am. Ceram. Soc. 71*, 770–775 (1988).
36. M. Bohmer and E. A. Almond, Mechanical properties and wear resistance of a whisker-reinforced zirconia-toughened alumina, *Mater. Sci. Eng. A105–A106*, 105–116 (1988).

37. C. S. Yust and C. E. DeVore, The friction and wear of lubricated $Si_3N_4/SiC_{(w)}$ composites, *Tribol. Trans. 34*, 497–504 (1991).
38. H. Ishigaki, R. Nagata, M. Iwasa, N. Tamari, and I. Kondo, Tribological properties of SiC whisker containing silicon nitride composite, *J. Tribol. 110*, 434–438 (1988).

10

Tribological Characteristics of Silicon Carbide Whisker-Reinforced Alumina at Elevated Temperatures

Christopher DellaCorte

Lewis Research Center, NASA
Cleveland, Ohio

ABSTRACT

The enhanced fracture toughness of whisker-reinforced ceramics makes them attractive candidates for sliding components of advanced heat engines. Examples include piston rings and valve stems for Stirling engines and other low heat rejection devices. However, the tribological behavior of whisker-reinforced ceramics is largely unknown. This is especially true for the applications described, where use temperatures can vary from below ambient to well over 1000 °C.

The chapter describes an experimental research program to identify the dominant wear mechanism(s) for a silicon carbide whisker-reinforced alumina composite (SiC_w-Al_2O_3). In addition, a wear model is developed to explain and corroborate the experimental results and to provide insight for material improvement.

I. INTRODUCTION

A. Motivation

A major obstacle to the application of ceramics in machinery is the inherent brittleness of these materials and their tendency to fracture. This behavior is in contrast to metals, which yield in a plastic manner before fracture and usually avoid catastrophic failure. One way to improve ceramics is to enhance their toughness by the addition of secondary phases, such as whiskers and particles.

These secondary phases act to inhibit and deflect crack propagation and thereby improve toughness. One example of such a material is silicon carbide whisker-reinforced alumina (SiC_w-Al_2O_3). This composite consists of an Al_2O_3 matrix with approximately 8–40 vol % SiC reinforcing whiskers. The composite's fracture toughness, typically 8.0–8.5 $MPa \cdot m^{1/2}$, is about twice that of unreinforced Al_2O_3 [1]. Improved fracture toughness makes this ceramic composite material attractive for use in a number of advanced applications.

Since improved fracture toughness generally means improved wear resistance, one potential area of application for this composite is high temperature sliding components in advanced heat engines such as low heat rejection (LHR) diesels, Stirling engines, and aircraft turbine engines. Cylinder wall–piston ring sliding couples, valve stems, bushings, and seals are specific examples of applications for which reinforced ceramics are being considered as tribological elements [2].

The successful application of ceramics as triboelements depends on a full understanding of their wear behavior, especially at elevated temperatures. Recent tribological data on ceramics, both monolithic and reinforced types, indicate that in the absence of lubrication, friction and wear can be quite high [3]. To predict and possibly improve the friction and wear characteristics of ceramics at high temperatures, the wear mechanism(s) must be understood. For studying monolithic ceramics, brittle fracture theory can be used with satisfactory results. Reinforced ceramics like SiC_w-Al_2O_3 can behave in more complex ways.

B. Background

Researchers have studied the tribological behavior of SiC_w-Al_2O_3 composites. Sliney and Deadmore, for instance, examined the effect of temperature and whisker volume content of SiC_w-Al_2O_3 using a double-block-on-ring apparatus [4]. The blocks were made from SiC_w-Al_2O_3 composites and the ring material was the nickel-based superalloy Inconel 718. Their work indicated that the wear of the ceramic blocks decreased and the friction coefficient increased as the whisker content increased from 8 to 30 vol %. These investigators also pointed out that the transfer of lubricious metal oxides from the metal rings to the ceramic blocks has a pronounced effect on friction and wear: namely, high metal oxide transfer improves the tribological properties. However, the exact nature of the transfer film was unclear and complex, making specific conclusions from the data, especially regarding wear mechanisms, difficult.

In somewhat similar experiments, Fukuda et al. [5] studied the effect of SiC whisker content on the tribological properties of SiC_w-Al_2O_3 composites when sliding against heat-treated bearing steel with a Vickers hardness of 8.7 GPa. The tests were conducted at room temperature using a ceramic composite pin-on-steel disk test, and the results agreed with those described earlier by Sliney and Deadmore [4]. As the whisker content increases, the wear of the ceramic decreases and the wear of the metallic counterface increases. Fukuda et al. also detected significant metal oxide transfer to the ceramic and, like Sliney and Deadmore, concluded that the transfer films had a significant effect on friction and wear. This is also most likely the case for monolithic and composite ceramic counterfaces.

These experiments indicate that the tribological behavior of SiC_w-Al_2O_3 is not well understood. Furthermore, the tribological behavior is significantly complicated by the presence of transfer films. Research on SiC_w-Al_2O_3 composites, which are tested in sliding against themselves, has proven somewhat more helpful in terms of understanding wear mechanisms and behavior.

Bohmer and Almond studied the wear resistance of self-mated Sic_w-Al_2O_3-ZrO_2 using a pin-on-disk type of apparatus [6]. They found that the wear resistance increased as the whisker content was increased for both the pin and the disk. They attributed these results to improved toughness of the composite. Although they performed Scanning electron microscopic (SEM) analyses of the wear surfaces, they offered no explanation of the wear mechanism(s).

The most extensive research, by far, done on the wear of SiC_w-Al_2O_3 has been that by Yust et al. [7–9]. These studies have concentrated on understanding the wear processes and mechanisms of SiC_w(20 vol %)-Al_2O_3 sliding against itself from 20 to 800°C. Reference 7 describes tests conducted in a nitrogen atmosphere using a pin-on-disk configuration. The atmosphere was specifically chosen to avoid tribochemical reactions, to promote a better understanding of the tribomechanical wear process. The results of this work indicated that the material wears by fracture, since cracks in the wear tracks and faceted wear particles were observed. The authors did determine, in this work, that the wear of monolithic alumina was at least two and as much as four orders of magnitude greater than the 20 vol % SiC_w-Al_2O_3 composite. Despite acquiring appreciable wear data, the authors were unable to establish a relationship between the composite's microstructure and a wear mechanism.

Yust and Allard also tested their 20 vol % SiC_W-Al_2O_3 composite in an air atmosphere from 20 to 800°C [8]. They obtained results similar to their earlier results in nitrogen except that at the highest test temperature, 800°C, an oxide layer developed on the rubbing surface and reduced wear. Auger and energy-dispersive X-ray spectroscopic analysis indicated that the oxide layer was predominantly a mixture of Al_2O_3 (from the matrix) and SiO_2 (from oxidized SiC whiskers). Transmission electron microscope (TEM) analysis confirmed that the wear debris was made up of very fine (10–50 nm diameter) particles. These particles agglomerate into larger debris "areas" on the wear tracks, which macroscopically resemble plastically deformed areas. Further work by Yust and Allard indicates that subsurface dislocation movement may also play a role in high temperature wear behavior, although the mechanism is not exactly known [9].

It is clear from this review that the wear behavior of SiC_w-Al_2O_3 composites is not well understood. This is especially true when the wear behavior is complicated by a reactive test environment and high temperatures.

For this reason, a research program was conducted to further study the wear mechanisms of a SiC_w-Al_2O_3 composite as a function of test temperature and to determine the dominant wear mechanism(s).

In this program, pin-on-disk wear tests were conducted with a self-mated SiC_w-Al_2O_3 composite. Then the wear surfaces were subjected to SEM and TEM analysis. Some of the experimental results have been reported elsewhere [10]. Based on these analyses, the most probable dominant wear mechanism(s) were determined. Finally, to test the plausibility of the experimentally determined wear mechanism(s), an analytical model of the wear process was developed and applied.

II. EXPERIMENTAL PROCEDURE

A. Material

The SiC_w-Al_2O_3 composite material studied contains 75 vol % Al_2O_3 with 25 vol % SiC whiskers. Table 1 gives the material's detailed composition and manufacturer's strength–property data. Partially stabilized zirconia (ZrO_2) is included in the table for comparison of its bulk physical properties.

The composite is made by hot-pressing high purity alumina powder (> 99.9%

Table 1 Strength and Property Data

	Materials			
Property	Al_2O_3	SiC	Al_2O_3-SiC	ZrO_2
Density, g/cm^3	3.9	3.1	3.74	5.7
Young's modulus, GPa	386	406	393	200
Vickers hardness, kg/mm^2	2000	2800	2125	1050
Toughness, $MPa \cdot m^{1/2}$	4.2	3.8	8.8	8.2
Thermal expansion coefficient, $°C^{-1}$	8.0×10^{-6}	4.0×10^{-6}	6.0×10^{-6}	9.2×10^{-6}
Four-point bend strength, MPa at RT	344	448	641	630
Poisson's ratio	.23	.12	.23	.23
Thermal conductivity, $W\ m^{-1}\ °C^{-1}$	22	12.5	22.3	2.0
Thermal diffusivity, m^2/s	8.0×10^{-6}	6.0×10^{-2}	1.35×10^{-5}	7.0×10^{-7}

Source: ARCO Chemical Company, Green, SC; Carborundum Company, Niagara Falls, NY.

Al_2O_3 with traces of silicon and iron) mixed with SiC whiskers. During consolidation, most of the whiskers preferentially align themselves in a plane perpendicular to the pressing direction [11]. The pins and disks tested in this study have their rubbing surfaces parallel to the whisker planes (Fig. 1).

The whiskers are single-crystal SiC with lengths of 10–60 μm and diameters of ≈.75 μm. The matrix grain size is approximately 2 μm and the material is hot-pressed at approximately 1600°C.

Figure 2 shows TEM micrographs of the unworn or virgin material (i.e., after specimen preparation but before tribotesting: it can be seen that there is little residual porosity and that there is good contact between the matrix and the whiskers (i.e., no large voids between the whiskers and the matrix). Figure 2 shows a lengthwise cross-sectional view of a whisker.

Wear pins 0.48 cm in diameter and 2.5 cm long were made from the composite. Hemispheres of 2.54 cm radii were machined on the pin ends and were diamond polished to a surface finish of about 0.1 μm rms.

The wear disks are 6.35 cm in diameter and 1.25 cm thick. The faces were diamond polished to a surface finish of about 0.1 μm rms.

B. Apparatus and Procedure

The specimens were tested in a high temperature pin-on-disk tribometer (Fig. 3). With this apparatus, the pin is held in a torque tube and is loaded against a rotating disk, which is mounted on a ceramic spindle. The spindle penetrates a SiC glowbar furnace, which is capable of heating the specimens to 1200°C.

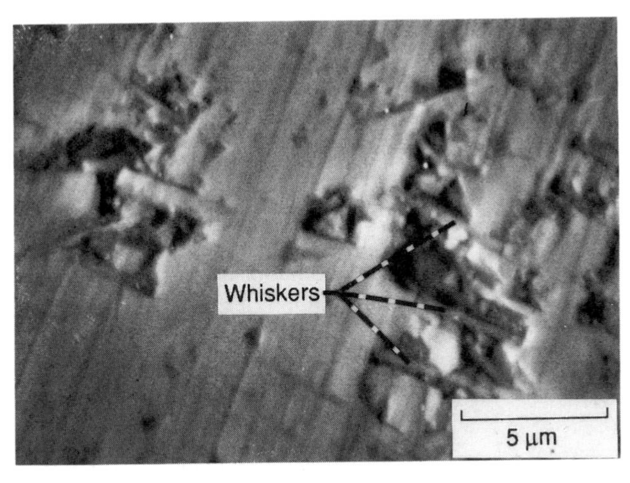

Figure 1 Scanning electron micrograph of ceramic surface showing orientation of whiskers in planes parallel to surface and sliding plane.

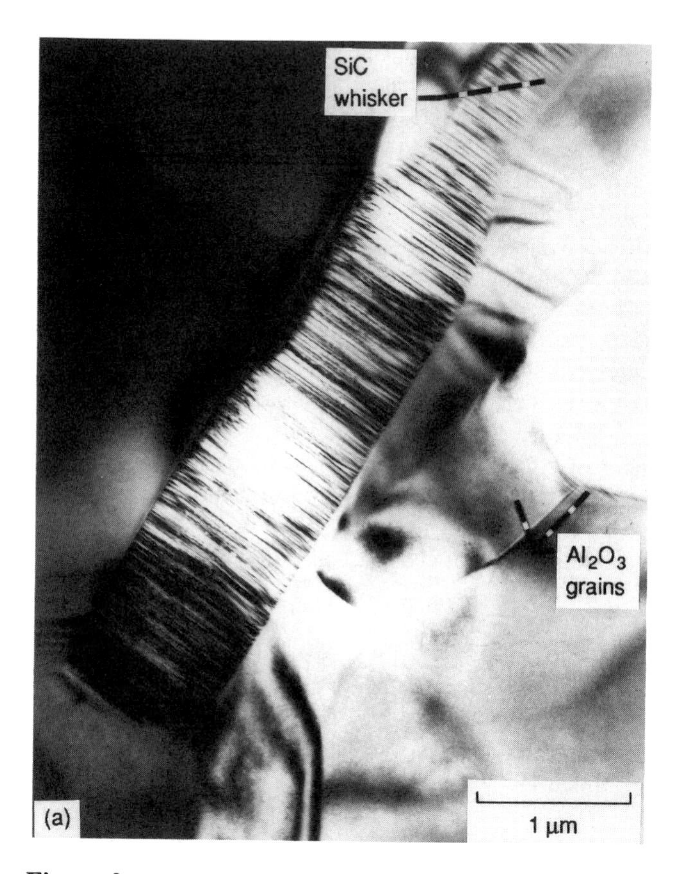

Figure 2 Transmission electron micrographs of virgin (unslid) material. (a) Interface between whiskers and matrix is free from inclusions, large-scale asperities, voids, etc. Alumina matrix grains visible. (b) Whisker cross section, showing some small enclosed pores, matrix grain structure.

Sliding velocity during these tests was 2.7 m/s (1000 rpm). The test atmosphere was ambient air with a relative humidity that ranged from 40 to 65% at 25°C. An air atmosphere was chosen to simulate conditions expected in future applications. Reference 12 gives a detailed description of the tribometer.

Prior to testing, the specimens were cleaned with pure ethyl alcohol, then rinsed with deionized water and dried before being mounted in the rig. To begin a test, the pin is slowly loaded against the rotating disk and data acquisition begins.

Since initial Hertzian contact stress for this material combination and geometry at the chosen test load of 26.5 N can be as high as 698 MPa, the first 30 seconds of sliding was at a much lower load, approximately 1 N. This reduced the initial

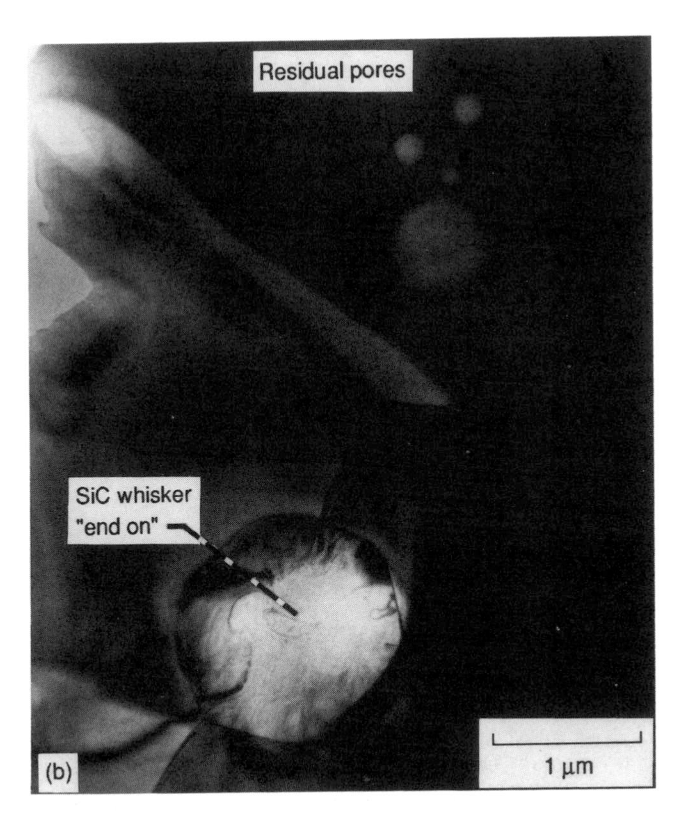

contact stress and allowed a wear scar to form on the pin so that at no time did the nominal contact stress exceed the material's compressive strength (≈ 500 MPa), which might have caused cracking due to overloading. The final test load, after the brief (≈ 30 s) run-in at 1 N was 26.5 N.

The tests were typically run for one hour for a total sliding distance of 9.7 km. The specimens were then unloaded and the furnace turned off and allowed to cool, whereupon the specimens were removed from the rig to make wear measurements and to analyze the wear surfaces. The pin wear measurements were made by using optical microscopy to measure the wear scar diameter and then calculate the wear volume. Disk wear was measured using stylus profilometry of the wear track to obtain an average track cross-sectional area. The disk wear volume was then calculated by multiplying the average cross-sectional area by the average track diameter.

To investigate the wear mechanisms of the SiC_W-Al_2O_3 ceramic composite, samples of the pin wear surfaces were prepared and analyzed using both SEM and TEM. For the SEM, the pin samples were coated with carbon to prevent charging and then analyzed. For the TEM, thin foils were prepared from the pin

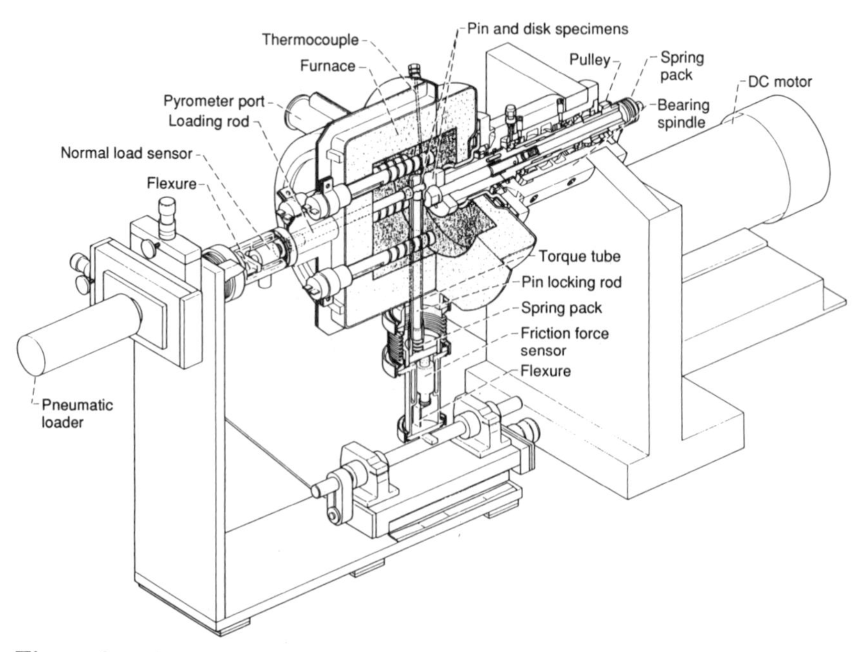

Figure 3 High temperature friction and wear apparatus.

wear scars themselves by slicing the worn tip from the pin, then ion-mill-thinning the foil from the unworn side until a hole was created in the center of the wear scar. Because it was very tedious to fabricate the TEM samples, only pin wear scars from the room temperature tests and the highest temperature tests (1200°C) were examined. Standard TEM procedures were then used to examine the pin wear scar. Disk surface specimens were not prepared; geometry complications made it too difficult to prepare thin foils. Also, because the pin surface is under continuous sliding, it suffers more frictional heating and severe wear conditions than the disk and may provide more information regarding wear mechanisms than the disk surface.

III. RESULTS AND DISCUSSION

A. Friction and Wear

The friction and wear data (for both the pins and the disks) are given in Table 2 and plotted as a function of temperature in Figures 4–6. The data indicate that only the pin wear increases as the test temperature is increased. Both the disk wear factor (defined as the wear volume divided by both the sliding distance and

Table 2 Friction and Wear Data Summary[a]

Test temperature (°C)	Coefficient of friction μ	Wear factors ($mm^3/N \cdot m$)	
		For pin	For disk
1200	0.58 ± .15	$(1.1 \pm 0.5) \times 10^{-6}$	$(5.1 \pm 2.0) \times 10^{-7}$
800	0.72 ± .22	$(6.1 \pm 1.0) \times 10^{-7}$	$(4.2 \pm 2.0) \times 10^{-7}$
600	0.60 ± .10	$(1.5 \pm 0.5) \times 10^{-6}$	$(7.0 \pm 2.0) \times 10^{-7}$
25	0.74 ± .10	$(2.4 \pm 0.5) \times 10^{-7}$	$(7.7 \pm 4.0) \times 10^{-7}$

[a]Uncertainties represent data scatter band.
Source: Reference 10.

the test load) and the friction coefficient remain relatively constant as the test temperature is increased from 25°C to 1200°C. The disk wear rate is relatively constant compared to the pin wear rate, which increases with temperature, probably because of the higher pin surface temperatures induced by frictional heating as previously indicated.

In general, average friction coefficients for the SiC_w-Al_2O_3 composite sliding against itself vary from a low of 0.58 to a high of 0.72 in the temperature range of 25–1200°C. Average disk wear factors for the alumina composites tested here are in the range of 4×10^{-7} to 9×10^{-7} $mm^3/N{\cdot}m$. Average pin wear factors show an increase with temperature from 2×10^{-7} at 25°C to 12×10^{-7} at 1200°C. The average pin wear factor at 600°C is highest, 15×10^{-7} $mm^3/N{\cdot}m$. This

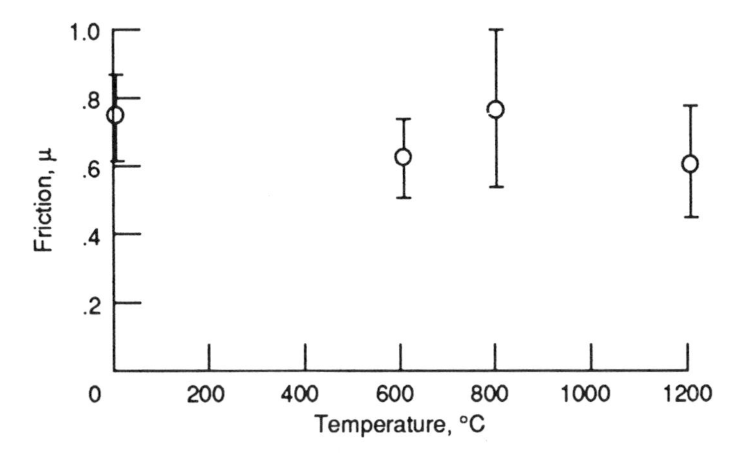

Figure 4 Friction coefficient versus test temperature for the Al_2O_3-SiC composite sliding against itself in air at 2.7 m/s, 26 N load. Error bars represent data scatter band.

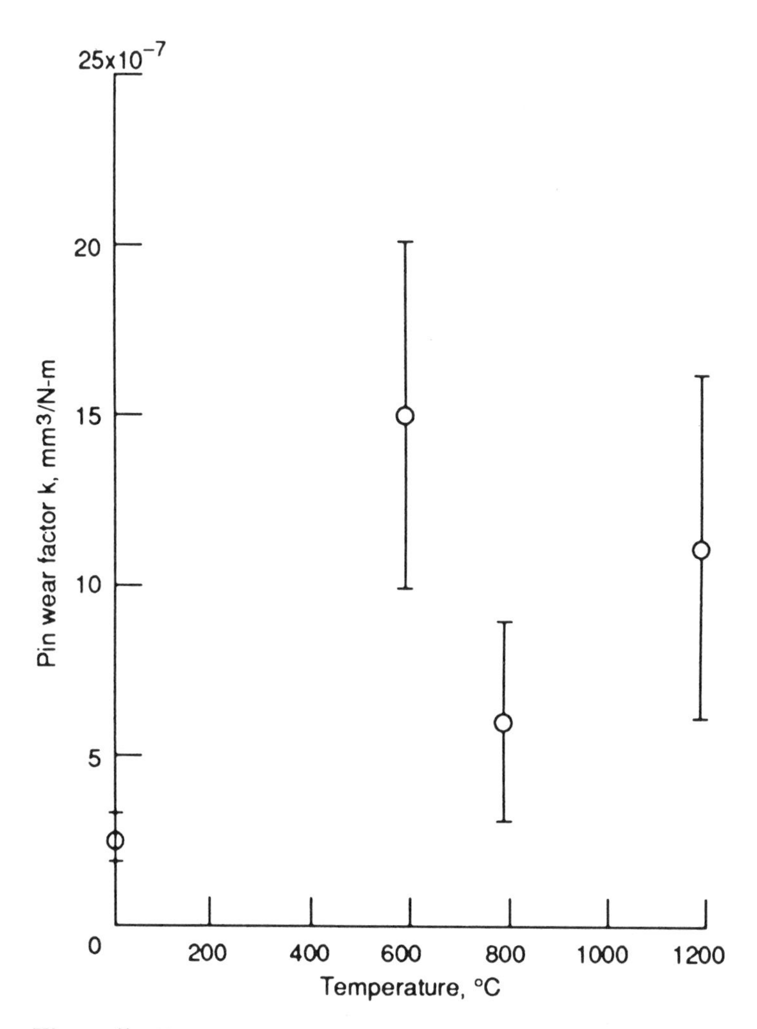

Figure 5 Pin wear factor k versus temperature for the Al_2O_3-SiC composite sliding against itself in air at 2.7 m/s, 26 N load. Error bars represent data scatter band.

may represent normal wear data scatter, since only two specimens were tested at this temperature, whereas usually three or more specimen sets were used at the other test temperatures. Although the friction coefficients are high, the wear is relatively low when compared to steel sliding against steel or monolithic alumina sliding against itself under similar conditions at room temperature, which have wear factors in the range of 10^{-3} to 10^{-4} $mm^3/N{\cdot}m$ (13 and 7).

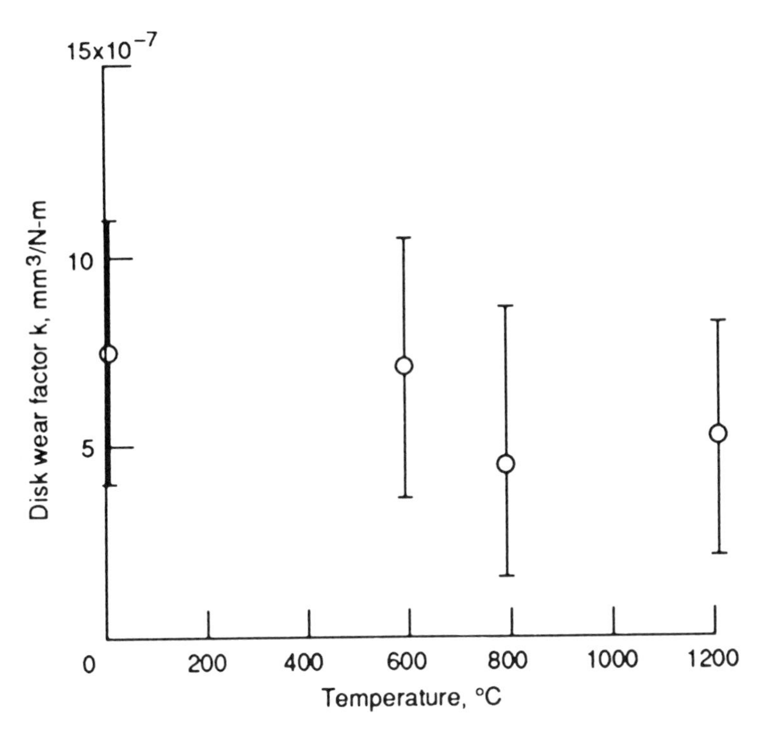

Figure 6 Disk wear factor k versus temperature for the Al_2O_3-SiC composite sliding against itself in air at 2.7 m/s, 26 N load. Error bars represent data scatter band.

B. Electron Microscopy

SEM analysis of the specimens from the room temperature tests indicated that the wear debris outside the pin wear scar was predominantly short, broken SiC fibers and Al_2O_3 matrix particles. Also present were large areas of compacted fine particles, which at lower magnifications look like plastically deformed areas (Fig. 7). In general, the room temperature wear surface indicates that the wear mode is primarily brittle fracture of both the matrix and the whiskers.

TEM analyses of the room temperature pin wear scar showed many cracks, as well as evidence of brittle fracture and individual wear particles. No evidence of plastic flow (i.e., dislocations) was seen. Typical TEM photomicrographs (Figs. 8a,b) suggest that under these test conditions, at room temperature, the wear mode is a generalized brittle fracture and subsequent removal of material.

SEM analyses of the pin wear surface and wear debris from the elevated temperature tests indicate a radically different wear mode. At 600°C, the pin wear scar shows evidence of whisker pullout. This can be seen as empty whisker

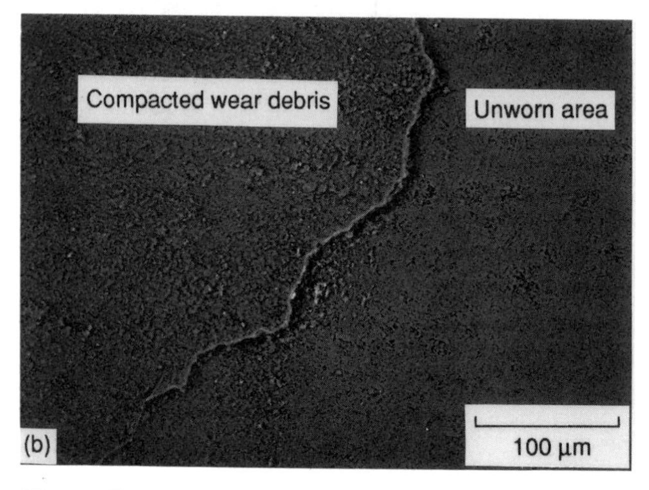

Figure 7 (a) and (b) SEM views of wear scar area from room temperature test. Wear debris is compacted into larger regions.

pockets or troughs on the pin wear scar (Fig. 9). Also detected at elevated temperatures were occasional wear debris particles that were in the form of rolls (Fig. 10). These rolls are probably the remnants of a glassy Al_2O_3/SiO_2 surface layer, which forms during sliding as the oxidation product of the SiC whiskers and the Al_2O_3 matrix. As wear takes place, this glassy layer debonds and is rolled upon itself or possibly around pulled-out whiskers, in the case of larger diameter rolls, to form the needlelike roll debris. This type of behavior with ceramics has been observed by other authors [14].

At 1200°C, large areas of long, unbroken whiskers, many as long as in the virgin material and devoid of Al_2O_3 matrix particles, are found outside the wear scar (Figs. 11 and 12). Near the scar, long whiskers and matrix particles are present. The long whiskers indicate whisker pullout. Since the whiskers are largely unbroken, it is plausible that they are somehow debonding from the matrix.

TEM analysis of pin wear scars from 1200°C tests also yielded features markedly different from room temperature specimens. Very few cracks were found, no wear particles were discovered, and a few dislocation regions were detected (Fig. 13). Clearly the wear behavior at 1200°C differs from that at 25°C.

TEM analysis of the virgin material indicates that it is free from voids, cracks, and dislocations. Therefore, changes in the material after testing can be attributed to effects from the sliding. Figure 14 shows a TEM micrograph of a whisker that has reacted with impurities possibly iron, when slid at 1200°C. The reaction product is identified by the TEM diffraction pattern as iron silicide. Since iron silicide is liquid at 1200°C [15], the remaining unreacted part of whisker may be debonding, hence easier to pull out. Thus, at elevated temperatures the wear mode seems to be whisker loosening, pullout, and breakup of the matrix, with possible whisker–matrix reactions.

C. Discussion of Experimental Results

The pin wear data indicate an increase of wear with test temperature. The reasons for this behavior are not clear, but a variety of factors, including the development of glassy surface layers and wear debris, may be cited. To better understand the wear process, pin wear surfaces from tests at room temperature and at 1200°C were examined using scanning and transmission electron microscopy.

The analyses indicate that at room temperature, the predominant failure mode is crack initiation and growth, followed by delamination and removal of the fractured particles. The analyses of high temperature specimens indicate that the predominant wear mode is by whisker pullout followed by increased matrix wear.

TEM analyses of the pin surface from the 1200°C tests indicated that dislocations in both SiC whiskers and the alumina matrix were present (Fig. 13), supporting the theory that plastic behavior may be playing an important role in high temperature wear. However, only a few dislocations were found. Thus, the theory that plastic deformation and subsequent particle removal dominate high temperature wear behavior may not hold up under these test conditions.

One strong argument for a whisker pullout wear mode is that the whiskers may be loosening at elevated temperatures because of differences in the thermal expansion coefficient between the Al_2O_3 matrix and the SiC whiskers. The thermal expansion coefficient for the alumina matrix is twice that of the SiC fibers (Table 1). Hence, as the material is heated, the whiskers loosen.

During hot consolidation (i.e., the initial production of the composite) at about

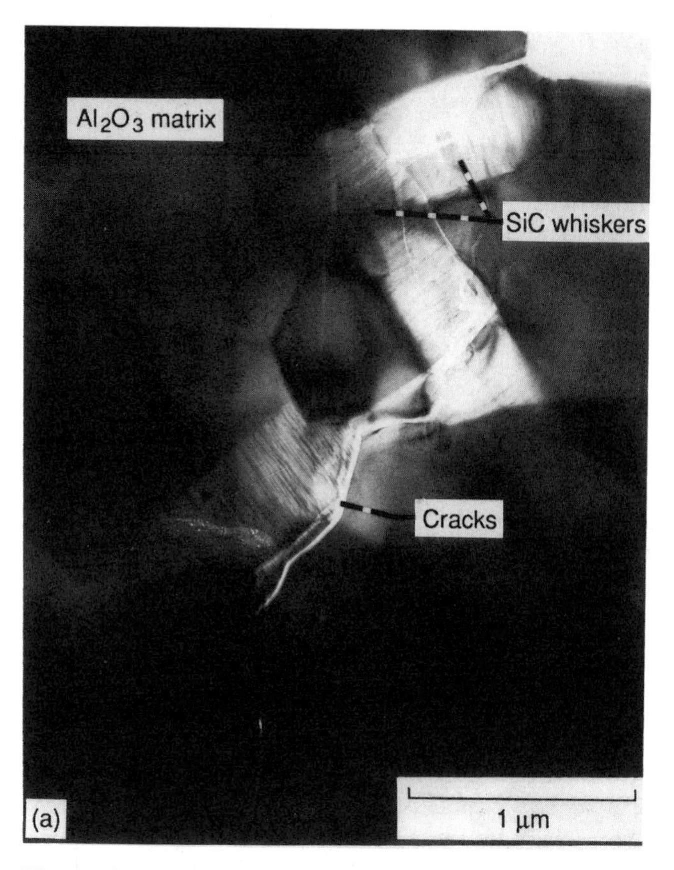

Figure 8 Room temperature TEM micrographs of pin wear scar showing cracks (a) and fractured wear debris particles (b,c).

1600°C, there is no thermal stress between the whiskers and the matrix. Upon cooling, however, the matrix contracts more than the whiskers, thus "clamping" the whiskers in compression. At temperatures lower than the consolidation temperature, the whiskers are in compression and the alumina matrix is in tension.

When sliding occurs, the friction force may pull or tug at the whiskers in the matrix. At low temperatures the whiskers are mechanically held in the matrix by the thermal compression. However, during elevated temperature testing or during high speed and load tests, which exhibit high frictional heating, the difference in thermal expansion lowers the "clamping" force on the whiskers. This allows them to be pulled out of the matrix more easily. Then the matrix, which is riddled with empty whisker pockets, cracks up and wears easily.

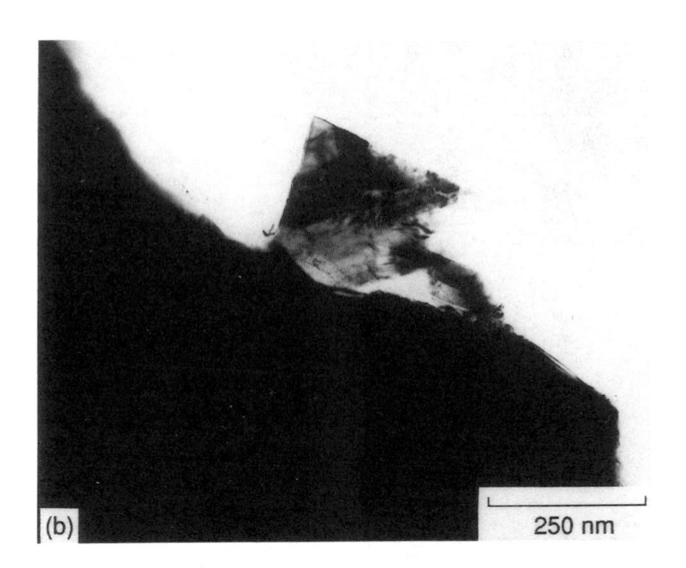
(b)
250 nm

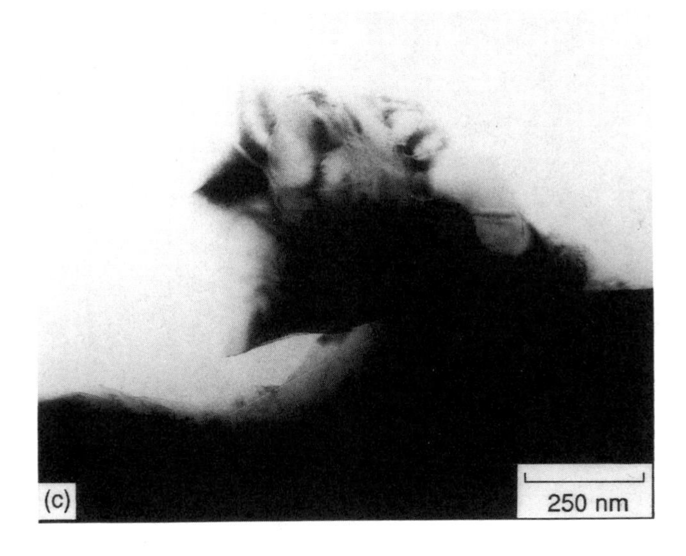
(c)
250 nm

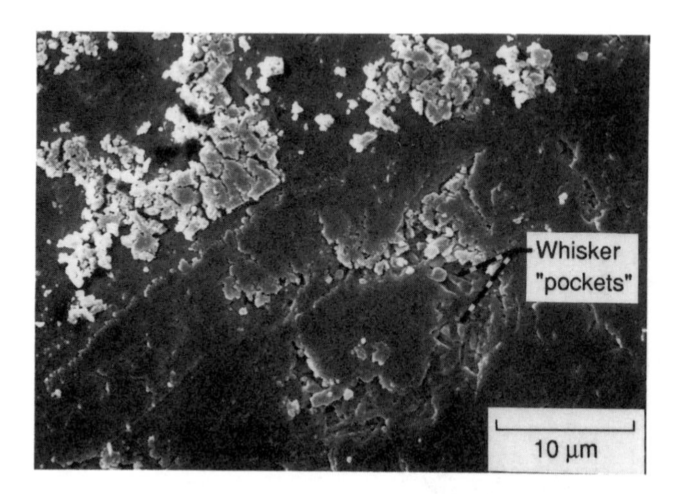

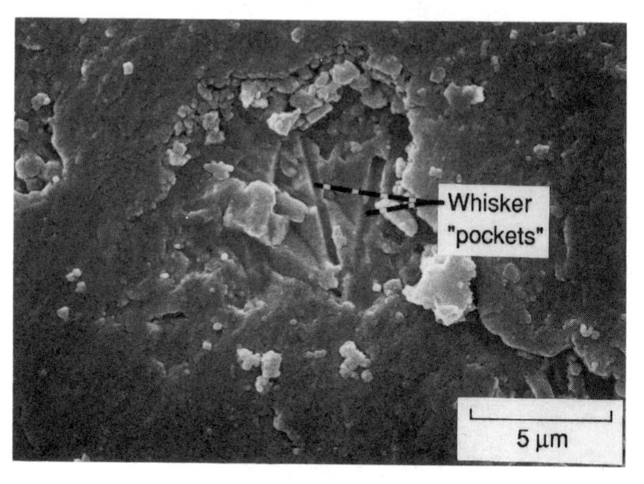

Figure 9 Whisker pockets left behind by pulled-out whiskers on pin wear surface from 600°C test.

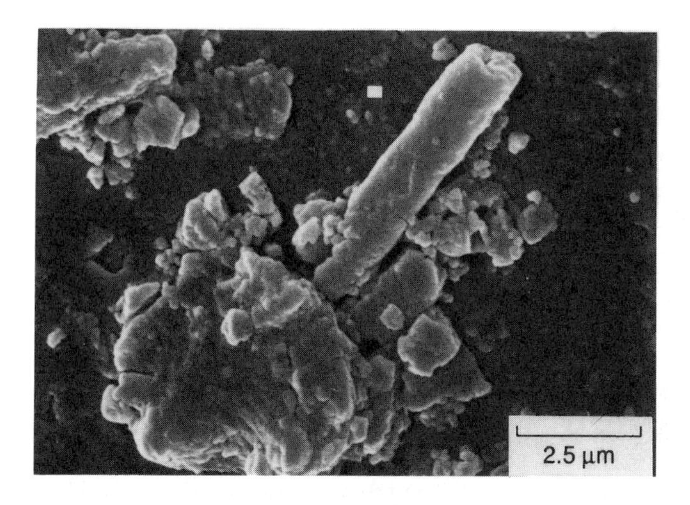

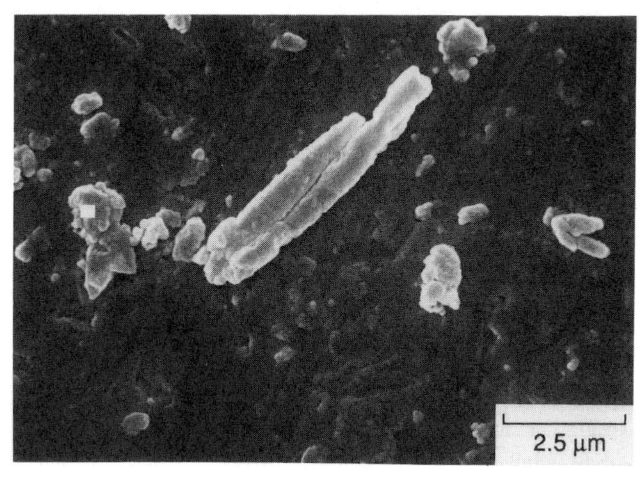

Figure 10 Wear debris "rolls" or needles from 600°C test specimens. Seam along needles and small diameters (≈ 0.4 μm) distinguish these formations from whiskers.

Figure 11 Pulled-out whiskers from 1200°C test. Uniform whisker diameters and lack of seams distinguish whiskers from debris "rolls."

Evidence for this wear mode is found by examining the large numbers of unbroken, long whiskers outside the wear scar after testing at 1200°C. Wear debris from tests at 25°C shows only short whisker pieces and matrix, presumably because at lower temperatures the whiskers are strongly held by the matrix and are being broken by the wear process rather than being pulled out.

TEM diffraction analyses indicate that at 1200°C some reaction between whisker and matrix may be occurring. The reaction products may have an effect on the friction and wear, especially since there appears to be a glassy layer forming at the sliding surface as evidenced by the roll debris observed. Also, perhaps a low shear strength reaction product such as mullite or iron silicide, due to iron impurities in the whiskers or Al_2O_3, is forming at the matrix–whisker interface. If the reaction products are liquid at the sliding temperatures, they may be allowing easier whisker pullout. Analyses by way of energy-dispersive spectroscopy (EDS) were inconclusive for this material because the analysis area was larger than the whisker diameter and the spacing between whiskers.

Because the SiC_w-Al_2O_3 composite exhibits a dual wear mode, any modeling of wear mechanisms must explain the reason for the two wear modes. The following analysis attempts to do just that. By combining a thermal stress analysis with a tribothermal analysis, a model is developed to describe the dual wear mode observed experimentally. In addition to determining why this dual wear mode exists, the analysis points out which material and test parameters have significant effects on tribological performance. With this information, steps toward material improvements can be made.

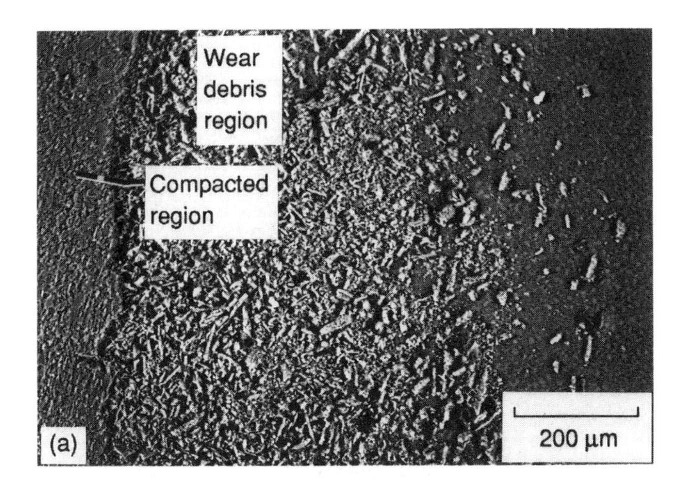

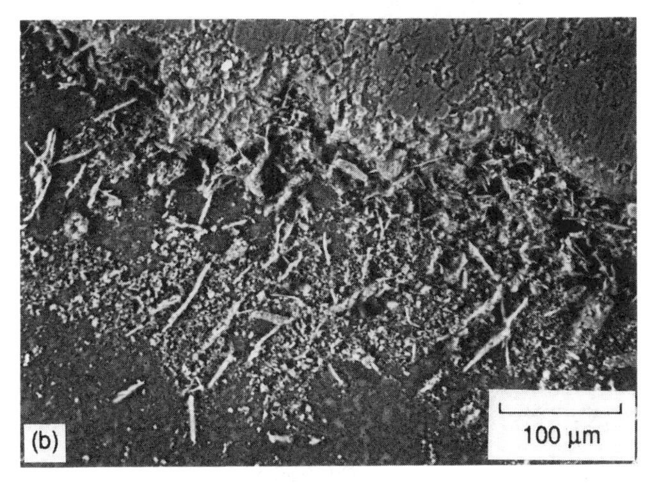

Figure 12 SEM view, of "pulled-out" whiskers outside wear scar from 1200°C sample; note that most of the whiskers are 10^{-7} μm in length. (a) Debris at 105× magnification and (b) debris at 200× magnification.

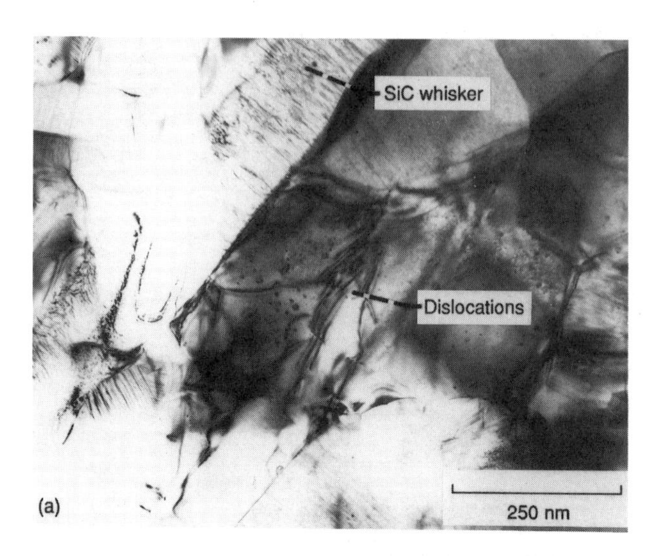

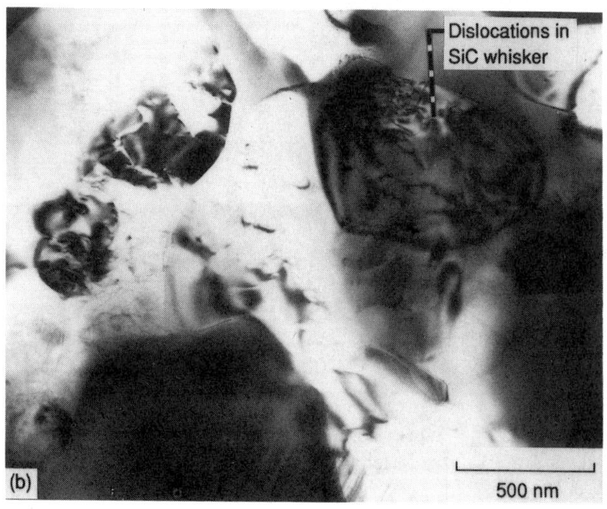

Figure 13 TEM views of pin wear surface region from 1200°C test sample. Note dislocations induced from sliding. No wear debris detected on wear surface area. Whiskers shown (a) sideways and (b) head on.

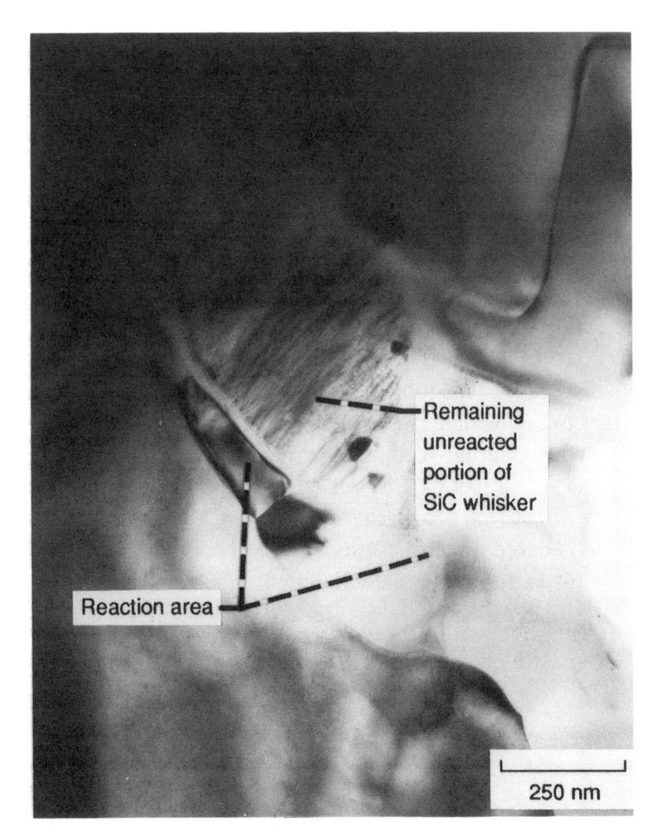

Figure 14 Transmission electron micrograph of pin wear scar from 1200°C test: this SiC whisker has partially reacted with impurities and matrix. Reaction products may be promoting whisker loosening and pullout.

IV. THERMOMECHANICAL WEAR ANALYSIS

A. Model

The following analysis models the wear behavior of the SiC_w-Al_2O_3 composite through a force balance on the whiskers coupled with the effects of bulk and frictional heating due to sliding. The model compares the sliding friction force F_T to the force required to fracture a whisker F_{break} and to the force required to pull out a whisker $F_{pullout}$ as a function of temperature, to explain the dual wear mode. The model includes the effects of material parameters (thermal expansion coefficients, thermal conductivity, whisker strength, etc.) as well as the effects on wear behavior of tribological and test parameters (such as friction coefficient, sliding speed, load, and temperature), and may aid in taking steps toward improving the composite.

Because the experimental program used a pin-on-disk configuration to generate wear conditions, the pin-on-disk configuration is modeled here. Figure 15 is a schematic representation of the wear specimen. It is assumed that the whiskers are pulled or pushed out of the matrix by the high frictional shear stresses present at the sliding interface or by a large counterface asperity, as shown in an exaggerated fashion in Figure 15.

The tangential forces present in the sliding contact, which can act to fracture or pull out whiskers, are appreciable and are approximated by the experimentally measured friction force (i.e. F_N μ). For our tests, the available friction force F_T is 26.5 N $\times$ $\approx$0.7 or about 18.6 N. This force F_T is much greater than the force required to break a whisker F_{break}, which can be estimated as the whisker tensile strength multiplied by the whisker cross-sectional area.

No data exist for the exact strength of the SiC whiskers used in this composite; however, Becher and Wei estimate the whisker tensile strength WTS to be about 7 GPa [16]. Their estimate is based on tensile tests of like diameter, longer SiC fibers and larger diameter (4 μm), but comparable length whiskers. Using this

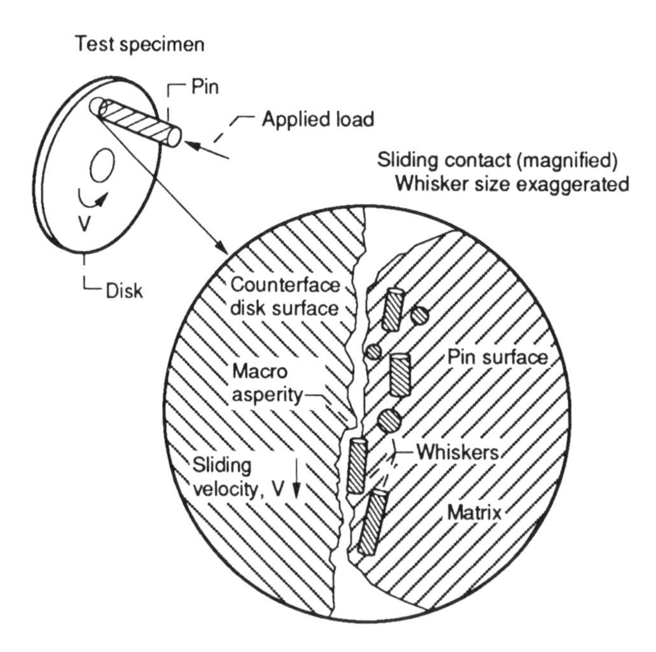

Figure 15 Pin-on-disk geometry of specimens modeled and an illustration of whisker–asperity interactions.

strength estimate (WTS = 7 GPa) and assuming that the average whisker is 0.75 μm in diameter, the force required to break a whisker F_{break} is about 3.1 mN. Although this type of calculation is approximate, it can be readily seen that the forces present in the sliding contact are appreciable compared to the strength of the whiskers; hence they generate wear debris.

When an asperity or simply the counterface contacts a whisker (as in Fig. 15), either the whisker can fracture, as is seen experimentally in low temperature sliding tests, or the whisker can be pulled out of the matrix, as is experimentally observed during sliding at higher temperatures. To summarize:

$$F_T > F_{break} \tag{1}$$

where F_T is the force available at the surface by the counterface to pull (or push) a whisker out of the matrix and F_{break} is the calculated force necessary to break an SiC whisker.

The force required to pull out a whisker, $F_{pullout}$, is equal to the compressive stresses acting on the whisker due to the matrix (σ_{csw}) multiplied by both the whisker surface area πDL and the friction coefficient between the matrix and the whisker, μ_{mw}. By comparing the forces required to pull out a whisker $F_{pullout}$ to the force required to fracture a whisker F_{break}, a prediction of the outcome of a whisker–counterface interaction can be made (i.e., the whisker is pulled out or broken).

For this analysis, it is assumed that the bond between the whiskers and the matrix is purely mechanical (i.e., arising from friction forces alone). Chemical bonding is neglected for this material system, and it if exists, it is probably small. This assumption is based on the widely held viewpoint that for a ceramic matrix composite to show improved toughness, as this composite does, the chemical bonding between the whiskers and the matrix must be small. Therefore, cracks are branched and deflected by the whiskers rather than propagating through them. Thus, the force required to remove a whisker from the surrounding matrix $F_{pullout}$ is equal to the frictional forces on the whiskers. For simplicity, μ_{mw} is considered to be about 1.0, which is typical for unlubricated ceramics; hence this term is dropped in the following equations.

Whiskers that are pulled from the matrix either by a counterface asperity or merely by the tangential friction stresses are necessarily exposed to the sliding surface. As such, they cannot be completely embedded in or surrounded by the matrix. Therefore, the actual surface area on which the compressive matrix stresses act on the whiskers is only a fraction of the total area, πDL. If a whisker were completely exposed to the sliding surface, the surface area acted on would be zero; if completely surrounded, the surface area would be equal to πDL. For the following analysis, we assume that the whiskers are only partially exposed and that the compressive matrix stresses act on about half the total whisker area or $\pi DL/2$.

At lower temperatures the compressive stresses on the whiskers due to the matrix are very high (on the order of 750 MPa), and thus the force required to pull out a whisker F_{pullout} is greater than the whisker fracture force F_{break}:

$$F_{\text{pullout}} > F_{\text{break}} \qquad (2)$$

$$\sigma_{\text{csw}} \frac{\pi DL}{2} > \text{WTS} \frac{\pi D^2}{4}$$

where WTS is the whisker tensile strength, D is the diameter, and L is the average whisker length.

Under this condition, which is characteristic of low temperatures, the whiskers are fractured rather than pulled out during sliding, and wear follows a brittle behavior.

At higher temperatures, thermal expansion of the matrix reduces the compressive stress on the whiskers and the inequality reverses, with the result that:

$$F_{\text{pullout}} < F_{\text{break}} \qquad (3)$$

$$\sigma_{\text{csw}} \frac{\pi DL}{2} < \text{WTS} \frac{\pi D^2}{4}$$

Under this condition, the whiskers are pulled out of the matrix during sliding. The second wear process, namely whisker pullout, which occurs at higher temperatures, leaves the matrix riddled with empty whisker pockets, which act as flaws or fracture initiation sites and lead to higher wear.

The main parameters affecting both the whisker strength and the compressive stresses on the whiskers is the near-surface temperature of the specimens (i.e., within a few whisker diameters from the surface). The temperature is, in turn, affected by the tribological and test parameters such as friction coefficient, load, and velocity, as well as material parameters such as thermal conductivity and diffusivity.

B. Thermoelastic Stress Analysis

Since brittle materials, such as SiC_w-Al_2O_3, behave elastically up to the fracture point, we can determine the stresses in the material using elasticity theory. The magnitudes of the compressive stresses on the whiskers and on the matrix have been experimentally measured and analytically modeled [17–19]. The models are based on the application of Hooke's law in three dimensions, where the residual thermal strains are calculated by elasticity theory.

In general, the whiskers are modeled as being completely embedded in an infinite isotropic ceramic matrix. Although in reality the elastic constants for the whiskers are not isotropic, when isotropic values are used, the results obtained are reasonable; that is, they agree with other more rigorous tests such as experimental stress analysis. The analytical method and results presented here are loosely based on just such an analysis by Eshelby [20]. To compensate for whisker interactions and nonisotropic elastic constants, a self-consistent approach was used: that is, the model is of a single whisker completely embedded in an infinite matrix, which has the elastic properties of the SiC_w-Al_2O_3 composite.

The analysis method, first outlined by Mori and Tanaka [21], was applied to this SiC_w-Al_2O_3 material by Majumdar and Kupperman [18]. The analysis consists of setting up a three-dimensional matrix of Hooke's law and following the stress and strain fields in a SiC whisker as temperature is changed.

Majumdar and Kupperman [18] applied this approach to the SiC_w-Al_2O_3 material system, and their results are shown in Figure 16, which clearly indicates that the compressive stresses on the whisker are highest at room temperature and decrease linearly with temperature. These analytical results are also in excellent agreement with experimental stress analysis results obtained by neutron diffraction techniques. This congruence helps increase confidence in the stress analysis method cited.

The results of the analytical stress analysis from Reference 18 can be summarized as follows. During cooling from the initial consolidation temperatures ($\approx$ 1650 °C), differences in the coefficient of thermal expansion (CTE)

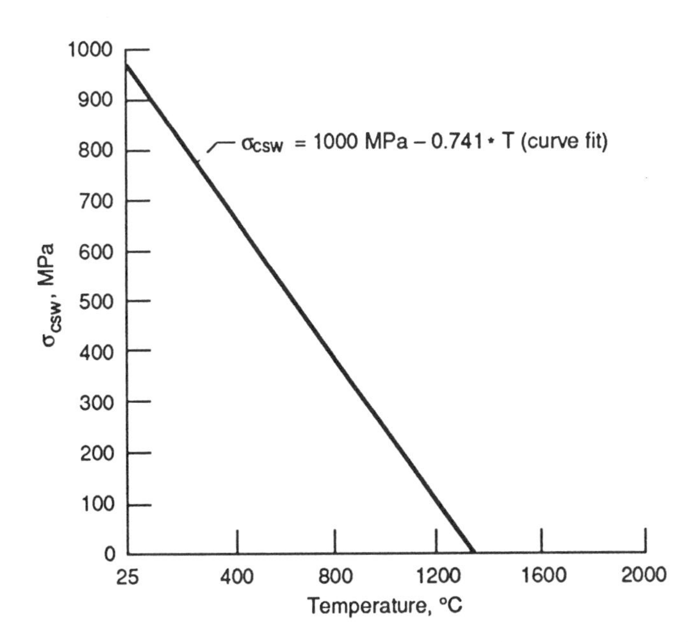

Figure 16 Variation of maximum residual hoop stresses at whisker surface with temperature. (From Ref. 15 for an 18 vol % SiC_w-Al_2O_3 composite.)

between the SiC whiskers and the Al_2O_3 matrix cause thermal stresses to form, which are relieved by matrix creep. At temperatures below about 1350°C, however, the matrix is stiffened and creep is no longer a dominant factor in stress relief; thus thermal stresses develop. Because the CTE for the whiskers is about half that of the matrix, the whiskers are placed in compression. As the temperature further decreases, the residual stresses at the whisker–matrix interface continue to rise in more or less linear fashion. It is this compressive stress on the whiskers that helps to hold them in the matrix during sliding.

The following curve-fitting equation describes the variation of whisker compressive stresses with temperature:

$$\sigma_{csw} = (1000 \text{ MPa} - 0.741 \text{ MPa/°C}) \tag{4}$$

for a composite containing 18% by volume SiC whiskers.

Majumdar et al. [19] extended their analyses to include the effect of whisker volume percent. The average strains (hence stresses) were found to decrease with SiC content in more or less linear fashion as shown in Figure 17. The decrease in stress was attributed to a dilution effect of the matrix stress effect by the SiC whiskers.

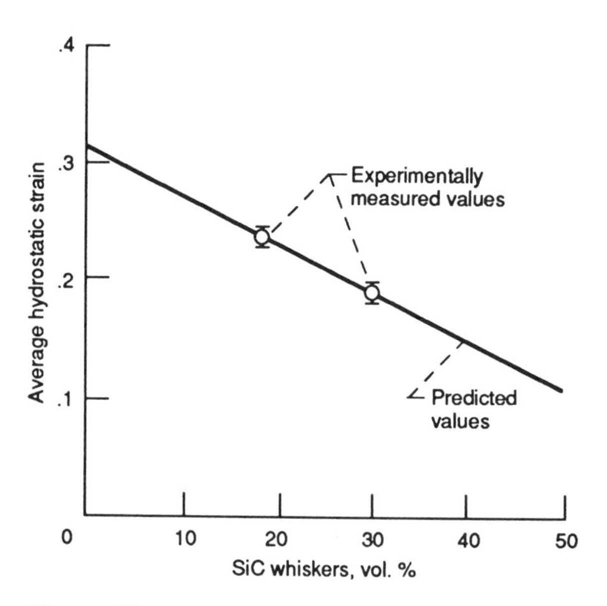

Figure 17 Compressive strain on SiC whiskers at room temperature as a function of whisker volume percent from Reference 15. Error bars represent one standard deviation of measured data.

A curve fit for this effect, taken from Figure 17, is given by the dimensionless coefficient as follows:

$$C_{vol} = [1 - 0.01449\ (\%\ \text{SiC whiskers} - 18\%)] \tag{5}$$

Equation (5) can be considered to be accurate for compositions ranging from about 15 to 40 vol % SiC whiskers.

By combining Equations (4) and (5) with the effect of bulk temperature, we get an equation that relates the compressive stresses on a fully embedded whisker with temperature and composition:

$$\sigma_{net}(T,\ \%\text{SiC}_w) = \sigma_{csw} \times C_{vol} \tag{6}$$

$$\sigma_{net}(T,\ \%\text{SiC}_w) = (1000\ \text{MPa} - 0.741\ \text{MPa/°C} \times \text{T}) \times [1 - 0.01449 \times (\%\text{SiC} - 18\%)]$$

where σ_{net} represents the stress on a whisker due to the matrix as a function of temperature and composite whisker content. To use Equation (6) to calculate $F_{pullout}$, Equation (2) can be employed, substituting σ_{net} for σ_{csw}.

From a tribological point of view, the key aspect of Equation (6) is that the compressive stresses on the whiskers decrease linearly with temperature and with volume percent SiC whiskers.

C. Tribothermal Analysis

If the composite were used in a static situation (i.e., without sliding) the analysis above would be sufficient to describe the stress state of the whiskers, assuming the sample to be in thermal equilibrium with the ambient temperature. However, during sliding, frictional heating can greatly increase the near-surface temperatures, significantly altering the whisker stresses, hence, potentially, the tribological properties of the material.

Many researchers have studied the problem of determining surface temperatures and near-surface-region temperature rise which occur as a consequence of frictional heating [13,22]. Although specific details vary, it is generally accepted that the temperature rise is a function of such variables as load, speed, friction coefficient, thermal conductivity, and diffusivity, as well as the type of environment.

To describe the temperature rise occurring for pin specimens studied here, an analysis by Ashby [22] has been found to be useful. Figure 18 shows the physical situation described by Ashby's model, which is based on the assumption that the frictional heating is conducted away from the sliding contact into the pin and its holder and also into the disk. Convection is neglected, and the mean heat diffusion distance (the near-surface region of the sliding contact) has been approximated

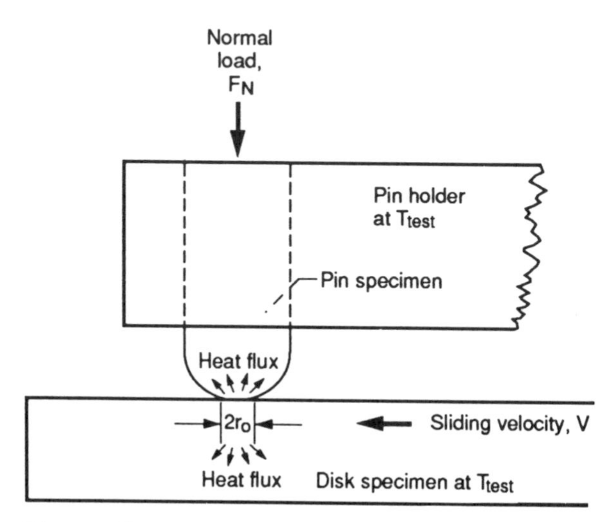

Figure 18 Sketch of physical situation modeled by heat transfer analysis. Pin and disk specimens have equivalent thermal properties.

by 1.6 times the contact radius r_0, as suggested by Ashby [22]. The following equation mathematically describes the "bulk" heating of the specimens, that is, the surface temperature of the pin in the region near the sliding interface T_S:

$$T_s = T_{test} + \frac{\mu F_N V}{2\sqrt{\pi}\, K r_0} \times \left[\frac{1 + \pi}{[2 \tan^{-1} \left(\sqrt{\frac{2\pi a}{V r_0}} \right)} \right]^{-1} \tag{7}$$

where

μ = friction coefficient
F_N = normal load force
K = thermal conductivity
V = sliding velocity
T_{test} = ambient test temperature
r_0 = wear scar radius
a = thermal diffusivity of the material

This equation considers both material parameters and test parameters. For the analysis presented here, the material parameters are essentially constant. Therefore, the important variables are load, speed, friction coefficient, and T_{test}. As these are increased, the near-surface-region temperature increases and thus the compressive stresses on the whiskers decrease. From these considerations, it is possible to have a room temperature test that exhibits whisker pullout because high loads and/or sliding speeds lead to high near-surface temperatures.

The magnitude of the frictional heating effects can be seen by examining Table 3, which shows the test conditions and calculated surface temperatures for the tests conducted in the experimental work [9]. It can be seen that even at moderate load and sliding velocity, the frictional temperatures rise above the ambient by about 400–500°C. This heating can greatly influence the wear mode of the materials by changing the stresses on the whiskers in the region of sliding.

D. Discussion of Analytical Results

The results of the preceding analyses and the implications for wear behavior are illustrated in Figure 19, which plots both the pullout force on the whiskers and the whisker fracture force versus the near-surface temperatures for a 25 vol % SiC_w-Al_2O_3. The inequality reversal or crossover point between the whisker fracture force F_{break} and the compressive forces on the whiskers $F_{pullout}$, shown graphically in Figure 19 as the "transition region" and described by Equations (2) and (3), gives a plausible reason for the existence of dual wear mechanisms (namely whisker pullout or generalized fracture) for the composite material. As described previously, when the whisker fracture force is less than the compressive forces holding the whiskers in the matrix, interactions with the sliding counterface will fracture the whiskers. This behavior occurs at temperatures below 1200°C. At temperatures above about 1200°C, the compressive forces holding the whiskers in the matrix drop below the whisker fracture force, and interactions with the counterface will result in whisker pullout rather than fracture.

The width of the transition region, which is based on the scatter of friction

Table 3 Test Conditions and Calculated Pin Surface Temperatures for Data Obtained by Yust and Allard [9]

T_{test} (°C)	F_N	V (m/s)	μ	T_s, calculated (°C)[a]
25	26.5	2.7	0.74	605
600	26.5	2.7	0.60	1071
800	26.5	2.7	0.72	1361
1200	26.5	2.7	0.58	1655

[a]Calculations done for 25 vol % SiC whisker-reinforced Al_2O_3 composite. Wear scar, radius; 1 mm; thermal diffusivity, 1.35×10^{-5} m/s; thermal conductivity, 22.3 W m^{-1} °C^{-1}. Calculated surface temperature uncertainties are ≈ ± 100 °C.

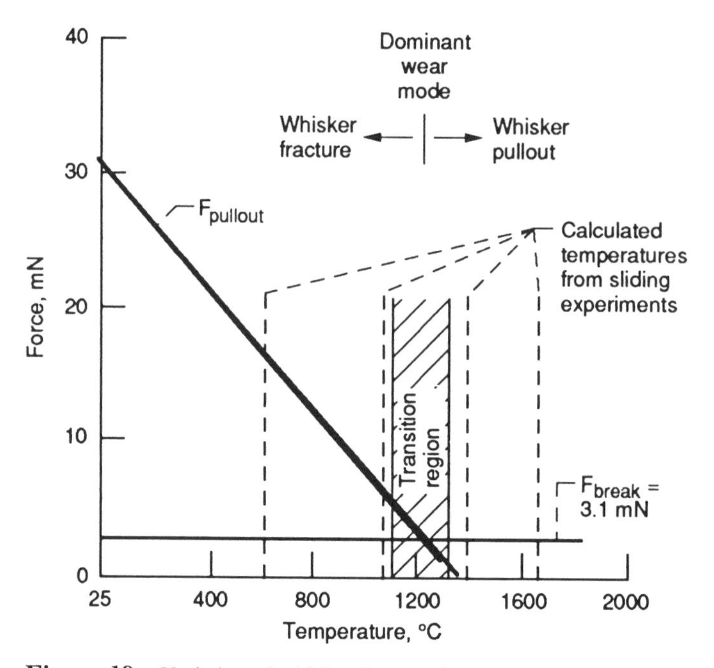

Figure 19 Variation of whisker fracture force F_{break} and whisker pullout force F_{pullout} versus temperature. Wear mode transition occurs at crossover.

coefficient values and uncertainties in the analysis, is typically $\approx \pm 100°C$. One of the major contributors to a large transition region is the effect of whisker length. Its effect can be seen in Equations (2) and (3). Since the whiskers range in length from about 10 to 60 μm, the pullout force, hence the transition point, can vary significantly (Fig. 20). From Figure 20 it can be seen that the transition point changes from about 1000°C for short (10 μm) whiskers to about 1300°C for long (60 μm) whiskers. Thus one would expect composites made with predominantly shorter whiskers to suffer from whisker pullout at lower temperatures than composites made with longer whiskers. Composites made with longer whiskers, however, would not necessarily be best, since other mechanical properties such as toughness and processing ease can be degraded if whiskers are unduly long.

By examining the equations describing the stress conditions on the whiskers, some insight into the effects of variables on the potential wear behavior can be determined. For example, Equation (6) relates the stresses on the whiskers as a function of bulk temperature and composite whisker content. By varying the whisker content between 18 and 40%, the transition temperature can be made to decrease by approximately 350°C (Fig. 21). Therefore, the whisker content of

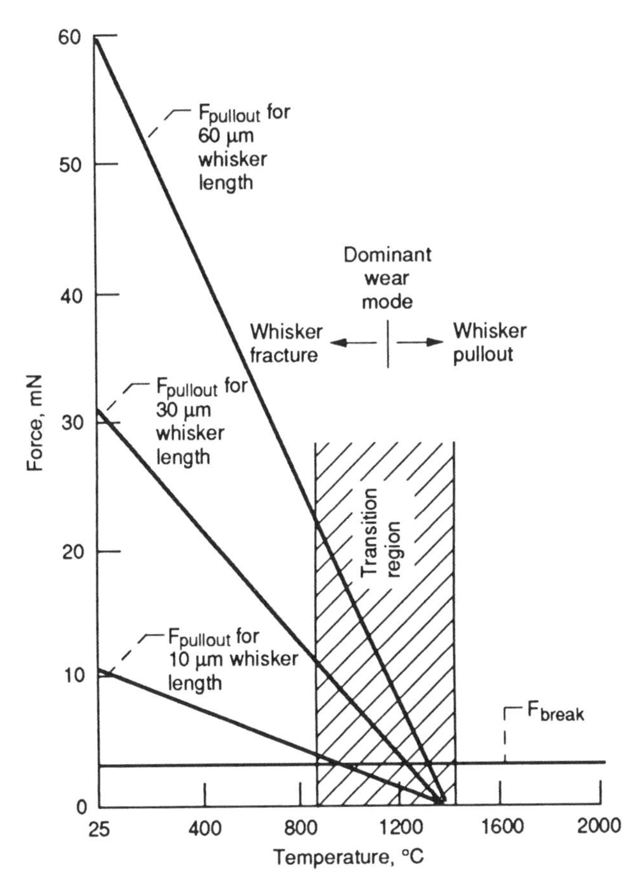

Figure 20 Effect of average whisker length on $F_{pullout}$ versus temperature. Transition is reduced from 1300°C to 950°C when whisker length is reduced from 60 μm to 10 μm.

the composite may have a significant effect in determining the wear mode. This effect needs to be verified experimentally because changes in whisker content can also affect the toughness in an inconsistent manner, thereby having an unknown effect tribological performance.

Changes in test conditions, such as load, velocity, and ambient test temperature, directly affect the calculated sliding temperature as in Equation (7). As these variables increase, the sliding temperatures also increase. Of course, increases in friction also increase the temperature, indicating that lubricating the sliding contact will reduce the temperature and may have a significant effect on the wear behavior.

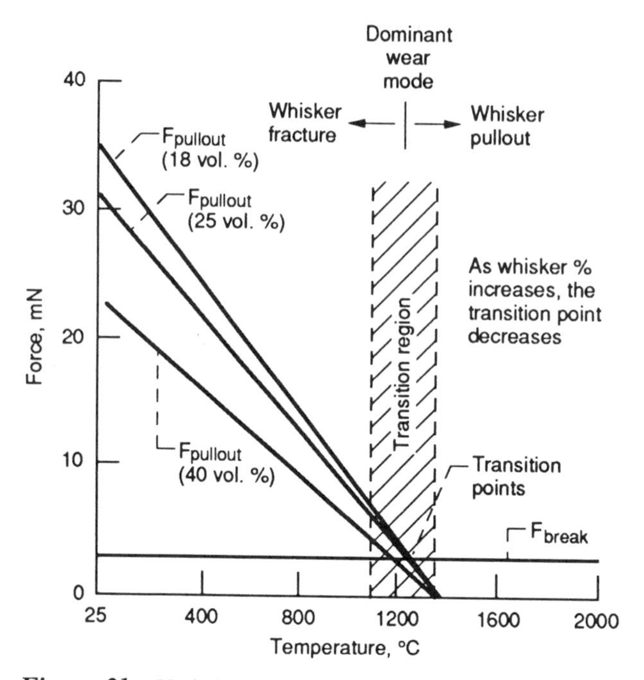

Figure 21 Variation of transition point as volume percent whisker content changes from 18 vol % to 40 vol % as a function of temperature.

E. Comparison to Experimental Data

Comparing the analytical results to experimental results is simply a matter of calculating the near-surface-region temperatures for tests conducted with this material and seeing whether the model predicts the experimentally determined wear mode.

Table 4 shows the calculated temperatures and experimentally observed wear modes for tests conducted at 25, 600, and 1200°C. Figure 19 also shows the experimental data points (using the calculated temperatures) plotted with the predicted curves. The model correlates very well with the observed wear behavior. Therefore, it may be useful in predicting the wear behavior of other similar materials under a wide variety of test conditions.

One interesting point is that the experimentally determined wear factors indicate a maximum at a calculated sliding temperature of 1071°C. This temperature is near the transition point of ≈ 1200°C. Other authors have also measured wear maxima with the alumina–silicon carbide composite [8]. There may be a relationship between the maximum wear rate and the transition in the wear behavior.

Table 4 Comparison of Predicted Wear Mode and Experimentally Determined Wear Mode[a]

Temperature (°C)		Wear mode	
T_{test}	T_s, calculated	Predicted	Observed
25	605	Whisker fracture	Whisker fracture
600	1071	Transition—mixed mode	Mixed mode
800	1365	Whisker pullout	Whisker pullout
1200	1655	Whisker pullout	Whisker pullout

[a]Test load 26.5 N; test velocity, 2.7 m/s. Calculated surface temperature uncertainties are ≈ ± 100 °C.

V. CONCLUDING REMARKS

The tribological behavior of whisker-reinforced ceramics is complex. The wear mechanisms can be affected by the environment, the sliding conditions, the counterface material, and the composition of the composite. Based on the experimental results and the model, an exhaustive series of experiments could be envisioned to further test and refine the analyses. These experiments might include tests in inert atmospheres at a wide range of loads, sliding speeds, temperatures, and whisker volume contents.

In addition, other ceramic composite systems could be tested. For example, testing of a zirconia-reinforced alumina, ZrO_2-Al_2O_3, could help to verify the effect of the coefficient of thermal expansion of the reinforcement phase on pin wear. With SiC_W-Al_2O_3, the pin wear increased with temperature because the differences in CTE between SiC and Al_2O_3 lead to a reduction in the compressive whisker stresses. However, with ZrO_2-Al_2O_3 the reverse would probably occur because the CTE of ZrO_2 is larger than that of Al_2O_3. In fact, research by Yust and Devore on ZrO_2-Al_2O_3 did display a reduction in pin wear with temperature [23]. Although their tests were not for a whisker-reinforced composite, the change in the ZrO_2-Al_2O_3 stress state with temperature may have played a role similar to that of SiC_W-Al_2O_3.

The results also indicate that the SiC_W-Al_2O_3 wear properties might be improved by inhibiting whisker pullout. This could be accomplished through the use of a high friction whisker coating or by using variable diameter whiskers, which may promote whisker–matrix interlocking. These techniques, however, may reduce toughness.

Furthermore, the model defines the limits or envelope of usable test conditions, beyond which the wear performance degrades. By using the analysis presented, the effects of a wide variety of variables can be more or less predicted.

Finally, much of what is learned by experimental testing probably never could be deduced. Careful experimental research coupled with a useful model, which helps explain the results, can be used for improving the tribological performance of these composites.

NOMENCLATURE

a	thermal diffusivity (m^2/s)
C_{vol}	compressive coefficient for the effect of whisker volume percent
D	whisker diameter
F_{break}	calculated fracture force for SiC whisker (mN)
F_N	applied normal test load (N)
$F_{pullout}$	force required to pull out a whisker at the sliding surface and only partially ($\approx$0.5 embedded) surrounded by the matrix
F_T	total friction force in sliding contact (N)
K	thermal conductivity (W/m °C)
L	whisker length
r_0	wear scar radius (m)
T_S	calculated surface region temperature (°C)
T_{test}	ambient test temperature (°C)
V	sliding velocity (m/s)
WTS	whisker tensile strength (GPa)
μ	sliding friction coefficient
μ_{mw}	static friction coefficient between whisker and matrix (assumed to be $\approx$ 1.0)
σ_{csw}	compressive stress on whisker completely embedded in matrix of an 18 vol % SiC_w-Al_2O_3 composite (MPa)
σ_{net}	compressive stress on an embedded whisker as a function of temperature and whisker volume percent (MPa)

REFERENCES

1. P. F. Becher and T. N. Tiegs, *Adv. Ceram. Mater. 3*, 148 (1988).
2. H. E. Helms and S. R. Thrasher, *Engineering Applications of Ceramics Materials*, American Society for Metals: Metals Park, OH, 1985.
3. K. C. Ludema and O. O. Ajayi, Wear mechanisms in ceramic materials engine applications, in *Proceedings of the 22nd Automotive Technology Development Contractors' Coordination Meeting*, Dearborn, MI, 1985, SAE Publication No. P-155, pp. 337–341.
4. H. E. Sliney and D. L. Deadmore, Friction and wear of oxide–ceramic sliding against IN-718 nickel base alloy at 25 to 800 °C in atmospheric air, NASA Technical Memorandum No. TM-102291 (August 1989).
5. K. Fukuda, Y. Sato, T. Sato, and M. Veki, Wear properties of SiC whisker

reinforced ceramics against bearing steel, Presented at the Conference on the Tribology of Composite Materials, Oak Ridge, TN, May 1–3, 1990.

6. M. Bohmer and E. A. Almond, *Mater. Sci. A105–A106*, 105 (1988).
7. C. S. Yust, J. M. Leitmaker, and C. E. DeVore, *Wear*, *122*, 151 (1988).
8. C. S. Yust and L. F. Allard, *STLE Trans. 32*, 331 (1989).
9. C. S. Yust and L. F. Allard, in *Ceramic Materials and Components for Engines*, V. J. Tennery (Ed.), American Ceramic Society: Westerville, OH, 1989, p. 1212.
10. C. DellaCorte, S. C. Farmer, and P. O. Book, Experimentally determined wear behavior of an Al_2O_3-SiC composite from 25 to 1200 C, NASA Technical Memorandum No. TM-102549 (1990).
11. J. Homeny and W. L. Vaughn, *Mater. Res. Soc. Bull. 12*, 66 (1987).
12. H. E. Sliney and C. DellaCorte, A new test machine for measuring friction and wear in controlled atmospheres to 1200 C, NASA Technical Memorandum No. TM-102405 (1989).
13. M. B. Peterson and W. O. Winer, *Wear Control Handbook*, American Society of Mechanical Engineers: New York, 1980, p. 475.
14. P. Boch, F. Platon, and G. Kapelski, Effect of temperature and environment on wear and friction of Al_2O_3 and SiC ceramics, in *Proceedings of the Fifth International Congress on Tribology*, Helsinki, Finland, K. Holmberg and I. Nieminen (Eds.), 1989, pp. 114–119.
15. E. M. Levin, C. R. Robbins, and K. F. McMurdie, *Phase Diagrams for Ceramists*, American Ceramic Society: Westerville, OH, 1964, p. 59, Fig. 81.
16. P. F. Becher and G. C. Wei, *Am. Ceram. Soc. Commun. 67*, C-267 (1984).
17. Z. Li and R. C. Bradt, *J. Am. Ceram. Soc. 72*, 70 (1989).
18. S. Majumdar and D. Kupperman, *J. Am. Ceram. Soc. 72*, 312 (1989).
19. S. Majumdar, D. Kupperman, and J. Singh, *J. Am. Ceram. Soc. 71*, 858 (1988).
20. J. D. Eshelby, *Proc. R. Soc. London, Ser. A*, *241*, 376 (1957).
21. T. Mori and K. Tanaka, *Acta Metall. 21*, 571 (1973).
22. M. F. Ashby, The development of wear mechanism maps, in *Proceedings of the International Workshop on Wear Modelling*, Argonne National Laboratory, F. A. Nichols, A. E. Michael, and L. A. Northcutt, (Eds.), 1988, pp. 25–59.
23. C. S. Yust and C. E. DeVore, *STLE Trans. 33*, 573 (1990).

11

Microstructure and Wear Resistance of Silicon Nitride Composites

Steven F. Wayne

Valenite Inc.
Madison Heights, Michigan

S. T. Buljan

Norton-St. Gobain
Worcester, Massachusetts

ABSTRACT

Wear of machine components can take many forms, depending on the application. The design of wear-resistant materials must account for the factors that control wear in each specific instance. This chapter discusses the wear resistance of silicon nitride–titanium carbide composites subjected to abrasion or erosion, and through an example of cutting tool application examines a more complex problem of simultaneous action of abrasion and chemical wear processes.

In machining operations, both chemical and mechanical factors contribute to the wear of cutting tools. The dominant wear mode of the tool varies, depending on the workpiece material and the cutting conditions.

For ceramic composite cutting tools, the mechanical properties hardness (H) and fracture toughness (K_{IC}) expressed as a parameter $K_{IC}^{3/4}H^{1/2}$, which describes to a first approximation the abrasive wear resistance as determined in a laboratory test, predict performance of tools only in machining applications in which abrasion predominates. In the presence of significant chemical interaction between tool material and workpiece, any wear-resistance projections based solely on mechanical properties are not meaningful. In this context, the chapter elucidates the relationships between chemical and mechanical properties and wear resistance of silicon nitride–titanium carbide composite cutting tools designed for machining of ferrous alloys.

Isolated studies of either abrasion or erosion mechanisms suggest that predictions of wear resistance for ceramic composites based solely on hardness and fracture toughness, in the absence of additional information regarding microstructure, may be inaccurate. Wear resistance in these instances was found to be more reliably defined by including the mean free path of the dispersoid into predictive structure–property–performance relationships.

I. INTRODUCTION

Design of ceramic composite materials for enhanced wear resistance requires an improved understanding of the operative wear mechanisms and microstructure-property relationships. This point has been demonstrated with silicon nitride ceramics developed for wear applications [1–3]. As a result of the need to tailor material properties for specific applications, the development of ceramic wear-resistant materials has focused on composites, since this approach allows for tailoring of properties beyond those attainable with single-phase materials.

A significant portion of the reported studies that have examined abrasion and solid particle erosion of ceramics generally propose that resistance to wear is directly proportional to a material's resistance to penetration (hardness, H) and crack propagation (fracture toughness, K_{IC}). Based on indentation mechanics, wear resistance has been related to a product of hardness and fracture toughness ($K_{IC}^{m}\ H^{n}$), the exponents of which vary depending on the manner of material removal: $K_{IC}^{3/4}\ H^{1/2}$ for abrasion and $K_{IC}^{4/3}\ H^{1/4}$ for erosion [4,5]. Relationships such as these are useful and have been shown to be predictive, to a first approximation, especially for monolithic ceramics, which differ significantly in material properties [6]. Close examination of wear data, however, often reveals a considerable departure from predictions [7]. Such inconsistencies are particularly apparent in analysis of results obtained within a given material system [3]. Analyses and interpretation of wear data based on the assumption that wear resistance is fully describable by $K_{IC}^{m}\ H^{n}$ may lead to a wide range of exponents of little real physical meaning, thus indicating that a more relevant description requires additional definition of a material, specifically its microstructure. The uncertainties in ceramic material wear-resistance predictions for ceramics most likely stem from the fact that the aspects of material removal are not fully reflected by cumulative properties such as hardness and fracture toughness, but are also influenced by the microstructure and variations of mechanical properties on the scale of the wear process.

The wear of a cutting tool is a complex phenomenon involving many operative wear mechanisms. These can be generally classified as chemical (diffusion, oxidation, etc.) or mechanical (fracture, abrasion, etc.). The predominance of a particular wear mechanism is dependent on the tool material–workpiece combination and the cutting conditions. The cutting edge of a tool is exposed to high stress as well as to elevated temperatures produced by primary shear of the metal ahead of the tool and friction between the chip and the tool as the latter slides over the tool surface. Depending on the material being cut and the cutting speed, temperatures at the cutting edge may be as high as 1200°C. It follows then that the thermal shock resistance of the cutting tool material plays an important role in determining tool life [8]. This is particularly important if it is considered that tool use, more often than not, entails repeated interruption of cutting and sometimes requires use of coolants to prevent annealing of the metal surface. Ceramic cutting tools made from aluminum oxide, while proven successful in some operations, have been unable to make a significant impact on the cutting tool market because of their limited fracture toughness and thermal shock resistance. In recent years, this void has provided an impetus for the application of monolithic Si_3N_4 ceramics [9–12] and the more recent development of SiC whisker-reinforced alumina (Al_2O_3) and Si_3N_4 composites [13–16]. Fracture resistance of these composites is twice that of monolithic materials and represents a major improvement in reliability and abrasive wear resistance of ceramic cutting tools. The development of Si_3N_4 and Si_3N_4-composite cutting tools has improved productivity considerably. Today it is common in the auto industry to reliably machine cast iron at speeds exceeding 1400 m/min using Si_3N_4 cutting tools in turning or milling operations where the tool is subjected to intermittent impact cutting cycles. Although Si_3N_4-based cutting tools show outstanding wear resistance in machining of cast iron, they have been mostly unsuccessful in machining of steels because of the chemical interactions that occur at the cutting interface.

Extensive machinability testing of ceramic cutting tools, including Si_3N_4-based composites, shows that ceramic tools used in machining of gray cast iron wear primarily by abrasion [14] (i.e., mechanical wear). In these circumstances, wear resistance can be related directly to the mechanical properties of the tool. This behavior is contrary to observation of Si_3N_4 cutting tool performance in machining of steel, where chemical wear dominates and determines tool life. It follows, then, that the options of tailoring both mechanical and chemical characteristics of Si_3N_4-based composites must be considered for the development of an effective cutting tool material for machining of both steel and cast iron. Compared to simple binary or "monolithic" ceramic systems, the composite allows tailoring of chemical and mechanical properties through the introduction of selected dispersoids, as well as modifications through sintering aid additions and microstructural control through processing [15].

Through specific examples of Si_3N_4-TiC composites, this chapter discusses the principles leading to a successful design of cutting tools for a broad range of applications. Furthermore, the relationship between microstructure, mechanical properties, and wear resistance is described for Si_3N_4-ceramic composites subjected to abrasion or particle erosion.

II. MATERIALS AND METHODS

If processing-related options (i.e., sintering aids and microstructure) were not considered, the basic elements of material design would be those shown in Figure 1. Silicon nitride based materials are modified by adding sintering aids or by introducing dispersoids to obtain specific mechanical and chemical properties to meet the design criteria. Another option is to modify the surface of Si_3N_4 cutting tools by applying coatings [17] that reduce friction, increase near-surface hardness, or reduce chemical interaction between the tool and the workpiece. In this study, we discuss bulk property modifications obtained through the addition of sintering aids (Y_2O_3, CeO_2, Al_2O_3) and dispersoids (TiC) to Si_3N_4-based materials, which were consolidated by hot-pressing in an inert atmosphere. Hardness and fracture toughness of the consolidated composites were determined by standard indentation techniques [18]. Table 1 lists Si_3N_4 composites that were milled with Al_2O_3 media and exclusively contain 6 wt% Y_2O_3 and TiC as additives. These materials were thoroughly characterized in terms of their microstructure and tested in abrasion and erosion. Table 2 lists the Si_3N_4-based compositions and the associated properties of the consolidated materials used as cutting tool materials.

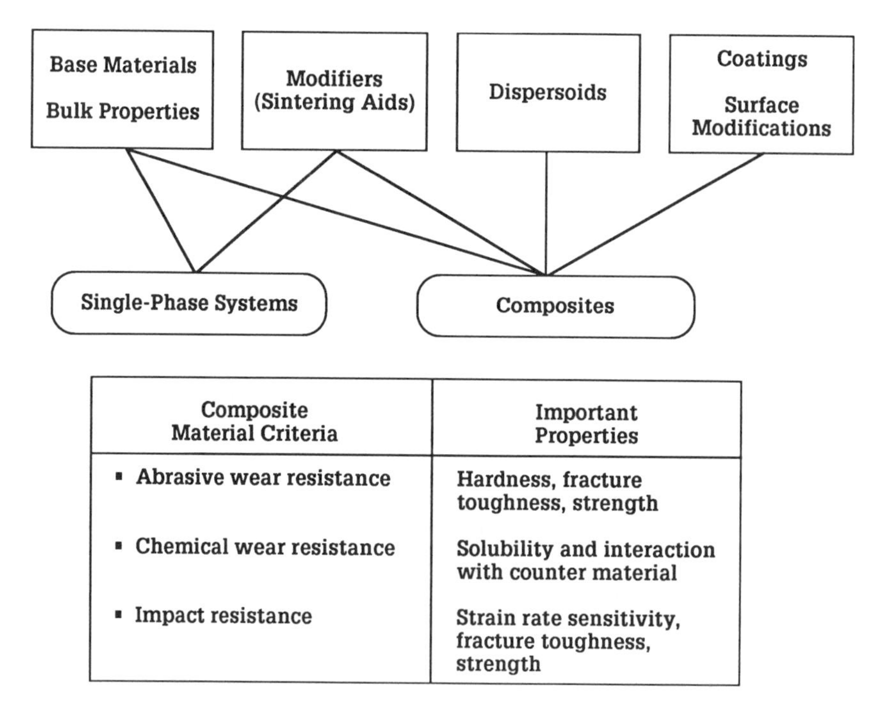

Composite Material Criteria	Important Properties
▪ Abrasive wear resistance	Hardness, fracture toughness, strength
▪ Chemical wear resistance	Solubility and interaction with counter material
▪ Impact resistance	Strain rate sensitivity, fracture toughness, strength

Figure 1 Basic elements of ceramic composite materials design.

Table 1 Mechanical Properties and Microstructural Parameters[a] of Test Materials

Material	$\bar{D}$ (μm)	$\bar{\lambda}$ (μm)	Indentation fracture toughness ($MPa \cdot m^{1/2}$)	Knoop hardness (GPa)
Silicon nitride (AY6)[b]			3.5 ± 0.1	13.8 ± 0.4
AY6 + 20 vol% TiC	0.4	1.1	3.7 ± 0.1	15.5 ± 0.4
	1.5	4.0	3.8 ± 0.1	14.9 ± 0.4
	3.4	9.1	4.1 ± 0.1	15.0 ± 0.4
AY6 + 10 vol% TiC	0.4	2.4	3.8 ± 0.1	15.0 ± 0.4
AY6 + 20 vol% TiC	0.4	1.1	3.7 ± 0.1	15.5 ± 0.4
AY6 + 30 vol% TiC	0.5	0.8	3.2 ± 0.1	15.9 ± 0.3
AY6 + 40 vol% TiC	0.5	0.5	2.9 ± 0.1	15.8 ± 0.5
AY6 + 10 vol% TiC	2.5	15.0	4.2 ± 0.1	14.7 ± 0.4
AY6 + 20 vol% TiC	3.4	9.1	4.1 ± 0.1	15.0 ± 0.4
AY6 + 30 vol% TiC	3.5	5.5	3.8 ± 0.1	16.0 ± 0.5
AY6 + 40 vol% TiC	3.2	3.2	3.7 ± 0.1	16.2 ± 0.4

[a] $\bar{D}$, average grain size of TiC particulates; $\bar{\lambda}$, mean free path of Si_3N_4 between TiC particulates.
[b] Designated "monolithic Si_3N_4" in this study.

Table 2 Calculated Chemical and Mechanical Wear Terms[a] for Selected of Ceramic Tools

Matrix composition (vol %)				
Si_3N_4	Oxide Additive[b]	Dispersoid (vol%)	Chemical term, $1/C_i$ (mol/mol)	Mechanical term, $K_{IC}^{3/4} H^{1/2}$
93.1	3.9Y + 3A		11.3	9.5
63.1	3.9Y + 3A	30 TiC	16.2	11.8
56.1	3.9Y + 10A	30 TiC	18.1	9.9
46.1	3.9Y + 20A	30 TiC	21.6	8.5
66.1	3.9Y + 20A	10 TiC	15.6	8.0
56.1	3.9Y + 20A	20 TiC	18.1	8.2
96.1	3.9C		10.9	8.4
66.1	3.9C	30 TiC	15.5	11.8
92.1	3.9C + 3A		11.3	9.0
63.1	3.9C + 3A	30 TiC	16.2	11.4
	70A	30 TiC	5.36×10^3	9.9

[a] $1/C_i$, calculated equilibrium solubility limits in α-Fe at 1400 K; $K_{IC}^{3/4} H^{1/2}$, K_{IC} = [$MPa \cdot m^{1/2}$]; H = [GPa].
[b] Y, Y_2O_3; A, Al_2O_3; C, CeO_2.

As indicated in Table 1, the materials that contained 20 vol% TiC differed in the average size of the TiC particulates, thus providing a means to examine the influence of dispersoid grain size on abrasive and erosive wear resistance. To study the influence of particle dispersion (mean free path), the volume fraction of TiC was varied from 10% to 40% for two TiC grain size groups, representing fine ($\leq$ 0.5 μm) and coarse ($\geq$ 1.5 μm). The mean free path ($\bar{\lambda}$) is defined as follows:

$$\bar{\lambda} = (\bar{D})\ \frac{2}{3}\left(\frac{1 - V_f}{V_f}\right) \tag{1}$$

where $\bar{D}$ is the average grain size of TiC, V_f is the volume fraction of dispersoid, and $\bar{\lambda}$ is the average spacing between TiC grains.

Wear testing focused on diamond abrasion or gas-jet Al_2O_3 particle erosion. The abrasive wear testing apparatus is the basic pin-on-disk type, in which the disk is a removable resin-bonded, 45 μm diamond surface. The pin test surface (a 6 mm/side square) is swept and rotated at 20 rpm under a 9.8 N static load against the spinning (40 rpm) abrasive disk. All tests were performed under dry abrasive conditions in a flowing air atmosphere [1].

The gas-jet Al_2O_3 particle erosion tests were configured to conform with ASTM-G-76. A 0.9 mm diameter nozzle is placed 1 mm from the polished ceramic surface (90° orientation), upon which 50 μm Al_2O_3 particles are impacted at a velocity of 115 m/s employing an argon gas carrier. All tests were performed at room temperature for 1 minute, corresponding to 6.3 g of Al_2O_3 abrasive being impinged on the sample. The samples were then ultrasonically cleaned in methanol, and the volume of the eroded material was calculated from depth gage readings and optical measurements of the diameter.

The machining performance of the ceramic tools in continuous turning was evaluated using sand-cast gray iron (class 40) and AISI 4340 steel workpieces, each 20 cm in diameter and 1 m long. All tests were performed without cutting fluid on a 60 hp lathe equipped with a 15° lead angle tool holder and SNG-432 inserts. The cutting parameters for steel were 3.55 m/s (213 m/min), 0.254 mm/rev, 1.27 mm depth of cut, and for gray cast iron, 12.7 m/s (762 m/min), 0.38 mm/rev, and 1.27 mm depth of cut.

Wear of the cutting tool was monitored with an air gage based system [19]. This technique is a direct means of measuring tool nose wear ΔR during turning operations. A nozzle placed adjacent to the cutting tool creates an air gap backpressure, which provides an analog signal proportional to tool wear. In addition, both light microscopy and scanning electron microscopy (SEM) were used to evaluate conventional nose, flank, and crater wear of the tool after the cutting test was terminated.

To elucidate the contribution of diffusion, static diffusion couples were made with the ceramic tool materials sandwiched between the gray cast iron and 4340 steel materials. These diffusion couples were then placed in a hot press at 15 MPa, 900°C, for 30 minutes in an argon atmosphere, which was intended to approximate the pressure and temperature encountered during metal turning operations [20,21]. The heat-treated diffusion couples were then cross-sectioned and metallographically prepared to reveal the chemical interactions across the metal–ceramic interfaces.

III. ABRASIVE WEAR

For brittle substances such as the ceramics and ceramic composites, material removal by fracture that occurs in abrasion can be assumed to take place when there is intersection of lateral cracks of adjacent indentations caused by penetration of sharp surface protrusions (or abrasive particles as in pin-on-disk testing) of the opposing surface [22]. The removed volume, per particle V_i, is then

$$V_i = r_i h_i \, l_i \tag{2}$$

where r_i is critical indentation separation, h_i is the depth of the indentation, and l_i is the sliding distance.

Considering the dependence of the size of the indentation and the length of the cracks emanating from such angular indentations on the hardness H and fracture toughness K_{IC}, respectively, the following expression for maximum volume removed by the system of indenters in a grinding operation, or abrasive wear, was derived by Evans and Wilshaw [22]:

$$V \,\alpha\, \frac{1}{K_{IC}^{3/4} H^{1/2}} \sum_{i=1}^{i=n} p^{5/4} = \frac{1}{K_{IC}^{3/4} H^{1/2}} N(\overline{P})^{5/4}\, \overline{\ell} \tag{3}$$

where N is the number of abrasive particles, P is a vertical force on the particle, and l_i is the sliding distance. Earlier experiments conducted on a series of monolithic ceramic and dispersoid phase containing composites have shown that their respective abrasive wear resistance follows a general relationship:

$$\frac{1}{V} \,\alpha\, K_{IC}^{3/4} H^{1/2} \tag{4}$$

where V is the abraded volume (Fig. 2) [6]. These measurements were obtained from a large number of specimens representing each group of materials: Al_2O_3 monoliths, Al_2O_3-ZrO_2 composites, Al_2O_3-TiC composites, and Si_3N_4 monoliths. The results are plotted as a linear function, with boxed regions representing the scatter of data observed within each material system.

Since deviation could be observed within each group of compositionally

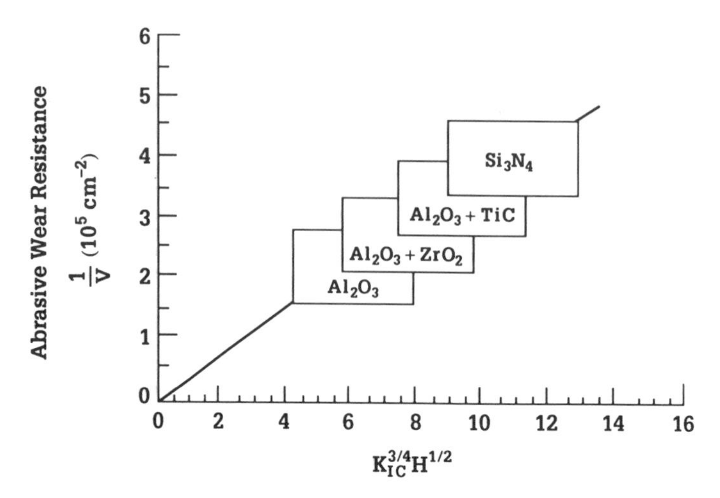

Figure 2 Abrasive wear resistance of ceramics and composites as a function of $K_{IC}^{3/4}H^{1/2}$. Relationship approximated by a straight line.

equivalent materials, it was speculated that microstructural differences might be responsible for the observed variation. It was therefore expected that composites, in which the content of dispersoid is held constant and average particulate size varied over a broad range, may provide an excellent example which could elucidate the origin of such discrepancies.

As can be seen from Table 1, composites with coarser TiC dispersoid have higher fracture toughness. The dispersoid size $\overline{D}$ has an inverse, although weaker, effect on the hardness of the composite. It would be expected from Equation (4) that addition of coarser dispersoids would lead to improved abrasive wear resistance. Measured wear resistances of Si_3N_4 + 20 Vol% TiC composites with different TiC grain size (Fig. 3), however, show an opposite trend. Silicon nitride composites containing finer TiC dispersoids exhibit higher wear resistance despite reduced fracture toughness.

The dependence of abrasive wear resistance on dispersoid size $\overline{D}$ can also be seen in Figure 4. Experimental results show that the abrasion resistance (Fig. 4a) for fine-grained TiC composites is higher than for the coarse-grained materials. This figure shows abrasion resistance versus volume content of TiC. Figure 4b is a prediction of the abrasive wear resistance of Si_3N_4-TiC composites based solely on hardness and fracture toughness according to Equation (4). For simplicity, TiC grain sizes of 0.5 μm and smaller are referred to as "fine" and those of 1.5 μm or greater as "coarse." The results show a trend opposite that of the experimental findings by predicting higher wear resistance for coarse TiC materials. Since the results in Figures 3 and 4 showed a dependence on

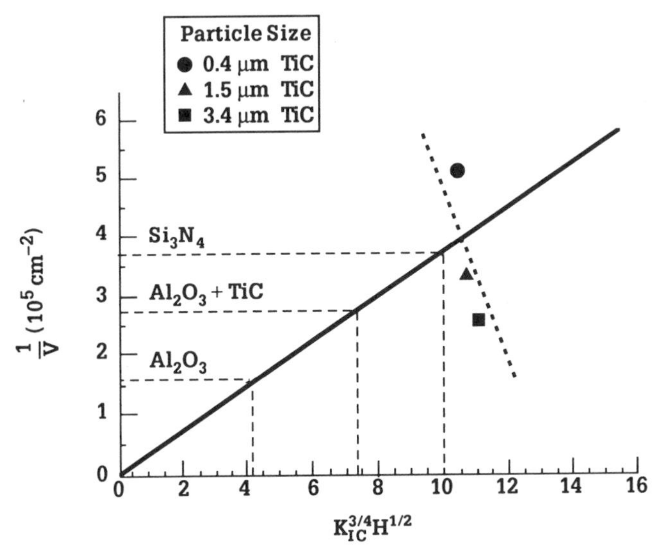

Figure 3 Abrasive wear resistance of Si_3N_4 + 20 vol% TiC composites versus $K_{IC}^{3/4}H^{1/2}$; note influence of TiC grain size (dashed line).

microstructure, it seemed appropriate to incorporate $\overline{\lambda}$ to account for both $\overline{D}$ and V_f of TiC. The abrasion resistance, calculated as follows

$$\frac{1}{V} \alpha K_{IC}^{3/4} H^{1/2} \left(\frac{1}{\overline{\lambda}} \right) \tag{5}$$

and plotted in Figure 4c, is in closer agreement with the experimental observations (Fig. 4a) insofar as finer TiC materials have higher wear resistance. Differences in slope between calculated and experimental results is due to the absence of the test specific proportionality constant implied in Equation (5).

The significance of $\overline{\lambda}$ in affecting the abrasion resistance of Si_3N_4-TiC composites becomes apparent when one considers the nature of material removal process as proposed by Evans and Wilshaw [22]. The abrasive particle penetrates the brittle surface, leading to a damaged zone formed by radial and lateral cracks, which can link up and lead to material removal. Thus, if $\overline{\lambda}$ is much less than the damage zone created by the abrader, the extent to which subsurface damage can occur is greatly reduced.

IV. PARTICLE EROSION

In many instances, ceramics and ceramic matrix composites are subjected to particle erosion. For successful wear-resistant material design, it is essential to

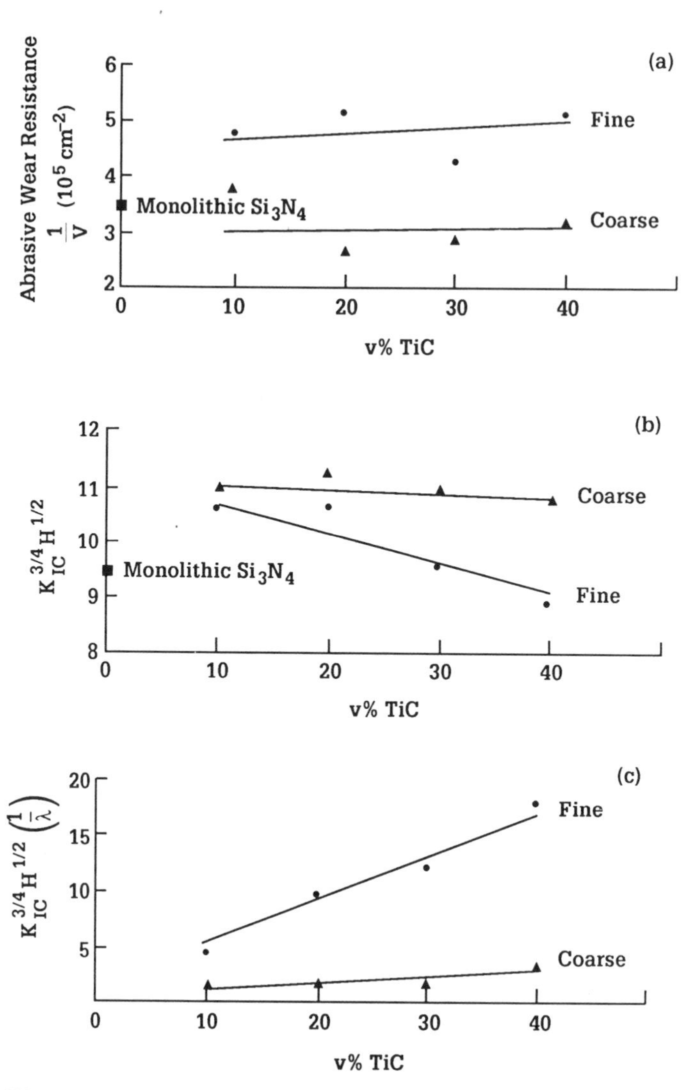

Figure 4 Abrasive wear resistance of Si_3N_4-TiC composites versus TiC content: (a) actual abrasive wear results, (b) prediction based on $K_{IC}^{3/4}H^{1/2}$, and (c) prediction based on combined mechanical properties–microstructural parameter.

understand the mechanism of material removal and identify critical material property–performance relationships. Composites in a series (Table 1) were subjected to gas-jet Al_2O_3-particle erosion. The eroded volume loss as a function of TiC content is plotted in Figure 5 (labeled Actual Results) and shows consistent increase in erosion resistance with increasing content of TiC. Coarser dispersoids are apparently less effective in reducing erosion than the fine-grained TiC at 10 and 20 vol% concentrations. Analyses of the erosion process for brittle materials have suggested that erosion resistance should also be expected to be proportional to a material's hardness and fracture toughness [23–26]. Evans et al. [5] have proposed that erosion of brittle materials can be predicted by the following relationship:

$$V \alpha \frac{1}{K_{IC}^{4/3} H^{1/4}} \tag{6}$$

where V is the eroded volume.

This relationship was used to predict the erosion wear of the Si_3N_4-TiC composites and is plotted in Figure 5 (labeled *Predicted Results;* Eq. 6). These data have been normalized to the measured values of eroded volume loss, K_{IC} and H of *monolithic* Si_3N_4, thus establishing a proportionality constant. The predicted results from Equation (6) clearly underestimate the erosion resistance of the Si_3N_4-TiC composites. Furthermore, the predicted results increase with the percentage of TiC, whereas experimental results show a decrease. The

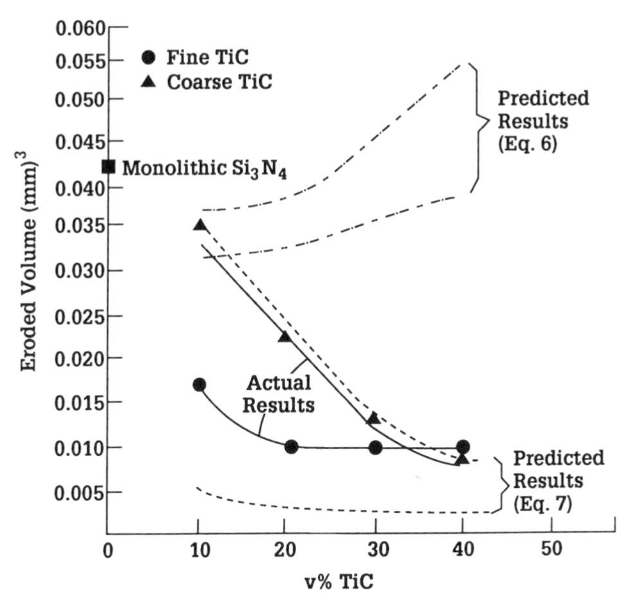

Figure 5 Eroded volume loss of Si_3N_4-TiC composites versus TiC content: actual results (solid curve), predicted results (Eq. 6) (dash–dot curve), and predicted results (Eq. 7) (dashed curve).

relationship between K_{IC},H, and erosion resistance has been established for monolithic ceramics systems and glasses, which develop radial and lateral cracking when impacted by solid particles [27–29]; composites were not considered. Inaccuracies in wear-resistance predictions of brittle materials have been attributed to the influence of microstructure [1], although only few systematic data exist. The extent of contact damage (cracks) during erosion has been estimated to be in the range of 0.1–10 μm for eroding particle sizes of 1 to 100 μm [30], which encompass the eroding particles used in this study. It has been postulated that resistance to crack extension must relate to cracks of this size, therefore any microstructural toughening should be on the order of the impact fracture size [4,30]. The significance of $\bar{\lambda}$ during particle erosion becomes evident when examining Figure 6, which depicts the relative size and interaction of the anticipated Al_2O_3-particle contact damage zone with the Si_3N_4-20 vol% TiC microstructural features.

The quantitative microstructural analyses, plotted as percentage of total TiC volume versus the TiC grain diameter, are the basis for the schematic microstructures illustrated in Figure 6. These quantitative results confirm the low volume fraction of small grains in the coarse dispersoid containing microstructure and the absence of large grains in the fine TiC containing material.

The microstructural comparison in Figure 6 depicts the increase in $\bar{\lambda}$ from 1.1 μm to 9.1 μm as the average TiC grain size $\bar{D}$ is increased from 0.4 μm to 3.4 μm, thus exposing the unreinforced Si_3N_4 matrix to impact damage. The erosive wear resistance of *monolithic* Si_3N_4 is lower than any of the Si_3N_4-TiC composites (Fig. 5), therefore the expectation is for lower wear resistance with increasing $\bar{\lambda}$. Therefore, the volume loss due to wear would be more appropriately described by

$$V \alpha \frac{1}{K_{IC}^{4/3} H^{1/4}} \bar{\lambda} \tag{7}$$

This implies that the highest values of $\bar{\lambda}$ would lead to higher erosive wear. The contribution of $\bar{\lambda}$ can be seen in Figure 5, which contains the curve labeled Predicted Results, Eq. (7). These results show that the actual erosive wear results obtained on Si_3N_4-TiC composites are more closely described by the combined mechanical property–microstructural parameter, thus establishing the significance of the microstructure in the prediction of erosive wear resistance.

Observations of the wear behavior of tungsten carbide/cobalt (WC-Co) systems have shown a strong dependence of abrasion and erosion wear resistance on $\bar{\lambda}$ (mean free path of cobalt binder) [31,32]. The softer cobalt phase was preferentially eroded as $\bar{\lambda}$ increased [32]. The Si_3N_4-TiC (brittle–brittle) system parallels, in some regards, the observations made for the erosion of WC-Co (brittle–ductile), particularly the observed reduction in erosion resistance with

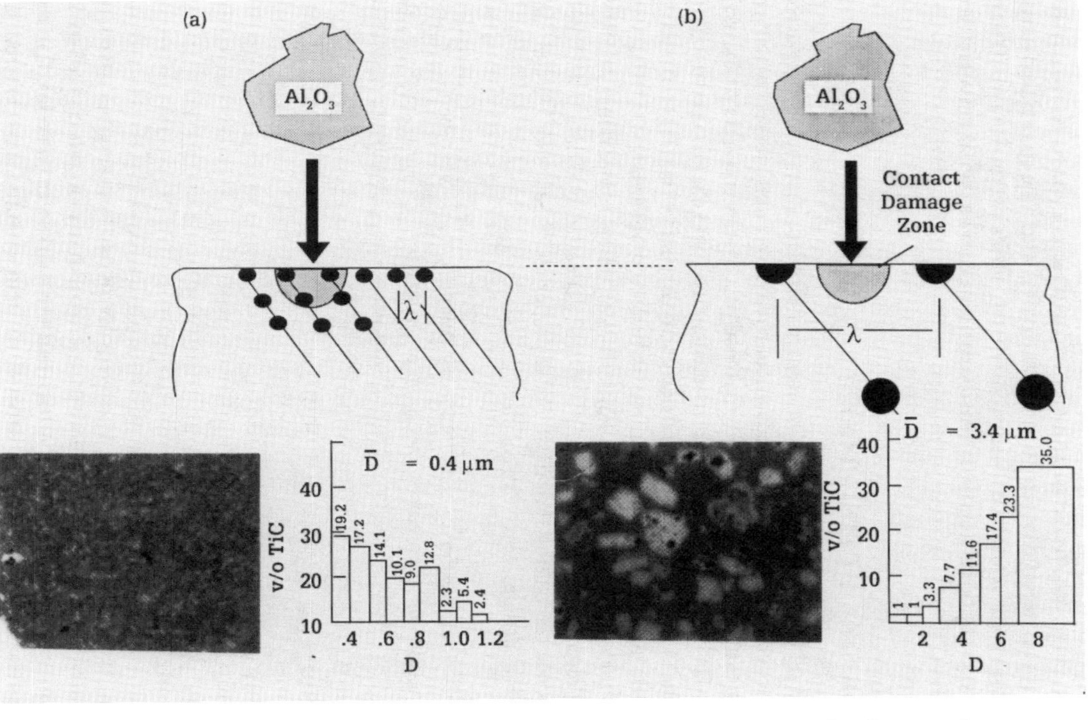

Figure 6 Schematic representation of Si_3N_4-20 vol% TiC composite microstructural and contact damage zone with photomicrograph of actual microstructure and TiC grain size distribution for (a) 0.4 µm TiC and (b) 3.4 µm TiC.

increasing $\bar{\lambda}$. However, the contribution of TiC grain size in the Si_3N_4-TiC system was found to be significant at 10 and 20 vol%, with fine-grained materials offering higher wear resistance. Reduction in TiC grain size at 30 and 40 vol% resulted in little additional benefit in erosion resistance, which qualitatively agrees with Equation (7).

V. WEAR OF CUTTING TOOLS

Wear of cutting tools represents a more complex problem. In machining, surface deterioration occurs as a result of mechanical and chemical processes, which can act in combination. The degree to which these processes contribute to the wear of cutting tools depends on the physical, mechanical, and chemical composition of the tool and workpiece materials. To design and optimize a cutting tool material for a specific application, considerable information regarding the tool wear mechanism under specific cutting conditions (i.e., force, temperature, environment) is a prerequisite. Elevated temperatures, generated by friction and metal shear, influence the cutting tool mechanical properties, create shock conditions, and provide a driving force for chemical interactions between the tool and workpiece.

Numerous attempts have been made to quantify tool wear behavior as a singular function of chemical or mechanical properties. Previous work [3] has shown that total wear W observed on cutting tools can generally be expressed as follows:

$$W_{total} = W_{mech} + W_{chem} \tag{8}$$

or in terms of wear resistance, R

$$W_{total} = \frac{1}{R_{mech}} + \frac{1}{R_{chem}} \tag{9}$$

where R_{mech} is related to $K_{\mathrm{IC}}^{3/4}\mathrm{H}^{1/2}$, primarily for abrasive wear resistance, and R_{chem} is a chemical interaction parameter (i.e., solubility of cutting tool in the workpiece material) [3] between the tool and the workpiece material. Clearly, these wear-resistance parameters depend on temperature, atmosphere, and pressure. However, in a specific cutting application, where the characteristics of the workpiece material and the machining conditions are constant, steady state tool wear is a reflection of the mechanical and chemical wear resistance of the cutting tool. This relationship is approximated by a parametric model, using arbitrary units, using Equation (9) (Fig. 7a).

The resulting three-dimensional surface constitutes a theoretical mapping of points whose coordinates are dictated by the intersection of mechanical and chemical wear-resistance components. As a first approximation, the surface is

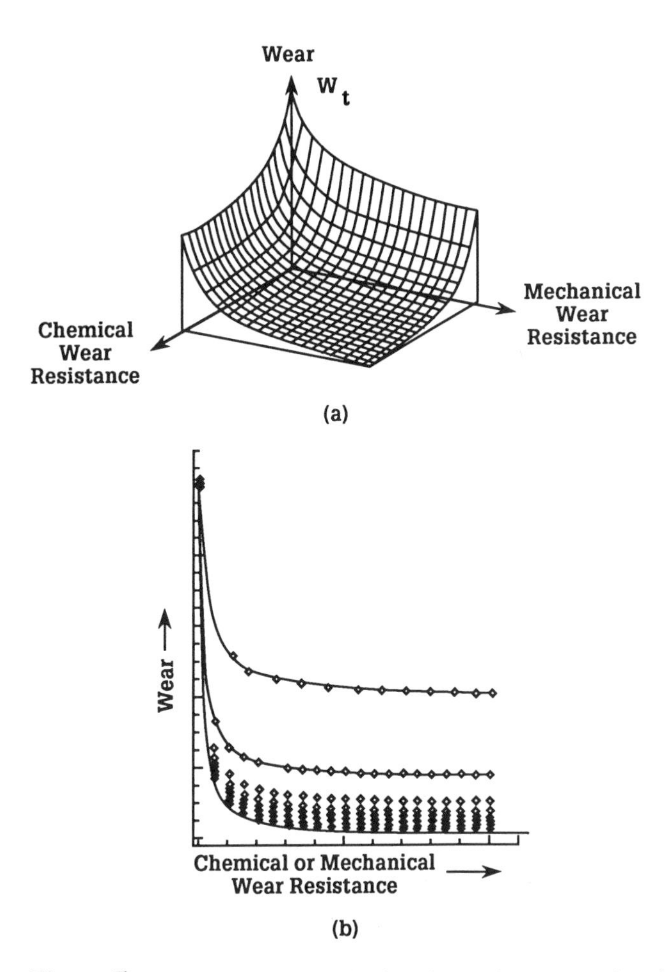

Figure 7 (a) Theoretical model of cutting tool wear as a function of mechanical and chemical wear parameters. (b) Two-dimensional projection showing dependence of wear on mechanical or chemical parameters.

consistent with anticipated tool wear insofar as (a) extremely low resistance to chemical or mechanical wear leads to highest total wear, (b) reductions in total wear can be achieved by improvements in chemical or mechanical tool material properties, and (c) minima in total wear are obtained when both property parameters are simultaneously increased to their limits.

Further insight into cutting tool wear can be extracted from Figure 7a by viewing the surface as a two-dimensional entity. Figure 7b shows the surface projections in the plane established by the W_{total}–R_{chem} axes, and similarly in the

W_{total}–R_{mech} plane. These views are instructive because they reveal the interdependence of the property parameters from a two-dimensional viewpoint, which is typically used for reporting wear results. This two-dimensional representation generates a bandwidth in wear behavior for both chemical and mechanical wear resistance. This bandwidth provides an explanation for the wide scatter in two-dimensional plots of cutting tool wear in which the interaction of either chemical or mechanical contributions was not considered. It is then apparent that observations of wear as an exclusive function of mechanical or chemical wear resistance could lead to erroneous conclusions if one or the other is not known and varies significantly (Fig. 7b).

To apply the mechanical wear principles addressed in this study to actual machining operations, two criteria must be met: (a) abrasive wear must be the dominant mode of tool failure for a given operation, and (b) the mechanical properties of the tool material must be known in the temperature range that the tool encounters in service. In the case of ceramic composite cutting tool materials, which tend to maintain their relative ranking of material properties over a broad temperature range, it is possible to relate the tool's wear resistance in machining of gray cast iron to the wear-resistance parameter based on room temperature fracture toughness and hardness [33].

Extensive observations of ceramic cutting tool performance in machining of cast iron have shown that tool life in continuous cutting is predictable by the relationship given in Equation (4) [34]. Silicon nitride materials, although lower in hardness than Al_2O_3-TiC composites, exhibit higher wear resistance and reliability because of increased fracture toughness, which also permits the tools to operate at higher feed rates. Extending the concept of performance prediction beyond wear to include behavior in interrupted cutting would require additional information and definition of properties, which reflect impact and thermal shock resistance of tool materials [8].

While Si_3N_4-based cutting tool materials, and Si_3N_4-TiC composites in particular, have shown an outstanding wear resistance in machining of cast iron, they are subjected to severe wear when used to machine AISI 4340 steel. A series of tests using Si_3N_4-based tools have shown that the predominant mode of tool wear is chemical when machining steels. Thus in cast iron machining tool wear is observed only on the nose and flank of the tool (Fig. 8a), while in steel machining crater formation on the rake face becomes a prominent wear feature (Fig. 8b). Crater formation indicates considerable contributions of chemical interaction between the tool and the chip generated from the workpiece.

To examine and define factors controlling wear, a series of composites in the system Si_3N_4-Y_2O_3-Al_2O_3-TiC were investigated. The material is composed of Si_3N_4 (2 wt% SiO_2) + 4 wt% Y_2O_3-based material with additions of up to 50 vol% of Al_2O_3 and up to 50 vol% of TiC.

As can be seen in Table 1, additions of TiC dispersoid prominently increase

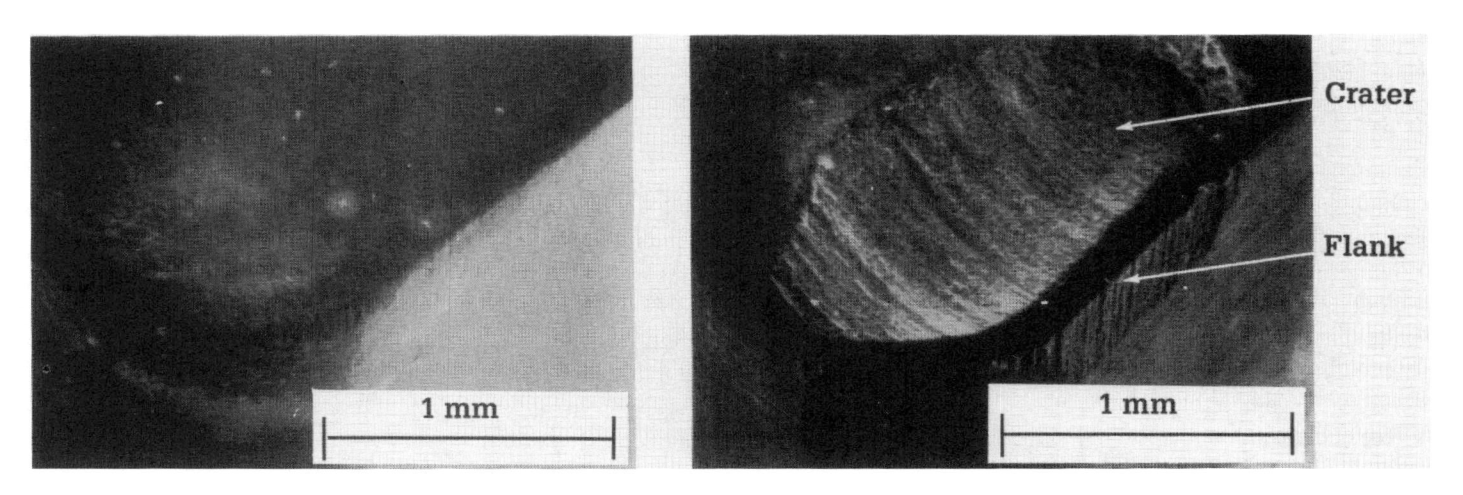

Figure 8 Apperrance of Si_3N_4-based cutting tool nose after machining: (a) gray cast iron at 7.1 m/s (426 m/min) for 12 minutes and (b) AISI 4340 steel at 3.55 m/s (213 m/min) for 1 minute.

the hardness (Knoop hardness number) of the composite and result in a relatively small decrease of fracture toughness measured by indentation. The room temperature modulus of rupture is only moderately affected, while additions of Al_2O_3 result in a decrease of the values for all three properties [36].

Figure 9a is a plot of wear resistance as a function of Al_2O_3 additions to a

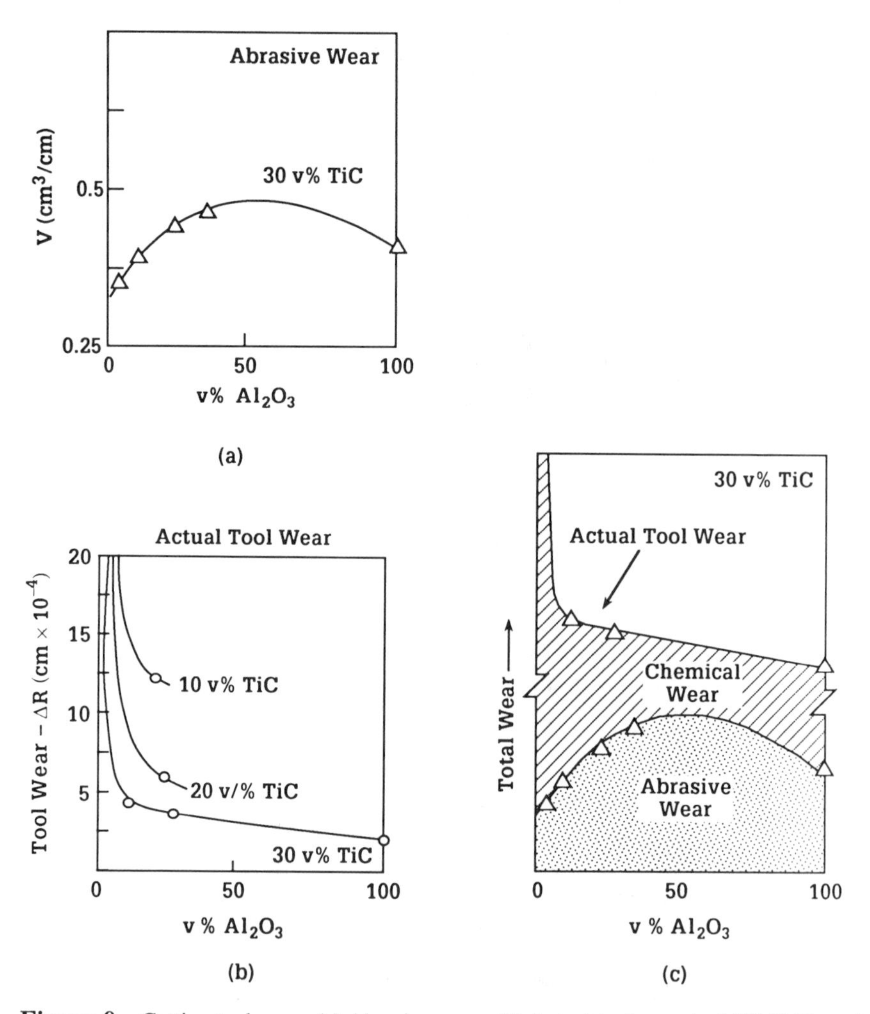

Figure 9 Cutting tool wear: (a) Abrasive wear. (b) Actual tool wear in AISI 4340 steel at 3.55 m/s (213 m/min), 0.25 mm/rev feed, and 1.25 mm depth of cut, 10 minute cutting time. (c) Composite plot of components of cutting tool wear.

Si_3N_4-TiC composite measured by pin-on-abrasive-disk testing. These data show that as the concentration of Al_2O_3 increases, so does the wear of the composite. Results on tool wear ΔR after 10 minutes of turning 4340 steel (Fig. 9b) indicate that wear is reduced by additions of Al_2O_3 and/or refractory carbide dispersoid. Changes in R are a consequence of tool nose wear, which in turn is a measure of tool wear resistance, both mechanical and chemical. Taking into account the known abrasive wear resistance of these materials, it becomes possible to illustrate the wear of these materials in steel machining (Fig. 9c) [35]. This representation considers actual tool wear with observed abrasion (Fig. 9a). Complete absence of similarity indicates an overwhelming contribution of chemical wear, which obliterates any improvements in mechanical (or abrasive) wear resistance. Under these circumstances, it is apparent that the outcome of efforts expended on further mechanical property optimization (such as microstructural tailoring for fracture toughness improvements) for increased wear resistance would be futile or limited at best.

The basis for understanding the chemical wear behavior of tool materials outside the metal cutting environment has been pursued by thermochemistry and diffusion studies. Based on the degree of adherence and materials transfer between the tool and the workpiece when in contact at 1000–1200°C, it has been reported that Si_3N_4 and TiC show greater reactivity with AISI 4340 steel than does Al_2O_3 [36]. These diffusion tests are indicative of the degree of possible chemical interaction at high machining speeds. It has also been suggested that the free energy of formation, diffusion, and solubility of the tool in workpiece materials can be used to predict tool wear [20,37,38]. Estimates of the solubility limit of single-phase materials in α-Fe have been used to predict performances of coatings in steel machining [39,40]. Taking into account the complex makeup of the aforementioned Si_3N_4-based composite tool materials, their solubility C_i (mol/mol) in α-Fe at 1400 K was estimated by considering fractional contributions V_f of individual microstructural phases:

$$C_{tool} = \sum_{i=l}^{i=n} V_f\, C_i \tag{10}$$

For the sake of simplicity, this estimate considers only the solubility of precursor materials and disregards their interaction in the densified composite.

While these thermodynamic estimates do not consider reaction kinetics and, therefore, may not be exactly and directly transferable to the dynamic situation of tool wear, they at least reflect the potential for chemical reaction. The resulting estimate of relative tool wear resistance for chemical wear (i.e., solubility limit) is shown in Figure 10. As can be seen, this simplified approach yields an approximation of actual tool performance in AISI 4340 steel (Fig. 9b), although it apparently fails to account for the strong effect of small additions of Al_2O_3 or

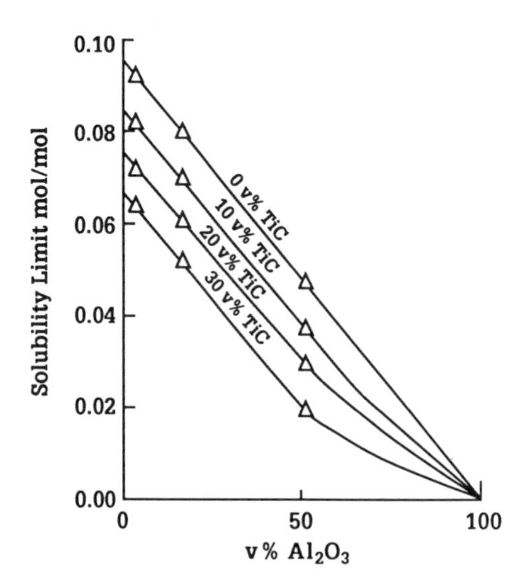

Figure 10 Estimate of composite tool solubility in α-Fe at 1400 K: (Si_3N_4 + 6 wt% Y_2O_3)-Al_2O_3-TiC system.

TiC. The absolute contribution of solubility, determined in this manner for cutting tool wear in ferrous alloy machining, varies depending on alloy characteristics (e.g., concentration, distribution of carbon, or alloying elements). Observations of wear behavior of silicon nitride cutting tools in high speed machining of cast iron, up to 30.4 m/s (1825 m/min) fail to reveal a substantial chemical wear component. Diffusion experiments indicate virtually no reaction between the Si_3N_4 composite and cast iron (Fig. 11a), but a considerable reaction with AISI 4340 steel (Fig. 11b). The change in workpiece composition strongly affects the tool–workpiece chemical interaction and requires more careful consideration in estimates of wear resistance. Furthermore, in a complex tool material consisting of two or more phases, reactions during densification may considerably change the phase assembly and render simple projections inaccurate.

Based on these observations, it appears acceptable to apply the predicted solubility (Eq. 10) in ceramic cutting tool design. Wear of ceramic cutting tools in AISI 4340 steel machining would, therefore, be proportional by a first approximation to selected parameters in the following relationship:

$$W \propto \frac{1}{K_{IC}^{3/4} H^{1/2}} + C_i \tag{11}$$

where K_{IC} is fracture toughness (MPa · $m^{1/2}$), H is hardness (GPa), and C_i is solubility (mol/mol).

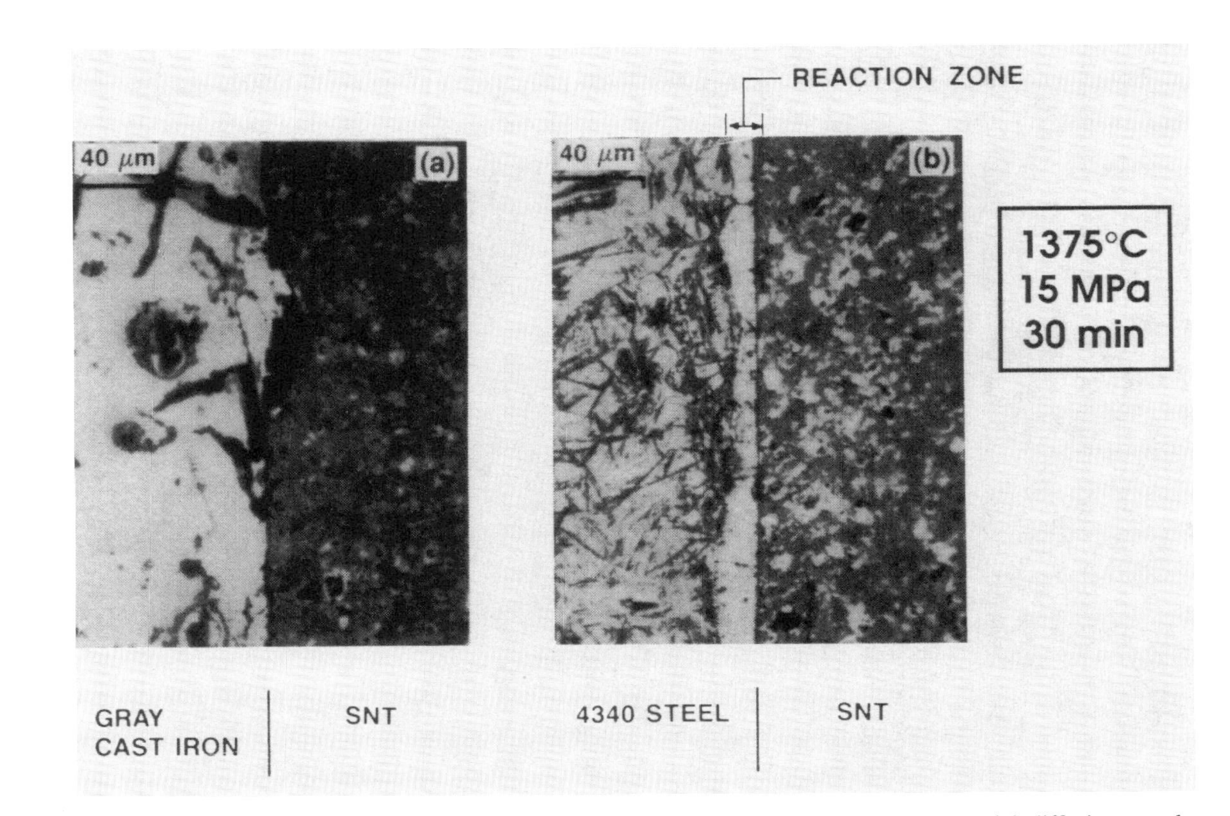

Figure 11 Photomicrographs of workpiece and Si_3N_4-TiC (SNT) composite tool material diffusion couple hot-pressed in argon at 1375°C, 15 MPa, for 30 minutes: (a) gray cast iron and (b) AISI 4340 steel.

The concept just discussed of cutting tool material design and established parameters were applied to Si_3N_4 composites. Chemical and mechanical properties of composites were tailored through modifications to the secondary glass phase in the ceramic matrix and dispersoid selection. Choice of components was governed by the estimated solubility and expectations of hardness and fracture toughness improvements. The composition, estimated chemical solubility, and mechanical wear parameters, calculated based on measured properties, were given in Table 2.

SNG432 cutting tools in a series selected from Table 2 were tested in AISI 4340 steel machining. Tool wear rate, $W = \Delta R/\Delta t$, which is plotted in Figure 12 as an inverse function of parameters, appears to be in agreement with the general relationship given in Equation (4). As anticipated, materials of the same or similar mechanical properties with reduced chemical solubility exhibit higher resistance to wear. Similarly, materials with equivalent solubility show improvement in wear resistance with increasing values of $K_{IC}^{3/4}H^{1/2}$. Since the range of mechanical property parameters is relatively narrow for the series of Si_3N_4 materials examined, the dependence of wear on reciprocal solubility is instructive, as shown in Figure 12. Conversely, a plot of wear versus mechanical property parameter is inappropriate and would be misleading because of the broad range of chemical parameter values.

VI. SUMMARY

Studies of composite silicon nitride ceramics have shown that predictive equations that are based solely on the hardness and fracture toughness only in part reflect the wear resistance in either erosion or abrasion. More precise predictions of ceramic composite wear behavior require additional information regarding the microstructures.

In more complex situations, such as cutting tool wear, the combined actions of two kinds of wear—chemical (i.e., solubility) and mechanical (i.e., abrasion)—must be considered. The studies have established material characteristics that reflect cutting tool resistance to chemical and mechanical wear. These parameters, considered simultaneously, have been used to predict successfully the wear resistance of a series of Si_3N_4-based cutting tools in steel machining.

While discussions here pertain more specifically to Si_3N_4-TiC composites, findings have a wider implication and will be applicable to other polyphase brittle materials and composites.

ACKNOWLEDGMENTS

The authors gratefully acknowledge the contribution of Mr. E. Geary, M. L. Huckabee, G. Zilberstein, J. G. Baldoni, the late Mr. R. Wentzell, and our

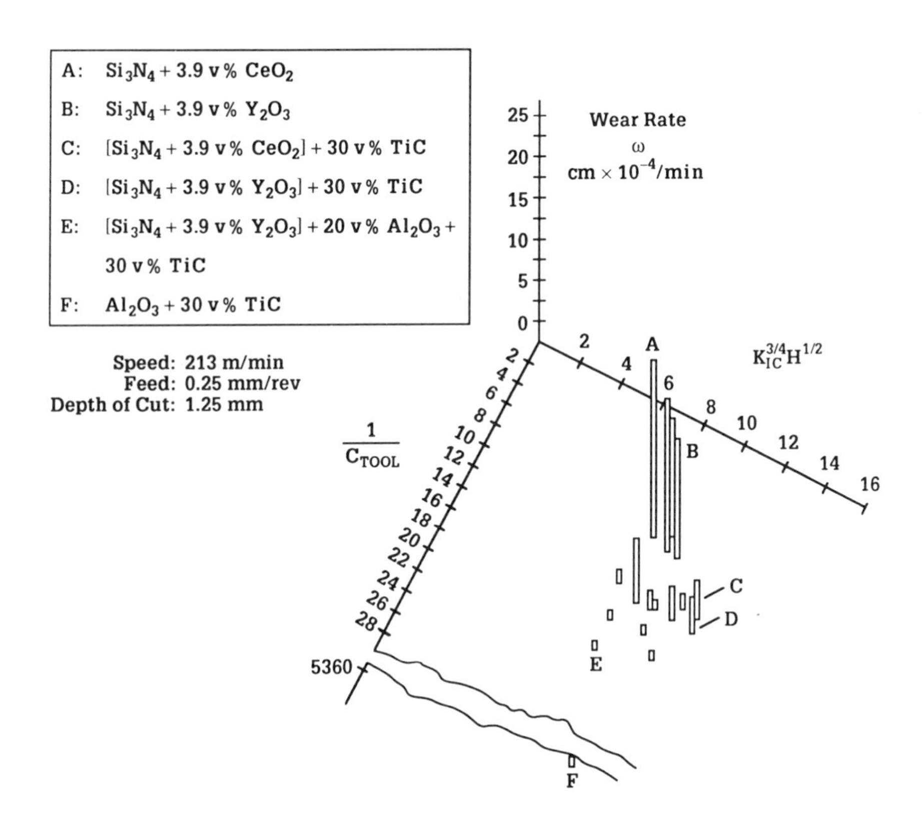

Figure 12 Selected Si_3N_4- and Al_2O_3-based cutting tool wear results plotted as a function of chemical and mechanical wear resistance parameters.

colleagues at GTE Laboratories, without whose assistance this work would not have been possible.

REFERENCES

1. J. G. Baldoni, S. F. Wayne, and S. T. Buljan, Cutting tool materials: Mechanical properties/wear resistance relationships, *ASLE Trans.* *29*(3), 347–352 (1986).
2. S. T. Buljan and S. F. Wayne, Si_3N_4 ceramic composite cutting tools: Material design criteria, *Adv. Ceram. Mater.* *2*(4), 813–816 (1987).
3. S. T. Buljan and S. F. Wayne, Wear and design of ceramic cutting tool material, *Wear*, *133*(2), 1–13 (1989).
4. A. G. Evans and D. B. Marshall, Wear mechanisms in ceramics, in *Fundamentals*

of Friction and Wear of Materials, D. A. Rigney (Ed.), ASM International: Metals Park, OH, 1981, pp. 439–450.
5. A. G. Evans, M. E. Gulden, and M. Rosenblatt, Impact damage in brittle materials in the elastic–plastic response regime, *Proc. R. Soc. London, Ser. A*, *361*, 343–365 (1978).
6. S. T. Buljan and V. K. Sarin, The future of silicon nitride cutting tools, *Carbide Tool J.* May/June 1985, pp. 4–7.
7. S. Wada and N. Watanabe, Solid particle erosion of brittle materials: Part 2. The relationship between erosive wear and α- or β-phase content of hot pressed Si_3N_4, *Yogy-Kyokai-Shi*, *95*(5), 468–471 (1987).
8. S. F. Wayne and S. T. Buljan, The role of thermal shock on tool life of selected ceramic cutting tool materials, *J. Am. Ceram. Soc. 72*(5), 754–760 (1989).
9. F. Heinrich, D. Munz, and G. Ziegler, Influence of microstructural variables on mechanical properties, creep, and thermal shock behavior of reaction-bonded silicon nitride, *Powder Metal. Int. 14*(3), 153–159 (1982).
10. S. K. Samanta and K. Subramanian, Hot-Pressed Si_3N_4 as a high-performance cutting-tool material, *Proceedings of the 13th North American Manufacturing Research Conference*: *Society of Manufacturing Engineers*, Dearborn, MI, 1985, pp. 401–407.
11. D. H. Jack, Ceramic cutting-tool materials, *Mater. Des. 7*(5), 267–273 (1986).
12. R. Wertheim and D. Agranov, Wear behavior of silicon nitride tools as a function of their specific properties, *Ann. CIRP*, *35*(1), 63–66 (1986).
13. K. H. Smith, Whisker-reinforced ceramic composite cutting tools, *Carbide Tool J. 18*(5), 8 (1986).
14. S. T. Buljan and V. K. Sarin, Silicon nitride-based composites, in *Sintered Metal Ceramic Composites*, G. S. Upadhyaya (Ed.), Elsevier: Amsterdam, 1984, pp. 455–468.
15. J. G. Baldoni, S. T. Buljan, and V. K. Sarin, Particulate TiC ceramic matrix composites, *Proceedings of the Institute of Physics Conferences*, Series No. 75, Adam Hilger: Bristol, England, 1986, pp. 427–438.
16. T. N. Tiegs and P. F. Becher, Thermal shock behavior of an alumina–SiC-whisker composite, *J. Am. Ceram. Soc. 70*(5), C-109–C-111 (1987).
17. V. K. Sarin and S.-T. Buljan, Coated ceramic cutting tools, *Proceedings of the ASM Conference on Advanced Metal Removal Techniques*, New Orleans, 1985.
18. A. G. Evans and E. A. Charles, Fracture toughness determination by indentation, *J. Am. Ceram. Soc. 59*(7–8), 179–182 (1976).
19. J. G. Baldoni, S. T. Buljan, and V. K. Sarin, Continuous tool wear monitoring in turning, in *Science of Hard Materials*, R. K. Viswanadham (Ed.), Plenum: New York, 1983.
20. E. M. Trent, *Metal Cutting*, Butterworths: London, 1978.
21. M. C. Shaw, *Metal Cutting Principles*, Massachusetts Institute of Technology: Cambridge, MA, 1968.
22. A. G. Evans and T. R. Wilshaw, Quasi-static solid particle damage in brittle solids. I. Observations, analysis and implications, *Acta Metal. 24*, 939–956 (1976).
23. A. G. Evans, Strength degradation by projectile impacts, *J. Am. Ceram. Soc. 56*(8), (1973).

24. A. G. Evans, Impact damage mechanisms: Solid projectiles, in *Treatise on Materials Science and Technology*, Vol. 16, C. M. Preece (Ed.), 1979, pp. 1–65.
25. M. E. Gulden, Solid particle erosion of Si_3N_4 materials, *Wear*, *69*, 115–129 (1981).
26. M. E. Gulden, Correlation of experimental erosion data with elastic–plastic impact models, *Commun. Am. Ceram. Soc.* March 1981, pp. 59–60.
27. D. B. Marshall, B. R. Lawn, and A. G. Evans, Elastic/plastic indentation damage in ceramics: The lateral crack system, *J. Am. Ceram. Soc.* *65*, 561 (1982).
28. B. R. Lawn, B. J. Hockey, and H. Richter, Indentation analysis: Applications in the strength and wear of brittle materials, *J. Microsc.* *130*, 295–308 (1983).
29. J. E. Ritter, Jr., P. Strzepa, K. Jakus, L. Rosenfeld, and K. J. Buckman, Erosion damage in glass and alumina, *J. Am. Ceram. Soc.* *67*(11), 769–774 (1984).
30. A. W. Ruff and S. M. Wiederhorn, Erosion by solid particle impact, in *Treatise on Materials Science and Technology*, Vol. 16, C. M. Preece (Ed.), 1979, pp. 69–124.
31. J. Larsen-Basse, Wear of hard-metals in rock drilling: A survey of the literature, *Powder Metal.* *16*(31), 1–31 (1973).
32. S. F. Wayne, J. G. Baldoni, and S. T. Buljan, Abrasion and erosion of WC-Co with controlled microstructures, *Tribol. Trans.* *33*(4), (October 1990).
33. S. T. Buljan and V. K. Sarin, Improved productivity through application of silicon nitride cutting tools, *Carbide J.* *14*(3), 40–46 (1982).
34. V. K. Sarin and S. T. Buljan, Advanced silicon nitride-based ceramics for cutting tools, presented at the First International Materials Removal Conference, SME Technical Paper No. MR83-1891-8 (1983).
35. S. T. Buljan and V. K. Sarin, Design and wear resistance of silicon nitride-based composites, *Inst. Phys. Conf. Ser.* *75*, 873–882 (1986).
36. E. D. Whitney, Ceramic cutting tools, *Powder Metal. Int.* *6*(2), 73–76 (1974).
37. E. M. Trent, Some factors affecting wear on cemented carbide tools, *Proc. Inst. Mech. Eng.* *166*, 467 (1952).
38. E. M. Trent, Wear processes on cemented carbide tools used in cutting steel, *Proc. R. Soc. London, Ser. A*, *212*, 467 (1952).
39. B. M. Kramer and N. P. Suh, Tool wear by solution: A quantitative understanding, *J. Eng. Ind.* *102*, 303–309 (1980).
40. B. M. Kramer, A comprehensive tool wear model, *Ann. CIRP*, *35*(1), 67–70 (1986).

IV

TRIBOLOGICAL APPLICATIONS

12

Ceramic Cutting Tools

Ranga Komanduri

Oklahoma State University
Stillwater, Oklahoma

ABSTRACT

Cutting tools made of advanced ceramics have the potential for high speed finish machining as well as for high removal rate machining of difficult-to-machine materials. The raw materials used in these ceramics are abundant, inexpensive, and free from strategic materials. In spite of this, solid or monolithic ceramic tools are currently used only to a limited extent partly as a result of certain limitations of these materials and partly because of the inadequacy of the machine tools used. The advances in ceramic materials and processing technology, the need to use materials that are increasingly more difficult to machine, increasing competition, and rapidly rising manufacturing costs have opened new vistas for ceramics in machining applications. The development of ceramic tool materials can be broadly categorized into three types: monolithic forms, thin coatings, and whisker-reinforced composites. Such a classification, which provides a totally new perspective on ceramic tool materials and broadens their scope considerably, is justified because it is the ceramic addition that makes the tool material more effective [1]. This chapter presents an overview of advanced ceramic tool materials.

I. INTRODUCTION

Ceramics are the newest upcoming class of tool materials with potential for a wide range of high speed finishing operations as well as for high removal rate machining of difficult-to-machine materials. Advances in the ceramic tool materials are partially a spinoff of the high temperature structural ceramics technology developed in the 1970s for automotive gas turbine and other high temperature structural applications. Of the advanced ceramics, namely, alumina, partially stabilized zirconia (PSZ), silicon carbide, and silicon nitride, PSZ in bulk form is not used as a cutting tool in machining because of its limited hardness (Knoop hardness ≈ 1500 kg/mm^2) and neither is silicon carbide, for its brittleness as well as its reactivity with ferrous work materials. However, all these materials as well as others are used in various forms for cutting tool applications, as is elaborated in this chapter. Other ceramic tool materials include the super-abrasives, namely, the single-crystal and polycrystalline diamond and the polycrystalline cubic boron nitride (CBN).

The development of ceramic tool materials for machining applications can be classified into the following three broad categories:

Monolithic forms
Thin coatings
Whisker-reinforced composites

Hot-pressed alumina and alumina–titanium carbide composite, SiAlON, silicon nitride, polycrystalline diamond, cubic boron nitride (CBN), and CBN-TiC come under the first category. Thin coatings (2–10 μm) of alumina, titanium carbide, diamond and titanium nitride, singly or in combination, on substrates of cemented carbides, high speed steels, or ceramics fall under the second category. Alumina and silicon nitride reinforced with silicon carbide whiskers belong to the third category.

II. FACTORS IN THE CHOICE OF CERAMIC TOOL MATERIALS

The following factors must be considered for the appropriateness of a ceramic tool material for machining [2]:

1. Which cutting processes can employ ceramic tools to advantage or, alternatively, which processes are unsuitable for ceramic tools?
2. Which work materials can or cannot be machined by ceramics?
3. What are the machine tool limitations with regard to the use of ceramic tools?
4. What are the strengths and weaknesses of ceramic tools? How can some of the drawbacks be overcome?

These questions are addressed in the following sections. Successful implementation of the solutions can lead to wider use of ceramic cutting tools.

A. Machining Process Limitations

Figure 1 is a pie chart showing the percentage cost distribution in the United States of various machining operations [3]. In grinding, polishing, and cutoff operations, ceramics (alumina, silicon carbide, diamond, and CBN abrasives) are invariably used. Of the other operations, drilling is generally not amenable to the use of ceramic tools because of small diameter holes and consequent lower cutting speeds at which ceramics are not efficient. Of the other operations, turning (≈22%) and milling (≈13%), together account for a substantial portion (≈one-third) of the cutting processes. The strength and fracture toughness limitations of current ceramics somewhat restrict their application to milling, although some recently introduced ceramics are being applied successfully in milling of selected work materials. Rapid advances in ceramic tool technology by way of toughened

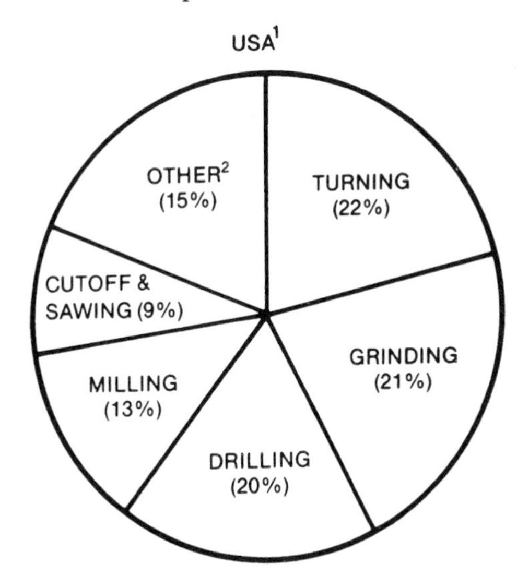

ANNUAL MACHINING COST: $115 BILLION

1 FROM STATISTICS BASED ON 1978 MACHINE TOOL INVENTORY

2 INCLUDES POLISHING AND BUFFING, BORING, GEAR CUTTING, TAPPING, SHAPING, HONING, BROACHING, THREADING, LAPPING AND PLANNING

Figure 1 Pie chart showing approximate percentage cost distribution of various machining processes used in the United States. In 1993, the annual machining costs can be as much as three times this figure or a third of a trillion dollars.

ceramics should, however, enable wider application of ceramics to turning operations.

B. Work Material Limitations

Not all materials can be machined presently with ceramics. Ceramic tools can be used successfully to machine most steels, cast iron, and nickel-based superalloys, even in their hardened condition, as well as many nonferrous alloys and composites. Titanium alloys, aluminum alloys, and some stainless steels cannot be machined with most ceramic tools available today because of the chemical interactions that occur between ceramic tools and these work materials.

In contrast, both high speed steels (HSS) and cemented carbides (e.g., those based on WC and TiC) are capable of machining a wide range of work materials. This is because both classes of materials were developed to exhibit a range of hardness and toughness for a variety of applications. They are, therefore, very versatile.

Unless ceramic tools can also be developed to broaden the field of applications, their use will be limited to niche areas. The challenge lies in the ability to develop ceramic tool materials for machining a wide range of work materials, at cutting speeds significantly higher than those currently used. It should, however, be realized that economics and machinability needs dictate the final choice of tool material for a given application.

C. Machine Tool Limitations

Ceramic tools, being brittle (low fracture toughness), require rigid, high precision machine tools for best performance. This is because even the smallest amplitude of vibration of the machine tool system can lead to chipping and failure of the tool, thus abruptly terminating its useful life. Many of the older machine tools were not designed for ceramic tool applications. As a result, they are inadequate in terms of cutting speed capability, power, precision, and rigidity requirements for this application. A new generation of machine tools with the foregoing requirements are required to take full advantage of ceramic tools.

D. Potential Strengths and Limitations of Ceramic Tools

Ceramic tools have great potential for the following applications:

High speed [> 5 m/s or 1000 ft/min] finish machining of many hard and difficult-to-machine materials

High removal rate [up to several thousand cubic centimeters (several hundred cubic inches) per minute] machining of some of the materials above with high horsepower, high rigidity machine tools using low to conventional cutting speeds

Machining of extremely difficult-to-machine materials such as hardened steels, alloy steels, and white cast irons, and nickel-based superalloys
Producing superior finish and accuracy of parts

This potential exists because ceramics have the following characteristics:

High room temperature and elevated temperature hardness
High strength in compression
Good chemical stability up to high temperatures

An additional advantage of ceramics is that the raw materials used in their synthesis include the elements Si, Al, O, N, and C, which are abundant in nature, inexpensive, and free from dependence on strategic materials. In spite of these advantages, monolithic or solid ceramic tools are currently used in industry only to a limited extent. This is partly because of certain limitations of current ceramic tools (e.g., fracture toughness, strength, and hardness, chemical stability, and reactivity with work materials), and partly because of the lack of adequate machine tools (e.g., rigidity, precision, and power). Also, there is a lack of training on the use and application of ceramic tools and on their proper industrial implementation. The inadequate strength and inconsistency in performance of some of the earlier ceramics did not help in promoting their use, especially when they were used in conjunction with low power, low speed, low pressure, and less rigid machine tools.

The main limitations of the ceramics as compared to cemented carbides are:

Lower transverse rupture strength (TRS).
Lower edge strength.
Lower fracture toughness.
Greater proneness to chipping, both micro- and macro-, and even gross chipping.
High processing and finishing costs of the ceramic cutting tools, even though raw materials are relatively inexpensive. If these costs can be reduced by technological advances, less expensive ceramic tools would result.
Lack of amenability to forming complex shapes in some cases.
Lack of consistency of the product and unpredictability in performance.
Proneness to notching at the depth-of-cut line and at the point at which the tool edge leaves the finish surface.
Abrupt ending of tool life without warning, compared to gradual tool wear at the end of tool life in the case of carbide tools.
More limited scope for machining of a variety of work materials or for use in different machining operations.
Competition among ceramic tools and consequent replacement of one for another.
Need for high power, high speed, and rigid machine tools.

While the cutting tool is a vital element of the machining system, there are other elements that affect the overall performance [4]. These elements include:

Work material characteristics (chemical and metallurgical state)
Part characteristics (geometry, accuracy, finish, and surface integrity requirements)
Machine tool characteristics (adequate rigidity with high horsepower and high speed capabilities)
Support systems (operator's ability, sensors, controls, method of lubrication, and chip control)
Cutting conditions and tool geometry

In the case of ceramics, improved machine tools (i.e., high speed, high power, and high rigidity devices) and the development of tougher ceramics, together with sensors to monitor the tool condition and give warning of tool failure, are significant areas of concern for increasing the efficient use of ceramics. Implementation of the former and research on the latter are crucial for extensive application of ceramic tools.

III. HISTORICAL BACKGROUND

The use of ceramic tools dates back to the dawn of civilization when man used stone as a tool to aid in achieving his basic needs of food, shelter, and protection [5]. Although ceramic tools were considered for certain machining applications as early as 1905, strength under the conditions of cutting was inadequate and the performance inconsistent [6]. In the mid-1950s, ceramic tools were reintroduced for high speed machining of gray cast iron for the automobile industry and for slow speed, high removal rate machining of extremely hard (and difficult-to-machine) cast or forged steel rolls in the steel industry. These materials were typically fine grain (<5 μm) alumina-based materials, either almost pure or alloyed with suboxides of titanium or chromium to form solid solutions, and containing small amounts of magnesia as a sintering aid and as grain growth inhibitor. For example, an alumina-TiO (TiO constitutes about 10% of the material) ceramic was made by cold-pressing and liquid phase sintering [7]. It is characterized by fine grain size (>3–5 μm) and uniform microstructure. A Rockwell hardness of 93–94 A and a transverse rupture strength exceeding 550 MPa($\approx$ 80 ksi) were achieved. However, limited strength and inconsistency made the applications of these materials extremely limited. The main success stories include the high removal rate machining of chill cast iron rolls on an extremely rigid, high power (up to 450 kW or 600 hp), high precision Binns superlathe (Fig. 2) [8], and the selective high speed finishing application on carbon steels and cast iron for automotive applications.

Several factors have recently rejuvenated interest in the development and

Figure 2 Photograph of an extremely rigid, high precision, high horsepower Binns superlathe. (Courtesy of J. Binns, Sr.)

application of ceramic cutting tools. These include the introduction of new ceramic materials, advances in ceramic processing technology, new opportunities (higher cutting speeds and removal rates) for reducing manufacturing costs, the need to use materials that are increasingly difficult to machine, advances in machining science and technology, rapidly rising manufacturing costs, and global competition.

IV. ADVANCED CERAMIC TOOL MATERIALS

Advanced ceramic tool materials, broadly classified into three categories, are discussed in this section.

A. Monolithic Forms

Over the past two decades, the following monolithic ceramic tool materials were developed:

1. Alumina-based ceramics (hot-pressed alumina and alumina-TiC)
2. Ultrahard ceramic tools (polycrystalline diamond or CBN and CBN-TiC)
3. Nitrogen-containing ceramics (silicon nitride and SiAlON)

1. Alumina-Based Ceramics

Advances in ceramic processing technology, such as hot-pressing and hot isostatic pressing (HIP) and the development of pure, fine-grained (micrometer size) ceramic powders led to the introduction of tougher [TRS ≈750 MPa (110 ksi)] ceramic tools in the early 1970s. The first tool material introduced was the hot-pressed, fully dense (>99.5%) fine-grained alumina, which is significantly tougher [*TRS* >685 MPa (100 ksi)] than the alumina-TiO tools [TRS ≈550 MPa (80 ksi)] [9]. This was soon followed by another hot-pressed ceramic tool material, alumina-TiC (alumina with 30% by volume of TiC and small amounts of yttria as the sintering aid) with enhanced mechanical and thermal shock resistance (Fig. 3). These two materials are used for high speed finish machining of steels and high temperature superalloys.

2. Ultrahard Ceramic Tools

The synthesis of diamond, the hardest material (8000 kg/mm^2) known, and the discovery of the cubic form of boron nitride (CBN), the second hardest (≈4700 kg/mm^2) material, by a high temperature and a high pressure process and in the presence of catalyst-solvents, led to the development of ultrahard ceramic tools, namely polycrystalline diamond and polycrystalline CBN [10].

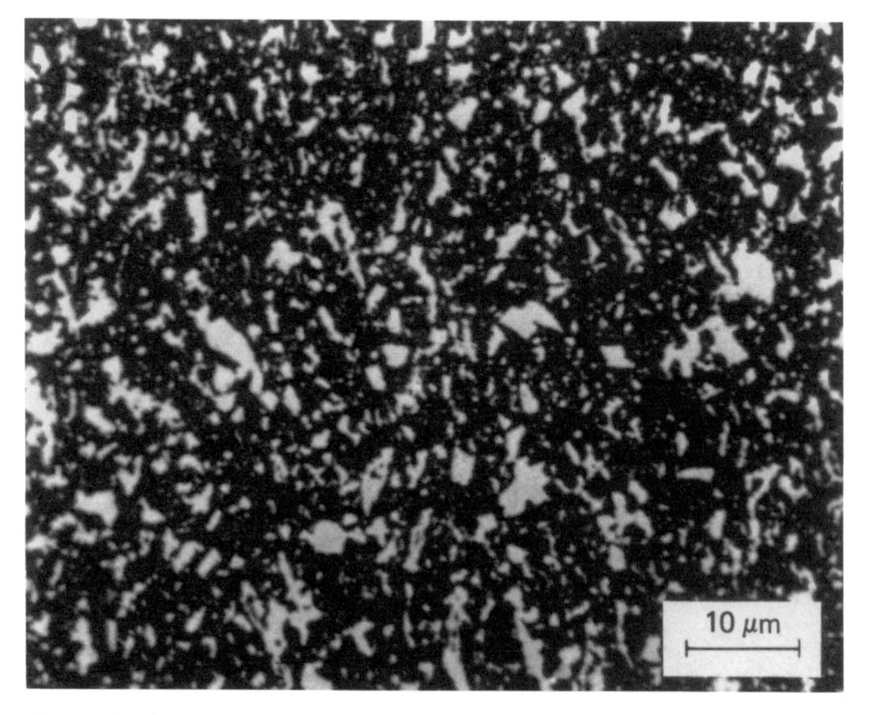

Figure 3 Optical micrograph of a hot-pressed alumina TiC ceramic tool material. (Courtesy of Carboloy.)

The polycrystalline diamond/CBN tools consist of a thin layer (0.5–1.5 mm) of fine-grained (1–30 μm) diamond/CBN powder sintered together and metallurgically bonded to a cemented carbide substrate (Fig. 4). The cemented carbide substrate provides the necessary elastic support for the hard and brittle diamond/CBN layer. At the appropriate conditions of pressure [≈5 MPa (50 kbar)] and temperature (≈1500C), full densification, extensive diamond-to-diamond bonding, and bonding between diamond/CBN and the tungsten carbide substrate takes place (Fig. 5). In addition, small amounts of binder phase, predominantly cobalt, present at the grain boundaries, infiltrate throughout the body. Because of this, both diamond and CBN tools can be thermally affected at temperatures above 700C. The resistance to thermal degradation can, however, be improved significantly by using aqua regia to leach the infiltrant [11]. The issue of thermal degradation assumes importance when the cutting temperatures are high, as in the machining of hardened steels, alloyed cast iron, and nickel–iron based superalloys at higher speeds.

Figure 4 Photomacrograph of a sintered polycrystalline diamond tool showing a dark, thin diamond layer (≈0.5 mm) on top, metallurgically bonded to a lighter colored cemented tungsten carbide substrate. Sintered polycrystalline CBN tools look similar to this tool. (Courtesy of General Electric Co.)

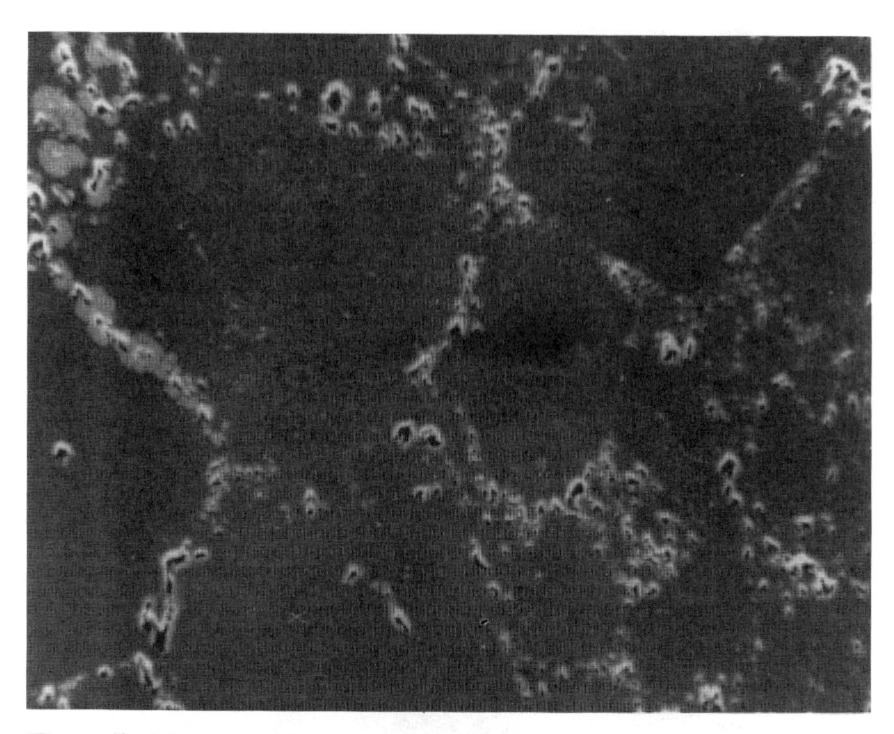

Figure 5 Micrograph of a polished polycrystalline sintered diamond tool showing diamond-to-diamond bonding, as well as the presence of binder phase, impurities, and voids at the grain boundaries. (Courtesy of General Electric Co.)

Polycrystalline diamond tools are used for machining of polymer and glass–epoxy composites, aluminum–silicon alloys, cast iron, and other nonferrous and nonmetallic materials. Because of their high reactivity with ferrous alloys, diamond tools are not recommended for machining carbon steels. Polycrystalline CBN tools, on the other hand, are less reactive with ferrous alloys. They are recommended for machining of hardened (>55 Rockwell C) steels, alloyed cast iron, and nickel–iron based superalloys.

To increase the toughness and to extend the fields of application of polycrystalline diamond/CBN tools, diamond or CBN is alloyed (with TiB_2, TiC, TiN, etc.), although this addition is at the expense of hardness [12,13]. Polycrystalline CBN tools mixed with TiC can be used for machining medium hard (≈40–50 Rockwell C) steels [14].

Sintered polycrystalline diamond or CBN tools can be fabricated in an assortment of shapes (squares, rounds, triangles, and sectors of a circle of different included angles) and sizes from round blanks. These tools are an order of magnitude or more expensive than the conventional cemented carbides or ceramic tools because of the high cost of both the processing techniques, which involve high temperature and high pressure sintering, final shaping, and the finishing methods. In spite of the higher cost, these tools are used for specific applications because they increase productivity, provide long tool life, and are economical on the basis of overall cost per part.

3. Nitrogen-Containing Ceramics

Silicon nitride and SiAlON are among the principal nitrogen-containing ceramics. Dense silicon nitride stabilized by yttria (or magnesia) is one of the more attractive ceramic tool materials recently introduced for the high speed machining [>25 m/s (5000 ft/min)] and/or high removal rate machining [several cubic centimeters per second thousand (several hundred to a thousand cubic inches per minute)] of gray cast iron for automotive and other applications, thus opening new opportunities for ceramic tool materials [15–17]. New machine tools with high power (up to 375 kW or 500 hp), high rpm [to provide cutting speeds of up to 50 m/s (10,000 ft/min)], high rigidity, and high precision are being built specifically for this purpose. Unfortunately, silicon nitride is not suitable for machining other ferrous materials or high temperature superalloys, thus limiting its application to the machining of cast iron.

Fully dense silicon nitride has a unique combination of properties, namely high strength and toughness, moderate hardness, good wear resistance, and low thermal expansion coefficient and consequent excellent thermal shock resistance. It is a covalently bonded material with a small self-diffusivity, hence cannot be easily densified by sintering without sintering aids. Dense material of β-Si_3N_4 is fabricated by hot-pressing α-Si_3N_4 powder with a suitable densification additive (e.g., yttria or MgO) at 30 MPa in a graphite die at 1700–1800°C. The Si_3N_4 powder is generally covered with a thin layer of silica. The additive reacts with this silica and a small amount of Si_3N_4 at the high hot-pressing temperature to form an oxynitride liquid in which α-Si_3N_4 dissolves and from which β-Si_3N_4 is precipitated.

Densification occurs by liquid phase mechanisms; but, unfortunately, the liquid that is necessary for densification cools to give secondary crystalline or vitreous phases, which affect the properties of the hot-pressed material at the glass-softening temperature (900°C). For example, with the yttria-stabilized β-Si_3N_4, yttrium–silicon oxynitride is formed at elevated temperatures (>900°C), with a pronounced change in specific volume that causes microcracking. It is somewhat difficult, if not impossible, to produce dense, pure, β-Si_3N_4 without any additive. Yttria- or MgO-stabilized silicon nitride material, produced by hot-pressing and subsequently fabricated into cutting tools by dicing and shaping with diamond grinding, is rather expensive. Need, therefore, exists for a less expensive near-net-shape processing technique for this material.

Since additives degrade the properties of Si_3N_4 at elevated temperatures, the alternative is to alloy Si_3N_4 with other inorganic compounds. This led to the development of another exciting, tough ceramic tool material, SiAlON, for machining of nickel–iron based superalloys [18–21]. This material is yttria-stabilized SiAlON, which is isostructural with, β-Si_3N_4. Because of the similarity of crystal structure, β′-SiAlON has physical and mechanical properties similar to those of silicon nitride; because of its composition, its chemical properties approach those of alumina [20].

To produce, β′-SiAlON, a mixture of alumina (≈13%), silicon nitride (≈77%), yttria (≈10%), and aluminum nitride is used as the starting material. Since this mixture produces, during sintering, a larger volume of lower viscosity liquid than in the synthesis of yttria-stabilized silicon nitride and its surface silica under similar conditions, β′-SiAlON can be fully densified by pressureless sintering. The powder mix for β′-SiAlON is first ball-milled, then preformed by cold isostatic pressing, and subsequently sintered at a maximum temperature of ≈1800°C under isothermal conditions for ≈1 hour before it is allowed to cool slowly. The microstructure of β′-SiAlON consists of β′ grains cemented by a glassy phase (Fig. 6) [22].

Figure 6 Micrograph of a SiAlON tool showing β′-SiAlON grains and the presence of glassy phase at the grain boundaries. (Courtesy of Dr. S. K. Bhattacharyya.)

SiAlON tools were originally developed by Joseph Lucas Industries Ltd. of the U.K. The material is marketed under licence by Kennametal, Inc., in the United States and by Sandvik in Sweden. It is much tougher than alumina-TiC and is used for rough machining of nickel–iron based superalloys. Further developments in microstructure and composition of SiAlON are likely to yield an even tougher material consisting of, β′-SiAlON and an intergranular phase of yttrium-aluminum-garnet (YAG) without any intergranular glassy phase. Similarly, addition of TiC is likely to yield a much harder and tougher SiAlON with enhanced mechanical and thermal shock resistance. The introduction of SiAlON tool material was for a time very exciting, since it led major tool manufacturers to concentrate on developing a similar material until another tool material, SiC whisker-reinforced alumina, a far tougher material, was introduced.

B. Thin Ceramic Coatings

An analysis of the cutting process indicates that the characteristics of the tool at or near the surface should be different from those of the bulk. To be abrasive resistant, the surface must be hard; to be chemically wear resistant, it must shield the constituents of the tool and the work material from each other under the conditions of cutting, especially at high cutting speeds. The bulk of the material, by contrast, should be tough enough to withstand high temperature plastic deformation and to resist breakage under the conditions of cutting. Since the surface and bulk requirements are markedly different, it is logical to consider engineered tool materials. This was the basis for the development of thin (2–10 μm) ceramic coatings on cemented carbide tools in the late 1960s and on HSS in the late 1970s, which are now used very extensively.

In cemented tungsten carbide cutting tools, cobalt metal is used as the binder. The cutting speed capability of these tools is limited primarily by the binder phase. Reducing the volume fraction of the binder phase increases the refractoriness of the tool material, at the expense of strength. It is somewhat difficult to develop a bulk material with these diverse characteristics. If, however, a binderless ceramic coating were deposited on a tough cemented carbide tool, such that the coating were thin enough and strongly bonded metallurgically to the substrate, such a coating might not affect the strength significantly; this property, in turn, could result in longer tool life and/or higher cutting speed capability. This idea has led to the development of single and multiple, hard, refractory ceramic coatings on cemented carbide tools by the chemical vapor deposition (CVD) technique [23].

To be effective, such coatings should be:

Hard
Refractory
Chemically stable

Chemically inert, to shield the constituents of the tool and the work material from interacting chemically under cutting conditions
Binder-free
Of fine grain size, with no porosity
Metallurgically bonded to the substrate, with a graded interface to match the properties of the coating and the substrate
Thick enough to prolong tool life, but thin enough to prevent brittleness
Free from the tendency of the chip to adhere (or seize) to the tool face
Easy to deposit in bulk quantities
Inexpensive

Several hard refractory ceramic coatings—including single coatings of TiC, Al_2O_3, TiN etc.; double coatings of Al_2O_3 and TiC, TiN and Al_2O_3, or TiC; and triple coatings of TiN, Al_2O_3, and TiC—were developed by the CVD process. To deposit a coating of TiC, the tools are heated to 1000C, and $TiCl_4$ and methane (CH_4) are reacted in hydrogen at a pressure of approximately one atmosphere. The hard reaction product, TiC, deposits on the tool and forms a strongly adherant metallurgical bond with the substrate at high temperatures. A coating thickness of 5 μm is found to be optimum from the standpoint of best performance and requires about 8 hours of deposition time.

Multiple coatings are developed with the following objectives:

To prolong the life of the tool by building up the thickness without encountering the deleterious effects of thick, hard coatings, such as delamination and chipping.
To form a stronger metallurgical bond with a graded interface.
To take advantage of the desirable features of each coating material without causing delamination.

Figure 7 is a micrograph of a representative multiple coating of Al_2O_3 and TiC on a cemented carbide tool. The microprobe traces of Al and Ti show clearly the graded interface between the substrate and TiC as well as between TiC and Al_2O_3. Currently, titanium nitride coating is deposited as the upper layer on most coated tools because it minimizes metal buildup and reduces friction.

At the cutting speeds at which HSS tools (e.g., drills, end mills) operate, metal buildup or periodic seizure between the chip and the tool is the main problem limiting tool life and productivity. To overcome this problem and to reduce friction, a titanium nitride ceramic coating has been developed successfully for HSS tools. However, the CVD process used for coating cemented carbide tools cannot be used for this material because the required substrate temperature of 1000C is too high, and results in alterations of the metallurgical structure of HSS. Hence, only coatings and processes that require heating below the HSS transformation temperature can be used. A coating of titanium nitride

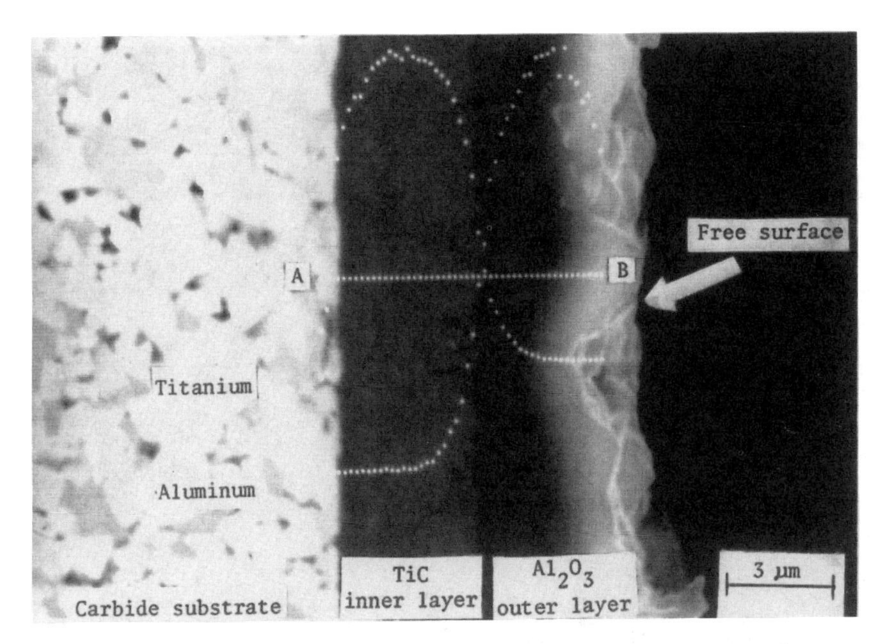

Figure 7 Micrograph of cross section of Al_2O_3-TiC coating on a cemented tungsten carbide substrate. (Courtesy of General Electric Co.)

by the physical vapor deposition (PVD) technique at low substrate temperature (≈400C) is increasingly used for these tools. The attractive golden color of the coating is an added aesthetic feature.

To take full advantage of the coating potential, substrate materials are being carefully matched with the coatings or appropriately altered to optimize properties, resulting in significant gains in productivity. Also, for strong bonding between the coating and the substrate, there should be good chemical and metallurgical compatibility between them. Since the mode of wear is different on the rake face and on the clearance face, coated-tool technology is advancing in the direction of selective compositions and/or modifications of the substrate in these areas to further prolong tool life and to make the coated tool more versatile for special or general purpose applications. An example in this direction is the recent development of a multi-layer-coated (TiC-TiCN-TiN or TiC-Al_2O_3-TiN) cemented carbide tool that has a cobalt-enriched layer (25 μm deep) on the rake face to provide superior edge strength, and a cobalt-depleted layer in the clearance face for high abrasion resistance and high temperature deformation resistance (Fig. 8) [24].

Ceramic coatings are also being developed on ceramic substrates mainly to limit chemical interactions between the tool and the work material. For example,

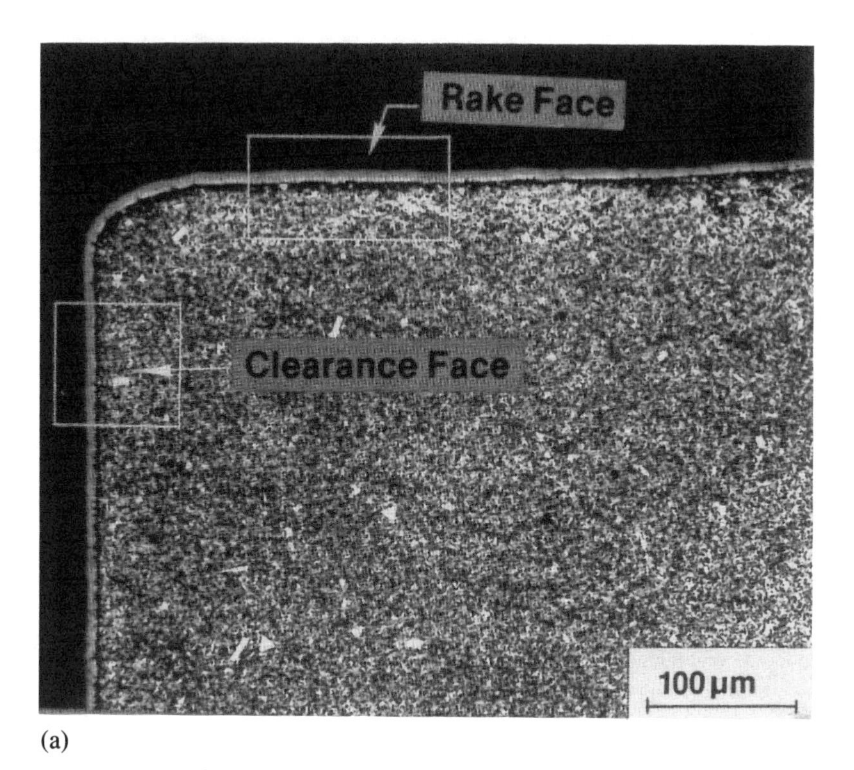

(a)

Figure 8 Micrograph of a multicoating of TiN-Ti (C, N)-TiC on an engineered cemented carbide substrate (note that the top layer is TiN). (a) Low magnification micrograph showing the areas around the rake face and the clearance face of the tool. (b) Micrograph at higher magnification of an area around the rake face, showing a thin cobalt-enriched straight WC layer near the rake face for increased toughness and edge strength. (c) Micrograph at higher magnification of an area around the clearance face showing a thin layer of cobalt-depleted multicarbide near the clearance face for increased wear resistance and high temperature deformation-resistance. (Courtesy of Dr. A. T. Santanam.)

silicon nitride tools are used successfully for high speed machining of cast iron. However, they react significantly with steels, hence cannot be used for this application as such. To take advantage of the high temperature deformation resistance of this material and to minimize chemical interactions when machining steels at high speeds, multiple coatings of TiC-TiN or Al_2O_3-TiC on silicon nitride and sialon substrates were developed, similar to those for cemented carbides [25–28]. Figure 9 is a cross section of multiple coatings of TiN and TiC on silicon nitride tool material. However, the extent to which such coated

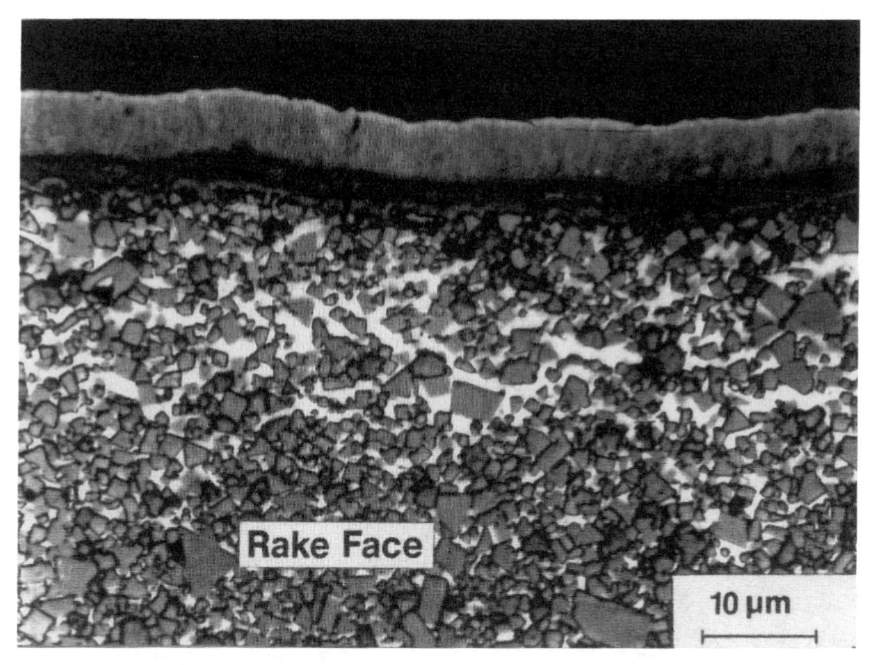

(b)

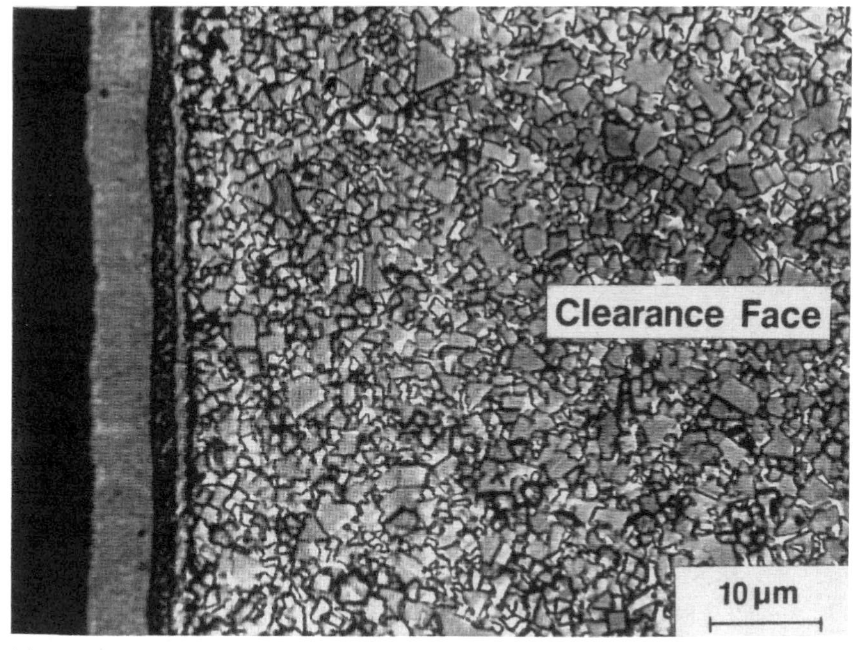

(c)

Figure 9 Micrograph of cross section of Al_2O_3-TiC coating on a silicon nitride based ceramic tool. (Courtesy of Dr. V. Sarin.)

ceramic tools will be used as compared to other competing materials for high speed machining of steels depends on the need and the economics of machining.

Recently, thick (1–1.5 mm) stand-alone layers of polycrystalline diamond grown by low pressure CVD techniques, such as microwave-assisted CVD and hot-filament CVD, are brazed onto the cemented tungsten carbide as an alternative to the polycrystalline diamond tools bonded on to the cemented tungsten carbide by the high pressure/high temperature techniques. These tools are still undergoing tests. Alternately, a thin diamond coating (up to 20 μm) on cutting tool is under development.

C. Fiber-Reinforced Composites

A fiber-reinforced ceramic composite material possessing improved fracture toughness (K_{IC} ≈9 MPa·$M^{1/2}$) was introduced in the mid-1980s as a cutting tool for machining of nickel–iron based superalloys used in aircraft engines [29,30]. This material is silicon carbide (SiC) whisker-reinforced alumina (Fig. 10). Single-crystal whiskers of SiC possess very high tensile strength (≈7 GPa). SiC

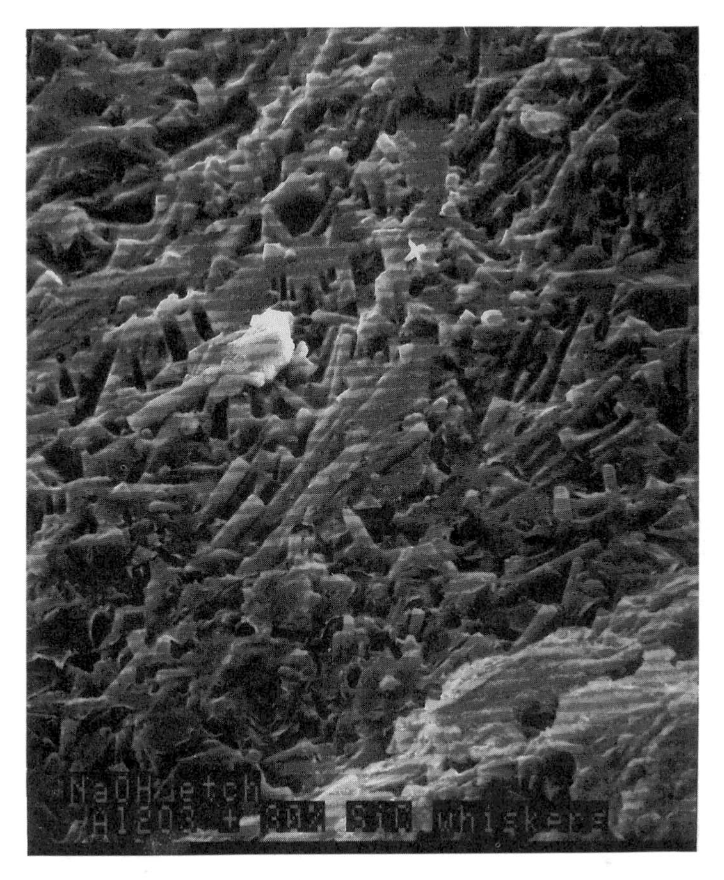

Figure 10 Fractograph of an SiC whisker-reinforced alumina tool.

also has a higher thermal conductivity and coefficient of thermal expansion than alumina. Consequently, the composite exhibits a higher strength and thermal shock resistance.

In the preparation of this composite material, SiC whiskers (0.5–1 μm in diameter and 10–80 μm long), about 20 vol %, are mixed homogeneously with micrometer-sized fine powder of alumina. The mixture is then hot-pressed to above 99% of the theoretical density at a pressure in the range of 28–70 MPa and temperature in the range of 1600–1950°C for pressing times varying from about 0.75 to 2.5 hours. The fracture toughness of this material is by far the highest among ceramic cutting tools, nearly twice that of its closest ceramics, Si_3N_4 and sialon. Although details of the micromechanisms for improved fracture toughness of this ceramic composite tool material are not clearly established, a plausible mechanism for the toughness is given in the following paragraph based on current knowledge of the behavior of composite materials in general [31].

As a crack propagates through the ceramic matrix, it is deflected by the whiskers; and the SiC whiskers, because of their inherently high strength (7 GPa), remain essentially intact. Consequently, whisker pullout occurs as a result of the separation of the matrix from the whiskers. Interfacial shear stress resisting the whisker pullout absorbs a substantial amount of this fracture energy and inhibits crack propagation. For improved fracture toughness, a strong metallurgical bond between the fibers and the matrix is not desirable, for it causes the whiskers to fail along with the matrix. This is the case with SiC whisker-reinforced alumina composites because the bond between the matrix and the fibers is not particularly strong. The bonding is weak because alumina and SiC are not metallurgically compatible. For best results, the fracture energy of the interface should not exceed one-tenth of the fracture energy of the matrix.

Although Si_3N_4 is a tough ceramic, its fracture toughness is not sufficiently high for interrupted cutting, such as milling of cast iron. So, using the foregoing approach, a tougher SiC whisker-reinforced silicon nitride was developed for high speed milling applications. When whiskers of other refractory materials become available, new tool materials may be developed with these whiskers.

The SiC whisker-reinforced alumina composite has opened new vistas for tool material development. One can consider this material to be a model engineered material and develop other composites for machining different work materials and for different applications. However, to accomplish this it is necessary to thoroughly analyze this material and its performance. For example, why is this tool not suitable for machining ferrous materials, such as steel or cast iron? What is the role of SiC whiskers? Can they be replaced with whiskers of other materials such as TiC, Al_2O_3, ZrO_2, Si_3N_4, and sialon? What are the optimum size (diameter and aspect ratio), volume fraction, and geometric distribution of the fibers? Can ceramic coatings be applied to shield the ingredients of the tool and work material from each other? Can one design a composite ceramic tool material based on the knowledge of the properties of the various ingredients of the tool and work materials as well as the cutting conditions for a given application?

V. CONCLUDING REMARKS

Significant innovations and technological advances were made in the last four decades in the adaptation of ceramic materials for many cutting tool applications. Thin ceramic coatings, synthesis of diamond and CBN, low pressure CVD diamond coatings, whisker-reinforced ceramic composites, processing techniques that produce ceramics close to the theoretical density, and toughened ceramics presented in this chapter are examples. Basically, many advanced ceramics are hard enough to machine a range of work materials with the possible exception of hard ceramics themselves. To extend the field of applications and to open up new opportunities, ceramic cutting tools must be developed even further: future applications are bound to be more challenging than at present. Some of the future directions in the development of ceramic tools include the following:

1. Ceramics with considerable increase in transverse rupture strength and fracture toughness than currently available.
2. A class of ceramics with a range of hardness and improved strength and fracture toughness characteristics, similar to the classes of HSS and cemented carbide tools.
3. Ceramics that are relatively inert chemically for machining a range of work materials at higher cutting speeds: for example, the development of a cutting tool material for high speed machining of titanium alloys. It may be pointed out that this is not an esoteric problem, since titanium is very reactive chemically with most tool materials at elevated temperatures. However, in the event that such a material is developed, it may revolutionize the way we make cutting tools in general, for this material may form the basis for the development of new materials.
4. Polycrystalline diamond and CBN coatings on cutting tools using the low pressure CVD techniques.
5. New ceramic tool materials developed on the basis of fundamental understanding on the nature and performance of current toughened ceramics, whisker-reinforced ceramics, etc.
6. New engineered ceramics with hierarchical structures, nanometer-sized grains, amorphous structures, and artificially layered structures.

ACKNOWLEDGMENTS

The author thanks Drs. A. T. Santanam, V. Sarin, and S. K. Bhattacharyya for providing micrographs presented in this chapter. Funds from MOST Chair, Eminent Scholars Program, the Society of Manufacturing Engineers (SME) Foundation grant, and Oklahoma Center for Integrated Design and Manufacturing (OCIDM) facilitated the preparation of this chapter.

REFERENCES

1. R. Komanduri, Advanced ceramic tool materials for machining, Parts 1 and 2, *Sadhana*, *13*, July 1988, pp. 119–137.
2. R. Komanduri and J. Desai, Tool materials, in *Encyclopedia of Chemical Techology*, Vol. 23, 3rd ed., Wiley: New York, 1983, pp. 273–309.
3. *Machinability Data Handbook*, Vol. 2, 3rd ed., Metcut Research Associates: Cincinnati, OH, 1980.

4. R. Komanduri, in *Encyclopedia of Materials Science and Engineering*, M. B. Bever (Ed.), Pergamon Press: Elmsford, NY, 1986, pp. 1003–1012.
5. N. P. Suh, *Wear*, 62, 1–20 (1980).
6. A. G. King, and W. M. Wheildon, *Ceramics in Machining Processes*, Academic Press: Orlando, FL, 1966.
7. E. W. Goliber, Ceramic and cermet compositions containing TiO, *Proceedings of the 66th Annual Meeting of the American Ceramic Society*, American Ceramic Society: Westerville, OH, 1960.
8. J. Binns, Cutting performance on rough turning and hogging cuts, ASTME (now SME) Paper No. 633, (1963), pp. 1–15.
9. R. L. Hatschek, *Am. Mach.* 1981, pp. 165–176.
10. R. H. Wentorf, Jr., R. C. DeVries, and F. P. Bundy, *Science*, *208*, 873–880 (1980).
11. H. P. Bovenkerk and P. D. Gigl, The temperature resistant abrasive compact and method for making same, U.S. Patent No. 4,224,380 (1980).
12. A. Hara and S. Yazu, Method of producing a sintered diamond compact, U.S. Patent No. 4,231,762 (1980).
13. A. Hara and S. Yazu, Sintered compact for a machining tool and a method of producing the compact, U.S. Patent No. 4,334,928 (1982).
14. N. Tabuchi, A. Hara, S. Yazu, Y. Kono, K. Asia, K. Tsuji, S. Nakaltani, T. Uchida, and Y. Mori *Sumitomo Electr. Tech. Rev. 18*, 57 (1978).
15. A. Ezia, S. K. Samanta, and K. Subramanian, Ceramic cutting tool formed from Si_3N_4-Y_2O_3-SiO_2 and method of making, U.S. Patent No. 4,401,617 (1983).
16. A. Ezia, S. K. Samanta, and K. Subramanian, Ceramic cutting tool formed from Si_3N_4-Y_2O_3-SiO_2 and method of making, U.S. Patent No. 4,401,238 (1984).
17. S. K. Samanta and K. Subramanian, Hot-pressed SiN as a high performance cutting tool material, *Proceedings of the North American Manufacturing Research Conference*, University of California, Berkeley, 1985.
18. Y. Oyama and O. Kamigaito, *Jpn. J. Appl. Phys. 10*, 1637 (1971).
19. K. H. Jack and W. I. Wilson, *Nature Phys. Sci. 238*, 28–29 (1972).
20. S. K. Bhattacharyya and A. Jawaid, *Int. J. Prod. Res. 19*, 589–594 (1981).
21. K. H. Jack, *Met. Technol. 9*, 297–301 (1982).
22. S. K. Bhattacharyya, A. Jawaid, and J. Wallbank, *Proceedings of the ASME High-Speed Machining Symposium PED,* Vol. 12, R. Komanduri, K. Subramanian, and B. F. von Turkovich (Eds.), American Society of Mechanical Engineers: New York, 1984.
23. S. J. Whalen, Vapor deposition of titanium carbide, ASTME (now SME) Technical Paper No. 690 (1971).
24. B. J. Nemeth, A. T. Santhanam, and G. P. Grab, *Proceedings of the 10th Plansee Seminar on Trends in Refractory Metals and Special Materials and their Technology*, H. Mortner (Ed.), Metallwork Plansee: Ruettle, Austria, 1981.
25. V. K. Sarin and S.-J. Buljan, Nitride coated silicon nitride cutting tools, U.S. Patent No. 4,406,668 (1983).
26. V. K. Sarin and S.-J. Buljan, Carbo-nitride coated composite modified silicon aluminum oxynitride cutting tools, U.S. Patent No. 4,406,669 (1983).
27. V. K. Sarin and S.-J. Buljan, Nitride coated composite modified aluminum oxynitride cutting tools, U.S. Patent No. 4,406,670 (1983).

28. V. K. Sarin and S.-J. Buljan, Alumina coated silicon nitride cutting tools, U.S. Patent No. 4,440,547 (1984).
29. G. C. Wei, Silicon carbide whisker-reinforced ceramic composites and method for making same, U.S. Patent No. 4,543,345 (1985).
30. H. K. Smith *Carbide Tool J. 18(5)*, 8–10 (1986).
31. C. F. Lewis, *Mater. Eng.* July 1986, pp. 31–35.

13

Ceramic Materials for Rolling Element Bearing Applications

R. Nathan Katz

Army Research Laboratory
Watertown, Massachusetts

ABSTRACT

Advanced ceramics, silicon nitride ceramics in particular, offer the potential for significant performance increases in rolling element bearings. This chapter reviews the properties of candidate ceramic materials relevant to rolling element bearing performance. The behavior of fully dense silicon nitride for a variety of bearing performance factors including rolling contact fatigue, lubrication behavior, and failure mode is reviewed. Processing advances are briefly described, and experience in actual applications is discussed.

I. INTRODUCTION

Rolling element bearings for use at very high speeds or in extreme environments are performance limited by the density, strength, corrosion resistance, wear resistance, and other constraints of traditional bearing materials. Modern high performance ceramics can provide the combination of material properties required to overcome these limitations. The present line of development of ceramic rolling element bearings was initiated in the late 1960s, when the availability of fully dense hot-pressed silicon nitride coincided with a growing awareness that new materials would be required to meet the evermore stringent demands imposed on aircraft gas turbine bearings [1]. Accordingly, in the 1970–1980 time period a variety of feasibility demonstrations and basic studies established that silicon nitride rolling elements could significantly increase bearing performance. However, the collective results of these efforts also pointed out that the best fully dense silicon nitride available at that time had insufficient reliability and reproducibility to meet high performance bearing requirements. Thus, during the 1980–1985 time period, efforts at various laboratories around the world focused on process modifications and controls to increase the reliability of silicon nitride rolling elements and races [2]. Since 1985 there have been considerable improvements in silicon nitride bearing materials and processes, which have led to substantial increases in bearing reliability. These have, in turn, led to commercial application of silicon nitride rolling element bearings and a growing experience base. This chapter reviews these developments.

II. POTENTIAL CERAMIC ROLLING ELEMENT BEARING MATERIALS

The properties that are desired of an advanced material for rolling element bearing application are listed in Table 1. Let us briefly review the importance of each property listed in the table and then examine the properties of various high performance ceramics to see which of them might have high potential as bearing materials.

Fracture toughness is often listed as a desired property for a bearing material. However, one must put this into the proper context. M-50 steel, the state-of-the-art high performance bearing steel, has a fracture toughness value of only about 15 MPa · $m^{1/2}$. As discussed later in this chapter, fully dense silicon nitride with an even lower fracture toughness of about 5 MPa · $m^{1/2}$ can provide significantly better bearing performance than M-50 steel. Thus, although if all other factors were equal one would chose a material with a higher fracture toughness, as one moves from one class of materials (steels) to another class of materials (silicon nitrides) it is evident that all factors are not equal. Nevertheless, within a given

Table 1 Desired Properties for Bearing Materials

Property	Range	Value
Fracture toughness, K_{IC}	High	>5 MN $m^{-3/2}$
Hardness	High	>1200 kg/mm^2
Elastic modulus	Low	<210 GPa
Density	Low	<4 Mg/m^3
Bend strength	High	>700 MPa
Corrosion resistance	High	
Upper use temperature	High	>800°C
Failure mode	"Steel-like" spallation	Small spalls

materials family (i.e., silicon nitrides), increasing fracture toughness probably would increase fatigue life and reduce wear, and therefore would be desired. Based on the foregoing reasoning, a fracture toughness value of at least 5MPa $\cdot$ m$^{1/2}$ is considered to be necessary for adequate bearing performance for ceramic materials.

Hardness is important for resistance to wear and abrasion. The elastic modulus determines the contact area and consequently the Hertzian stress at a given bearing load. The lower the elastic modulus, the lower the contact stress. If one is to use ceramic rolling elements with steel races (commonly referred to as a hybrid bearing), an elastic modulus lower than that of steel is desired. Since ceramics will have elastic modulii different from those of conventional bearing steels, modifications to bearing designs and to life prediction models are required.

Density (i.e., mass density) should be as low as possible to maintain low centrifugal loads, thus enabling the attainment of higher rotational speeds (or higher DN values, where DN is a rolling element bearing performance rating factor calculated by multiplying the bore diameter in millimeters by the shaft rpm). One must be careful that the low mass density is not obtained as a consequence of incorporating porosity into the microstructure. High strength and fatigue resistance require fully dense ceramics from a microstructural standpoint. It is well known that ceramic bearing life increases as porosity decreases [3]. The requirements for strength and corrosion resistance need no elaboration. The upper use temperature listed in Table 1 is far above that which any oil-lubricated bearing could tolerate. However, under lubrication starvation or dry lubricant conditions, such high temperature capability is to be desired.

Failure mode is not often listed as a material property. This is because the failure mode for a material is a function of the environment and the stress state. It may also stem from a natural tendency to design to avoid failure and therefore not to consider the consequences of failure. While designers and materials producers do their best to reduce the possibility of bearing element failure, they cannot entirely eliminate it. Therefore, when a bearing element does fail it must be in a relatively benign mode such as spalling, as opposed to catastrophic fracture. State-of-the-art M-50 steel fails by spallation. Any ceramic that is to be considered viable for rolling element bearing use must exhibit this failure mode. Additionally, if the candidate ceramic bearing material is to surpass M-50 steel's performance, spallation must occur at higher stresses and a higher number of fatigue cycles than in the case of this steel.

Table 2 lists the properties of several candidate bearing ceramics. How do these properties compare to the requirements just discussed? Only fully dense silicon nitride and transformation-toughened zirconia fail by spalling. Thus, other high performance ceramics such as alumina and silicon carbide are normally not used or advocated as rolling elements for bearings. Zirconia has a high fracture

Table 2 Properties of High Performance Structural Ceramics Versus M-50 Steel

Property	Silicon nitride: HIPed NBD-200	Silicon carbide: sintered	Alumina: fully dense sintered	Zironcia: sintered, transformation-toughened	M-50 steel: wrought ingot
Fracture toughness, K_{IC}, MN $m^{-3/2}$	5–6	4	5	8–10	12–16
Hardness, H, kg mm^{-2}	~1800–2000	~2800	~2000	~1300	~800
Elastic modulus, E, GPa	310	410	385	205	210
Density, Mg/m^3	3.2	3.1	4	5.6	8
Modulus of rupture, MPa	750	450	550	600–900	NA
Corrosion resistance	High	High	High	High	Moderate
Upper use temperature, °C[a]	1100	1400	1000^+	800–900	325
Failure mode	Spalling	Fracture	Fracture	Spalling	Spalling

[a]Based on temperature at which material loses so much hardness that wear behavior is likely to be affected.

toughness, high strength, and an elastic modulus very close to that of steel. Thus, one would expect that transformation-toughened zirconias would be utilized as ceramic rolling elements. However, early rolling contact fatigue tests of this material resulted in fatigue behavior considerably inferior to that of silicon nitride ceramics. For example, McLaughlin [4] performed comparative rolling contact fatigue (RCF) tests on several grades of silicon nitride, silicon carbide, and a transformation-toughened zirconia, all at a Hertzian stress of 5.9 GPa (860 ksi). The zirconia and the silicon carbides were both found to have average RCF lives two to three orders of magnitude less than any of the silicon nitrides. Thus, fully dense silicon nitride remains the preeminent ceramic for rolling element bearing application.

III. PERFORMANCE OF SILICON NITRIDE IN BEARING EVALUATION TESTS

A. Rolling Contact Fatigue

The increased rolling contact fatigue life of hot-pressed silicon nitride (HPSN) was demonstrated in the mid-1970s by Baumgartner [5] for rods and Dalal [6] for balls. Sibley [7] later documented the increased RCF life of HPSN balls compared to M-50 steel balls. The HPSN utilized in these early evaluations was NC-132, which contains about 1% MgO added as sintering aid. Improved versions of NC-132 have been developed utilizing hot isostatic pressing (HIP). The first of these, NBD-100, is essentially identical to NC-132 except that it is HIPed and consequently has less porosity and is isotropic. These improvements yield increased RCF performance, as shown in Figure 1 [8]. Figure 1 also illustrates one of the consequences of the higher modulus of silicon nitride as compared to steel, namely that at the same bearing load the ceramic will have a higher Hertzian stress (thus as shown in Fig. 1, silicon nitride is tested at a higher Hertz stress). A further improvement designated NBD-200 is processed similarly to NBD-100, except that extreme care is taken during powder processing to minimize tungsten carbide contamination due to milling. The benefits of this extra care are evident from the RCF performance of NBD-200 versus both M-50 steel and NBD-100, as shown in Figure 1.

The critical importance of surface finish to the RCF performance of HPSN was pointed out by Baumgartner [9], Dalal [6], and Sibley [7]. Dalal's data are reproduced in Table 3, which compares the fatigue lives of M-50 and NC-132 silicon nitride balls. The data show that the as-received silicon nitride balls have a significantly longer fatigue life than M-50 balls, but still fail by spallation. The data for the diamond-lapped (polished) silicon nitride balls indicate extremely long lives, and tests were terminated because the M-50 support balls failed, not the silicon nitride balls.

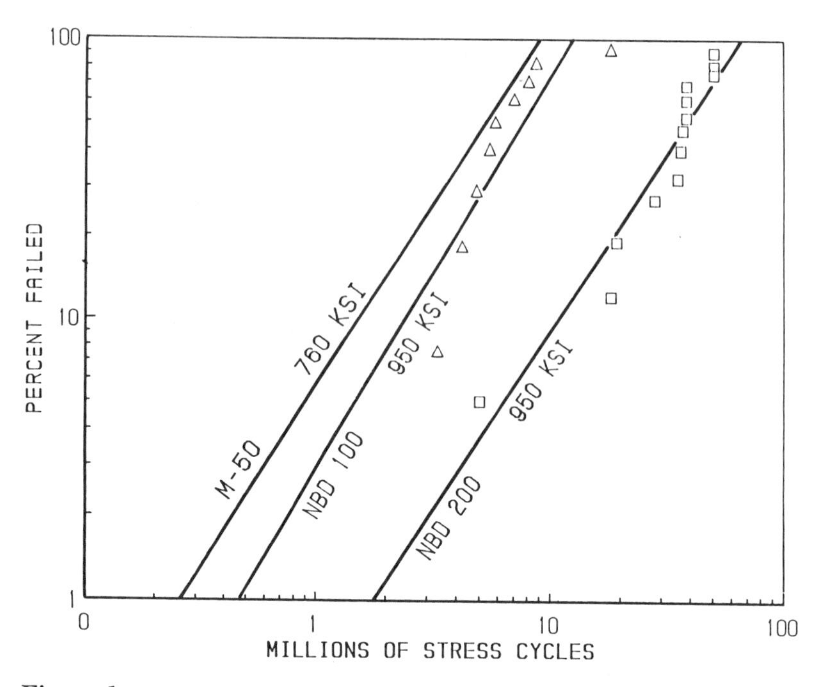

Figure 1 Rolling contact fatigue life of silicon nitride. (After Ref. 8.)

B. Lubrication Starvation Performance

Chui and Dalal [10] studied the effect of the elastohydrodynamic (EHD) lubrication on silicon nitride rolling element performance. Their work, as well as many subsequent tests, shows that HPSN gives EHD lubrication behavior essentially similar to that of steel with the same lubricants, inasmuch as:

1. It is wetted by hydrocarbon and ester lubricants "to a degree comparable to that for steel (based on contact angle measurements)."
2. In Hertzian contacts it forms "EHD lubricant films of thickness and traction properties very similar to those obtained with steel."

However, HPSN provides a unique capability, that is, lubrication starvation tolerance. Bersch [1] reports that a bearing with HPSN rollers survived a rig test for 117 hours with the lubricant shut off. The bearing condition at the end of this test was reported as "good"! Results of a lubrication starvation test of an all-NC-132 HPSN (except for the cage) roller bearing carried out in a small gas turbine engine have also been reported [2]. All components survived approximately one hour of full-speed testing in the engine.

The ability of HPSN to survive in very abusive environments was further

Table 3 Rolling Four-Ball Fatigue Life of Silicon Nitride Test Balls Stressed by M-50 Balls as a Function of Surface Finish

Spindle ball (test ball)	Spindle speed (rpm)	Maximum Hertzian stress [GPa (ksi)]	Number of M-50 support ball failures	Average test life ($\times 10^6$ rev)	Spindle ball condition post test
M-50: average of 6 tests	5,200	4.7 (680)	1 (in 6 tests)	13.2	6 of 6 spalled
As-received HPSN (NC-132)					
1 Test	5,200	4.7 (680)	2	117.3	Spalled
Average of 5 tests	10,000	5.5 (800)	4 (in 5 tests)	32.6	5 of 5 spalled
Diamond-lapped HPSN (NC-132): average of 6 tests	10,000	5.5 (800)	12 (in 6 tests)	170.9	5 of 6 intact (no spall)

Source: Reference 6.

substantiated by Hosang [11]. He reported two cases in which a hybrid bearing with HPSN balls and M-50 races, running at 93,000 rpm and a 200 N load, survived the complete destruction of the cage. Both the ceramic balls and the metallic races emerged in excellent condition.

Hanson and Ogden [12] have reported that ceramic rolling elements in metallic races significantly reduce lubricant breakdown. They attribute this advantage to the absence of microwelding between the ceramic elements and the races. This makes full ceramic or ceramic hybrid bearings attractive in situations characterized by the possible breakdown of an EHD film.

C. Failure Modes

Baumgartner [9] was the first to discuss the fatigue mechanism of HPSN, which he attributed to slow crack growth due to the nonelastic behavior of the material. The cracks ultimately link up and result in a spall. Lucek [13] has elaborated on the fatigue failure mechanism. In contrast to metals, where an accumulation of plastic deformation ultimately leads to cracking, in the fully dense silicon nitrides the mechanism is thought to be slow crack growth (akin to stress corrosion cracking). Alternatively, cyclic mechanical fatigue due to "far field" tensile stresses beyond the contact ellipse may be the cause of the progressive crack growth. Indeed, it has been shown by Beals and Bar-On [14] that for at least one variety of HPSN, cyclic mechanical fatigue cracks grow more rapidly than do static fatigue cracks (i.e., slow crack growth). Figure 2b shows a line of such cracks, which have coalesced to initiate the spall.

Whatever the precise details of the failure mechanism, the spalls exhibited in fully dense silicon nitride RCF test rods have remained remarkably constant in morphology over time. Figure 2a shows a spall from an RCF test of HPSN carried out by Baumgartner [5], and Figure 2b shows a spall in an RCF test of NBD-100 HIPed silicon nitride recently performed by Lucek [13]. The difference is that spalls in the improved material occur at either significantly higher Hertzian stresses or a higher number of stress cycles than earlier materials. In the absence of spallation, wear may eventually necessitate replacement of a bearing. Lucek [13] has evaluated the wear performance of several grades of fully dense silicon nitrides in RCF tests. Wear rates are a complex, not yet fully understood, function of the type and distribution of second phase, inclusion content, porosity, hardness, and fracture toughness. Table 4 shows the RCF and wear behavior of the silicon nitrides studied by Lucek. The L_{10} life listed in Table 4 is the fatigue life at which 10% of the rolling elements under test will fail.

It is important to reiterate that it is the existence of this noncatastrophic failure mode, identical to that of bearing steels, that differentiates silicon nitride from

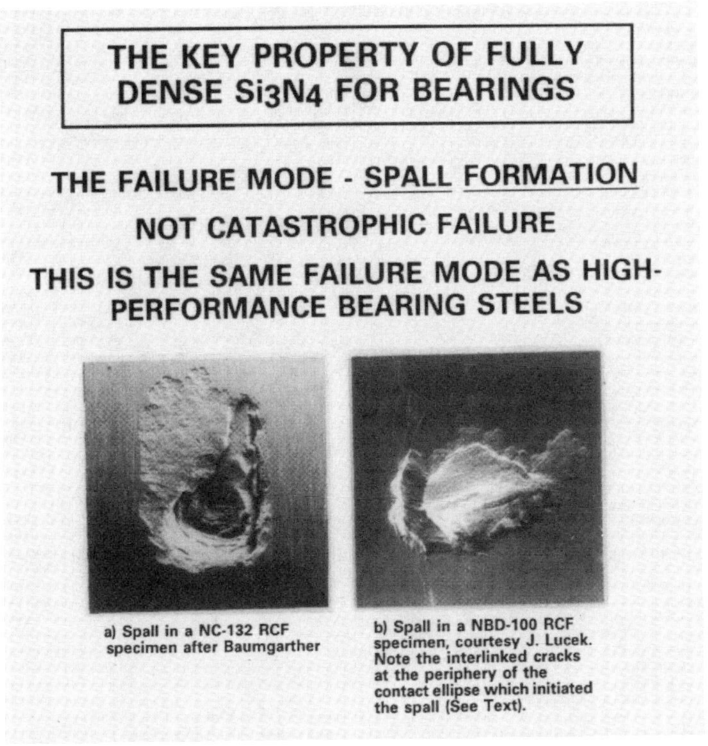

THE KEY PROPERTY OF FULLY DENSE Si_3N_4 FOR BEARINGS

THE FAILURE MODE - SPALL FORMATION

NOT CATASTROPHIC FAILURE

THIS IS THE SAME FAILURE MODE AS HIGH-PERFORMANCE BEARING STEELS

a) Spall in a NC-132 RCF specimen after Baumgarther

b) Spall in a NBD-100 RCF specimen, courtesy J. Lucek. Note the interlinked cracks at the periphery of the contact ellipse which initiated the spall (See Text).

Figure 2 Spall morphology of silicon nitride resulting from RCF testing (spalls are approximately 0.75 mm at their longest dimension).

other ceramics and makes its use in high performance bearings acceptable to designers.

D. DN Performance

The low density of fully dense silicon nitride ($\approx$3.2 g/cm^3) compared to that of bearing steels ($\approx$8 g/cm^3) significantly reduces the centrifugal loads on the bearing races and rolling elements at any speed. Thus, one would anticipate that an all–silicon nitride or a silicon nitride hybrid bearing could attain a higher speed or DN rating than an all-steel bearing. Tests of both all–silicon nitride and hybrid bearings have confirmed this expectation. In general it has been found that hybrid rolling element bearings can attain DN ratings approximately 50% higher than all-M-50 bearings. Figure 3 shows the calculated increase in DN ratings for a specific 50 mm bore hybrid bearing as a function of bearing lifetime [15].

Table 4 Effect of Composition on Silicon Nitride Rolling Element Bearing Performance[a]

Material	Process	Hertzian stress (GPa)	L_{10} (10^6 cyc)	Weibull (m)	Wear ($m^3 \times 10^{-10}$)
7% Y_2O_3 + 5% Al_2O_3 (sialon)	Sintering	13.5	1.66	1.09	≈1.6
5% Y_2O_3 + 2% Al_2O_3	Sintering and HIP	15.6	3.36	1.39	≈0.25
1% MgO	Hot-pressing	16.4	0.58	0.59	≈0.1
1% MgO	HIP (low WC)	16.5	>10.1	2.03	<0.1

[a]All RCF tests at 6.4 GPa; no information on grain size, morphology, starting powder, etc.; wear measured at 200 km element travel.
Source: Reference 13.

ratings for a specific 50 mm bore hybrid bearing as a function of bearing lifetime [15].

E. Heat Generation

Sibley [7] has pointed out that the heat generation in an all–silicon nitride or a hybrid silicon nitride–steel bearing will be lower than in an all steel bearing. Figure 4 shows the heat generation to be anticipated in two equivalent bearings: a conventional all-steel bearing and a hybrid with silicon nitride balls [15]. Two important points are made in this figure. First, at any given DN value the heat buildup of the ceramic element bearing is substantially less than the all-metal bearing. Second, confirming the performance described in the preceding paragraph, the DN limit is about 50% higher than that obtainable with the steel bearing. Recent studies by Aramoki et al. [16] relate the reduced temperature rise to the reduction of gyroscopic moments and centrifugal forces acting on the low mass density silicon nitride balls, which reduces frictional losses.

The paragraphs above have clearly shown the potential of fully dense silicon nitride rolling elements demonstrated in a variety of laboratory tests and in actual bearing tests. Given the high performance of these bearings, why have they seen

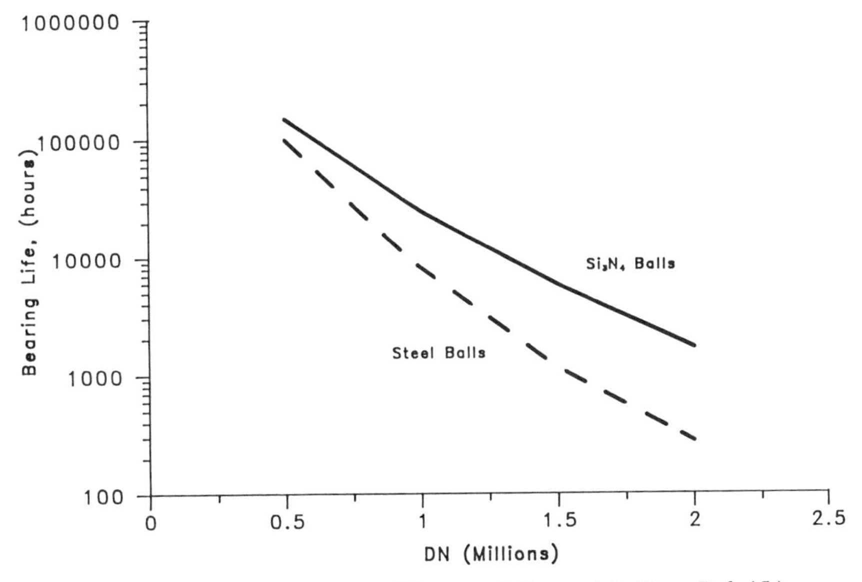

Figure 3 Calculated hybrid bearing life versus ball material. (From Ref. 15.)

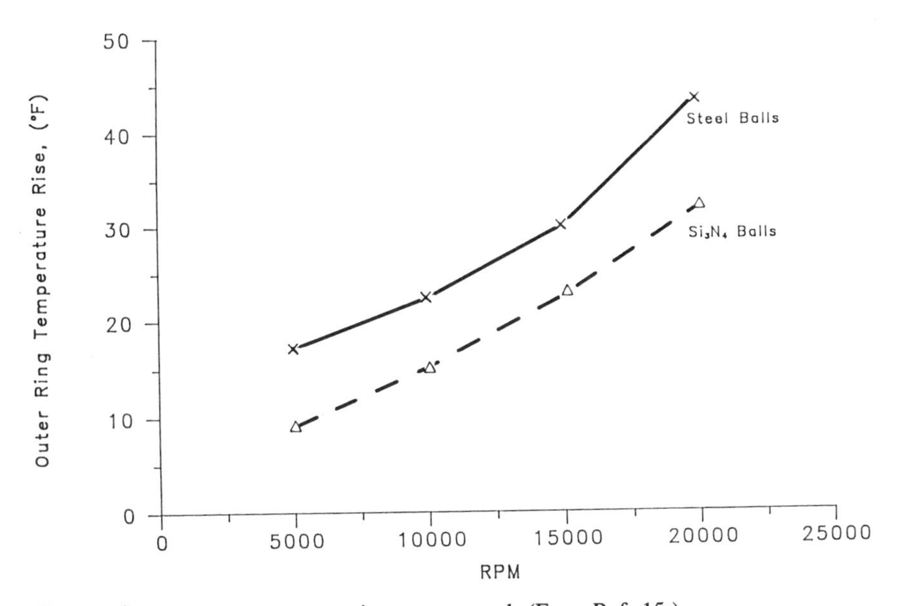

Figure 4 Bearing temperature rise versus speed. (From Ref. 15.)

such limited service to date? The two principal reasons are cost and the perception of inadequate reliability. Both these issues can be addressed and ameliorated by improved processing technology.

IV. PROCESSING AND RELIABILITY PROGRESS

Improved processing technology is the key to increasing the reliability and reducing the cost of silicon nitride rolling elements. Reliability ultimately translates into long life to failure, whether by spall formation or wear, and zero "infant mortality" or early failures. As discussed earlier by Katz and Hannoosh [2], reliability must be pursued by a total systems approach, which considers bearing design, processing, and quality assurance strategies as illustrated in Figure 5. Before the early to mid-1980s the standard silicon nitride rolling element material was hot-pressed with approximately 1% MgO densification aid (NC-132 type material). Because the material was hot-pressed, the strength, thermal conductivity, and several other properties were anisotropic. Processing runs were typically small in scale, and processing technology was rather immature. Although on the average early versions of NC-132 material outperformed M-50 steel, there were sufficient early failures to raise concerns about

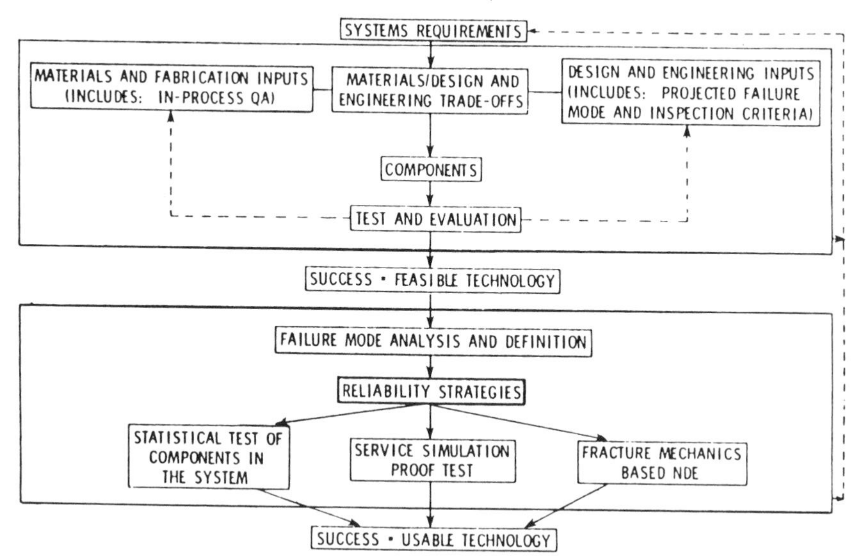

Figure 5 A manufacturing strategy for assured reliability.

reliability. Process improvements, which stressed cleanliness and process control, significantly reduced or eliminated this problem with NC-132. Nevertheless there was a need to further increase the performance (life and wear) of silicon nitride bearing elements and to simultaneously address the perception of low reliability. Hot isostatic pressing coupled with even more stringent emphasis on cleanliness has lead to the improved performance illustrated in Figure 1, combined with increased reliability, as measured by increased L_{10} life and wear resistance shown in Table 4.

Another aspect of processing, which as already shown in Table 3 has a profound effect on bearing performance, is surface finishing. The best current practice—the use of diamond grinding and lapping (polishing)—produces excellent surface finishes but is expensive. Recently a new method for finishing silicon nitride bearings utilizing magnetic fluid grinding has been developed [17]. While the surface finish attainable by this method is not yet as good as obtained by diamond grinding and polishing, a finished ball can be produced in 1/40 the time. If this method can be further developed, it offers the prospect of significant cost reductions for silicon nitride rolling elements.

Processing development and optimization of fully dense silicon nitride for bearing applications could benefit greatly from studies relating composition, microstructural features (grain size and morphology, α/β phase ratios, percentage and distribution of grain boundary phases, percentage and distribution of porosity, etc.) and processing parameters to bearing element performance. Katz [18] has pointed out that very little of such information is available in the literature. Accordingly, studies aimed at providing information elucidating such relationships are a research priority.

V. APPLICATIONS OF CERAMIC ROLLING ELEMENT BEARINGS

The principal applications of ceramic rolling element bearings today are in the chemical, food, biotechnology, instrumentation, and machine tool markets. In overwhelming predominance, these bearings are made of fully dense silicon nitride. Steinmann [8] provides an example of a bearing for a chemical processing facility in which an all–silicon nitride bearing eliminated the need for a mechanical seal, saving an estimated $6000. More importantly, bearings that previously had a one-month service life have been replaced with bearings giving at least 4 years of service.

Hannoosh [19] cites an application of hybrid silicon nitride bearings to enable a machine tool spindle to operate at one million DN with grease lubrication. The cost tradeoffs of using more expensive ceramic bearings versus cost savings in the lubrication system are shown in Table 5. More importantly than the initial systems savings ($700–$2200 per spindle) is the increased machining accuracy due to the higher stiffness and lower thermal expansion of the hybrid bearing.

Silicon nitride balls have just entered commercial service in turbomolecular pumps (Varian Turbo-V60). The bearings require no maintenance and contribute to the low vibration characteristics of this pump [20]. The potential of silicon nitride rolling elements to reduce noise levels in bearings for pumps and other machinery was recently reviewed by Philips [21].

An important exception to the general rule that high performance rolling element bearings are silicon nitride is the use of silicon carbide as rolling elements in pumps for offshore oil operations. The combination of resistance to abrasive wear and chemical resistance of SiC has made this material an attractive choice for relatively low DN (<500,000) centrifugal injection pumps lubricated only by seawater (which is highly abrasive due to entrained silica). Such pumps are reportedly in service in North Sea oil fields [22]. Hanson and Ogden [12] report the successful application of full ceramic bearings in tidal flow meters that operate immersed in seawater.

Future applications for full ceramic or hybrid ceramic bearings include bearings for use in strong magnetic fields or in marginaly lubricated applications. Hanson and Ogden [12] cite bearings for magnetic detectors for use in the superconducting supercollider as an example of the former and helicopter gearboxes as an example of the latter.

VI. SUMMARY

Advanced ceramic materials, principally silicon nitrides, possess combinations of material properties that make them the materials of choice for a variety of high performance rolling element bearing applications. During the past 10 years significant progress in processing technology has done much to reduce costs and

Table 5 Potential Cost Benefits of Ceramic Hybrid Bearings: A Case History

Application:
- Machine tool spindle bearing
- Silicon nitride bearing allows grease (vs. oil) lubrication at DN of 1×10^6 (vs. 0.6×10^6 for steel)
- DN of 1×10^6 was required

Cost:
- Silicon nitride hybrid bearing, added $800 per spindle
- Elimination of oil lubrication system saved $1500–$3000 per spindle
- Net savings: $700–$2200 per spindle

Source: Reference 19.

increase the reliability of ceramic rolling elements. Indeed these processing advances have been the basis for what is today a global market of approximately $10 million with growth potential to exceed $100 million by the end of the decade. It is important to note that this progress has been based on improved processing of silicon nitrides developed for hot components of gas turbine or other engines, not compositions developed specifically for bearing use. Future advances in this field will likely depend on the development of ceramic compositions and microstructures specifically designed for bearing application. The fundamental understanding of the composition/microstructure/processing/performance relationships that will enable such application-specific materials design is largely lacking.

REFERENCES

1. C. F. Bersch, Overview of ceramic bearing technology, in *Ceramics for High Performance Applications*, Vol. II, J. J. Burke, E. S. Lenoe, and R. N. Katz (Eds.), Brook Hill Publishing: Chestnut Hill, MA, 1978, pp. 397–405.
2. R. N. Katz and J. G. Hannoosh, Ceramics for high performance rolling element bearings: A review and assessment, *Int. J. High Technol. Ceram. 1*, 69–79 (1985).
3. E. V. Zaretsky, Ceramic bearings for use in gas turbine engines, *J. Tribol. ASME*, *111*, 146–157 (1989).
4. J. J. McLaughlin, Alternate ceramic roller bearing program, Final Report, Contract No. N00019-81-C-0284 (November 1983).
5. H. R. Baumgartner, Evaluation of roller bearings containing hot-pressed silicon nitride rolling elements, in *Ceramics for High Performance Applications*, J. J. Burke, A. E. Gorum, and R. N. Katz (Eds.), Brook Hill Publishing: Chestnut Hill, MA, 1974, pp. 713–718.
6. H. Dalal, Machining bearings for turbine application, in *Ceramics for High Performance Applications*, Vol. II, J. J. Burke, E. S. Lenoe, and R. N. Katz (Eds.), Brook Hill Publishing: Chestnut Hill, MA, 1978, pp. 407–422.
7. L. B. Sibley, Silicon nitride bearing elements for high-speed high-temperature applications, in *Problems in Bearings and Lubrication*, AGARD Conference Proceedings No. 323, AGARD, Neuilly sur Seine, France, Chapter 5, 1982.
8. D. Steinmann, Silicon nitride ceramics for balls and ball bearings, *Ceram. Forum Int. 67*, 584–588 (1990).
9. H. R. Baumgartner, Ceramics bearings for turbine applications, in *Ceramics for High Performance Applications*, Vol. 2, J. J. Burke, E. S. Lenoe, and R. N. Katz (Eds.), Brook Hill Publishing: Chestnut Hill, MA, 1978, pp. 423–443.
10. Y. P. Chiu and H. Dalal, Lubricant interaction with silicon nitride in rolling contact interactions, in *Ceramics for High Performance Applications*, J. J. Burke, A. E. Gorum, and R. N. Katz (Eds.), Brook Hill Publishing: Chestnut Hill, MA, 1974, pp. 589–607.
11. G. W. Hosang, Results and design techniques from the application of ceramic ball bearings to the MERADCOM 10 kW turbine, presented at the AIAA/SAE/ASME/ASEE 23rd Joint Propulsion Conference, AIAA Paper No. 87-1444 (1987).

12. R. A. Hanson and W. P. Ogden, Ceramic bearings: The methods, the materials, the marketplace, *Split Ballbearing Features*, Vol. VII, Part 1, Split Ballbearing Division of MPB Corp. Lebanon, NH: 1991, pp. 1–7.
13. J. W. Lucek, Rolling wear of silicon nitride bearing materials, ASME Paper No. 90-GT-165 (1990).
14. J. T. Beals and I. Bar-On, Fracture toughness and fatigue crack propagation of silicon nitride with two different microstructures, *Ceram. Eng. Sci. Proc. 11*(7–8), 1061–1071 (1990).
15. Cerbec Ceramic Bearing Co., East Granby, CT, product literature, 1990.
16. H. Aramoki, Y. Shoda, Y. Morishita, and T. Sawamoto, The performance of ball bearings with silicon nitride balls in high speed spindles for machine tools, *J. Tribol. ASME, 110*, 693–697 (1988).
17. K. Kato, Tribology of ceramics, *Wear, 136*, 117–133 (1990).
18. R. N. Katz, presentation at the NIST/DARPA Ceramic Bearing Technology Workshop, Rockville, MD, April 17, 1991.
19. J. G. Hannoosh, *Design News*, Nov. 23, 1988.
20. Varian Vacuum Products, Lexington, MA, company product literature, 1991.
21. G. Phillips, presentation at the NIST/DARPA Ceramic Bearing Technology Workshop, Rockville, MD, April 17, 1991.
22. Anon., *Eng. Mater. Design*, March 1988, p. 11.

14

Tribological Performance of Ceramic Cam Roller Followers

Arup Gangopadhyay
Ford Motor Company
Dearborn, Michigan

H. S. Cheng
Northwestern University
Evanston, Illinois

Joseph F. Braza
Torrington Company
Torrington, Connecticut

Stewart Harman
Chrysler Corporation
Detroit, Michigan

J. M. Corwin
Unique Solutions, Inc.
Royal Oak, Michigan

ABSTRACT

Friction and wear tests were conducted on three types of ceramic: alumina–titanium carbide composite, partially stabilized zirconia, and silicon nitride in lubricated rolling contact against nodular cast iron. In the case of silicon nitride, three variations in processing techniques were evaluated: pressureless sintered, reacted, and hot-pressed. The friction coefficients of the different ceramic materials were approximately the same, but they had different wear rates. The wear rate differences were found to be related to their wear modes. The wear of the ceramic rollers in a 100-hour test was very small (of the order of 0.1 mg) except for the partially stabilized zirconia rollers, whose wear was of the order of 1 mg. Full-scale motorized valve train tests were conducted on these ceramics as cam roller followers in contact with a nodular cast iron camshaft and compared to steel roller followers. The reduction in the largest diameter of the nodular cast iron cam lobes increased in the first 400 hours as a result of initial run-in, but no further reduction in diameter was observed after that period. The ceramic rollers exhibited negligible wear. The wear modes of the ceramic rollers and the cam lobes were analyzed by scanning electron microscopy. The results indicate that ceramic materials can replace steel roller followers.

I. INTRODUCTION

Turbine rotors, dies, and bearing components are currently being fabricated from advanced structural ceramic materials. In the future, these materials could potentially replace current metallic components in automobile engines [1]. Ceramic materials are promising because of their low density, low thermal expansion coefficient, high melting points, and high oxidation and corrosion resistance [2]. The major disadvantage of the ceramic materials is their low fracture toughness. Pores, flaws, impurities, and other defects in the material act as potential sites for crack nucleation. However, advances in processing techniques such as hot isostatic pressing [3] have reduced defects of these types and enhanced their mechanical properties.

One type of ceramic material, silicon nitride (Si_3N_4), exhibited exceptionally good rolling contact fatigue performance in hybrid ball bearings [4,5] in which the rolling elements are made from Si_3N_4 and the raceways are fabricated from various hardened steels. Dalal et al. [6] conducted the first comprehensive study on the rolling and sliding contact behavior of hot-pressed silicon nitride and observed that Si_3N_4 was adequately wetted by hydrocarbon- and ester-based lubricants to form an acceptable elastohydrodynamic (EHD) film. In a later complementary study, Bhushan and Sibley [7] investigated the effect of grinding damage on Si_3N_4 in relationship to the rolling contact fatigue life. They observed that poor finishing practices reduced the rolling contact fatigue life. In some applications, however, ceramic components are not subjected to such high localized Hertzian contact stresses on the microasperity scale (as in ball bearings) and therefore lower cost finishing practices can be utilized.

Ceramic materials have been generally aimed at elevated temperature applications, yet there exists a strong potential for using ceramic materials in low temperature, wear-resistant applications, such as cam roller followers in automobile or diesel engines [8,9]. In the past, the majority of the fundamental ceramic wear studies concentrated on the sliding or rolling wear behavior of silicon nitride, silicon carbide (SiC), partially stabilized zirconia (PSZ), and alumina (Al_2O_3) ceramics [10–14]. With the recent interest in using ceramic materials in automobile engines, fundamental investigations have been conducted to evaluate the lubricated rolling contact performance of various ceramics [15,16]. These controlled simulated tests produce knowledge and understanding of their lubricated rolling performance. The results from these tests complement and augment full valve train analyses [8]. This chapter synthesizes the rolling contact investigations with the full valve train test analyses. The ceramic rollers, made out of alumina–titanium carbide, PSZ, and several types of Si_3N_4, were tested against nodular cast iron cam lobes with a mineral oil based lubricant. Friction coefficients, wear rates, scanning electron microscopy results, and surface damage analysis are presented to show the performance of these ceramics in lubricated rolling contact tests.

II. ROLLING FRICTION AND WEAR TESTS

A. Materials and Test Apparatus

The physical and mechanical properties of the ceramic materials are listed in Table 1. The silicon nitride materials were obtained from three different commercial sources, each was processed by a different technique: hot-pressed, pressureless sintered, and reacted. Consequently, the microstructures of the materials were different. The surface roughness of the rollers was 0.15 μm R_a. The hot-pressed silicon nitride consisted essentially of β-Si_3N_4 grains with less than 5% silicon oxynitride and MgO as a sintering additive. The grains were approximately 0.7 μm in size. This material also contained up to 3 wt % tungsten carbide inclusions dispersed in the matrix. The primary source of tungsten carbide particles was the tungsten carbide balls used for milling silicon nitride powders. The size of the tungsten carbide particles varied from submicrometer to 3 μm. The grains of the pressureless sintered silicon nitride were needle shaped, with diameters approximately 0.5 μm and lengths up to 5 μm. The pore sizes were up to 5 μm. The reacted silicon nitride was extremely dense, with little surface porosity. This material is different from conventional reaction-bonded silicon nitrides in the sense that in the conventional reaction-bonded material the starting powder is silicon, whereas in reacted silicon nitride the starting material is liquid organochlorosilane. The organochlorosilane is first converted into polysilazane, which is later converted into silicon nitride using supercritical fluid extraction

Table 1 Physical Properties of the Ceramic Materials

	Si_3N_4				
	Hot-pressed	Sintered	Reacted	Al_2O_3-TiC	PSZ
Density, g/cm^3	3.29	3.2	3.24	4.23	5.75
Elastic modulus, GPa	310	270	310	421	200
Poisson's ratio	0.26	0.27	0.27	0.25	0.22
Knoop hardness, GPa[a]	17.9	15.9	17.0	18.1	9.1
Fracture toughness, MPa · m$^{1/2}$	5.4	6–7	6.0	3.0	8–12
Thermal expansion, × 10^{-6}/°C	3.5	3.4	3.4	7.8	10.1
Average grain size, μm	0.7	1–5	0.4	1–2	60

[a]Measured by Knoop indentation, 500 g load.

Table 2 Mechanical Properties of the Nodular Cast Iron

Tensile strength	550 MPa
Yield strength	380 MPa
Modulus of elasticity	145 MPa
Hardness	5 Rockwell C

process [17]. The alumina–titanium carbide composite contained approximately 45 wt % titanium carbide particles in the alumina matrix. The partially stabilized zirconia was MgO stabilized and consisted of tetragonal precipitates dispersed within large cubic grains that were approximately 60 μm in size. These ceramics were tested against nodular cast iron, which was austenitized, quenched in oil, and tempered. The mechanical properties of the cast iron are listed in Table 2.

The specimen configuration is shown schematically in Figure 1. The apparatus essentially consists of a ceramic roller, which is rotated against a nodular cast iron roller. The ceramic roller is supported on a steel pin, which constitutes a journal bearing. The cast iron roller is mounted onto a spindle shaft, and its speed is controlled by a variable speed motor drive system. The normal load is applied through the stationary steel pin, forcing the ceramic roller (which is free to roll) down onto the cast iron roller (the driving component). The specimens are lubricated by a jet of mineral oil, the properties of which are listed in Table 3. The paraffinic mineral oil contained 10–20% naphthenic fraction. The oil did

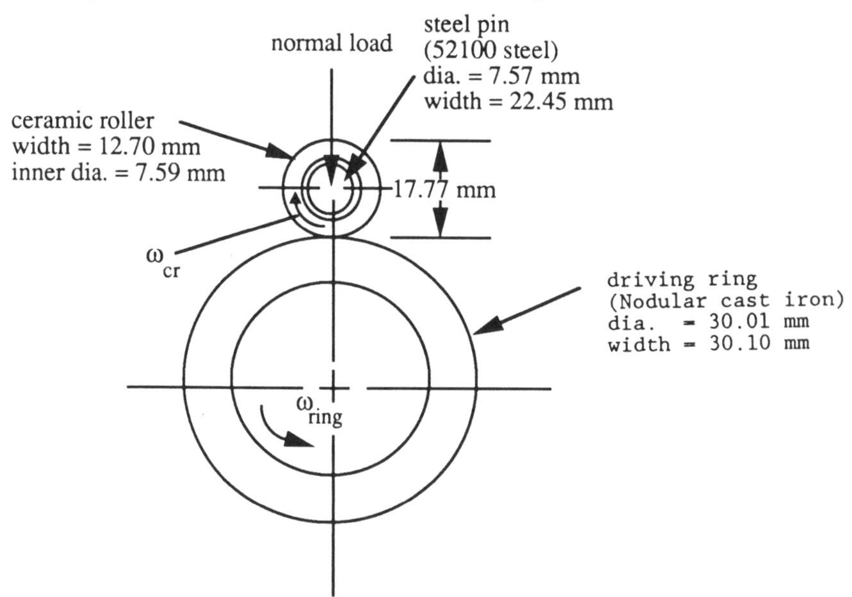

Figure 1 Schematic diagram of the contact geometry.

Table 3 Physical Properties of the Lubricant

Mass volume at 21°C, g/mL	0.866
Kinematic viscosity, m^2/s	
at 37°C	37.5×10^{-6}
at 100°C	5.94×10^{-6}

not contain any additives. The friction force was continuously measured by a load cell and displayed on a strip chart recorder. A detailed description of the apparatus can be found elsewhere [15].

B. Experimental Procedure

The experiments were conducted with a normal load of 1112 N, which corresponds to a maximum Hertzian contact pressure ranging from 1.19 GPa to 1.48 GPa for the various ceramics. This is the typical contact stress experienced in cam follower applications. The rotational speed of the cast iron roller was 1500 rpm, producing a rolling velocity of 3.12 m/s. Each test interval was 10 hours. At the end of each test interval, the ceramic roller was removed, rinsed with acetone, and weighed to the nearest 0.02 mg. The total length of the test was 100 hours.

C. Results

1. Friction and Wear Data

The friction coefficient of the different ceramic materials varied very little and was found to be approximately 0.005–0.006. This value is dependent on the operating conditions (i.e., normal load, speed, and temperature) that affect the lubricant film thickness in the journal contact zone. The wear data for the ceramic materials are shown in Figure 2 as a function of test length. The wear constant k was calculated from the weight loss measurements according to the following relationship:

$$k = \frac{W}{\rho P L}$$

where W is the weight loss, ρ is the density of silicon nitride, P is the applied load, and L is the sliding distance. This wear constant relationship is derived for sliding contacts but is also useful in rolling contact for comparison of the wear behavior of different ceramic materials. It was observed that little to no wear occurred on the steel pin, or on the inner diameter surface of the ceramic roller. If the journal bearing is properly designed, the lubricant film will support the load in the contact area. Consequently, the weight loss of the rollers represents the material loss from the outer diameter surface only.

The initial and the steady state wear of PSZ (Fig. 2a) is substantially higher than those of the other two ceramic materials, while the hot-pressed silicon nitride

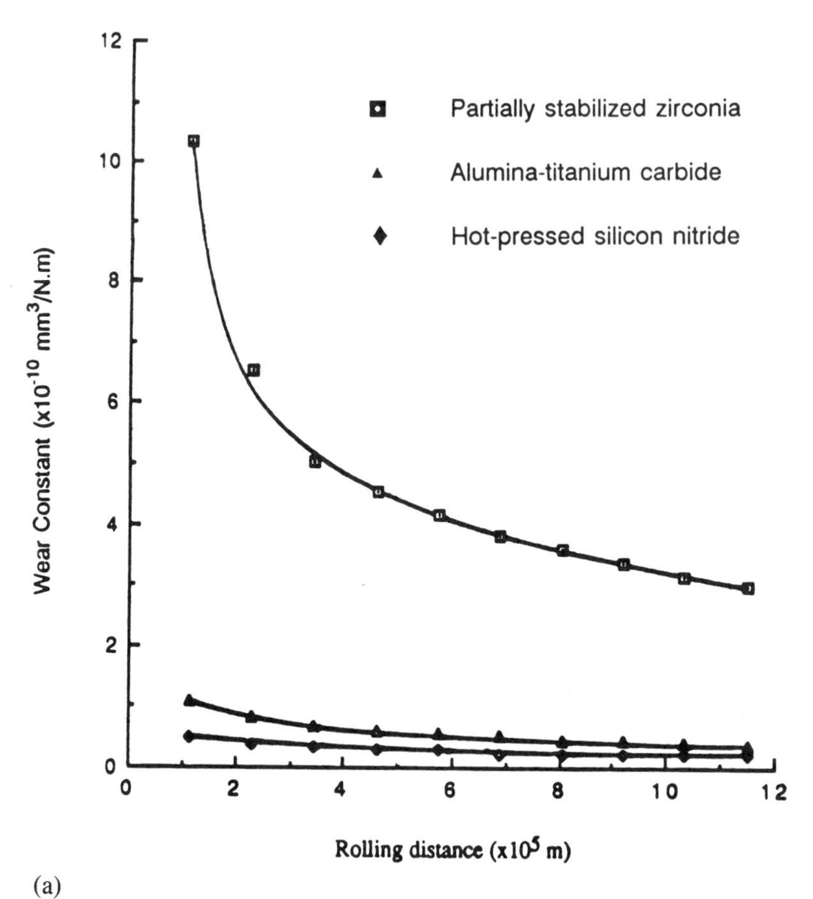

(a)

Figure 2 Wear rates of different ceramic materials.

is the lowest. The alumina–titanium carbide composite has a higher initial wear than the silicon nitride, even though its steady state wear approaches that of silicon nitride. To show the effect of the processing techniques on the wear of silicon nitride, the wear constants of the three silicon nitride ceramics are plotted separately in Figure 2b. (Note that the curve for the hot-pressed silicon nitride is the same in Figures 2a and b.) The reacted silicon nitride exhibited the largest initial and steady state wear compared to pressureless sintered and hot-pressed silicon nitride ceramics.

2. Scanning Electron Microscopy

The wear surfaces of the ceramics were examined under the scanning electron microscope in an attempt to explain their wear behavior. The original unworn

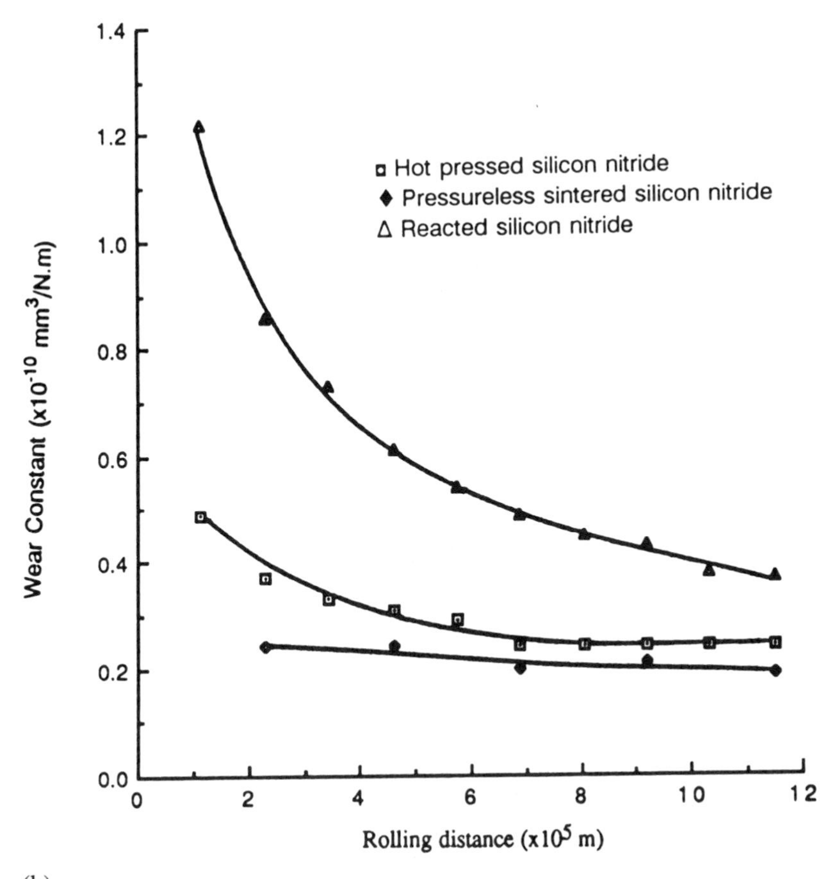

(b)

surface of PSZ in Figure 3a contains surface defects as large as 100 μm. These surface defects are sites for large weight loss, since the sides of the pores cave in and become larger over time, as shown in Figure 3b, which is the same region after 100 hours of testing. The wear mode of PSZ is classified as microchipping, as defined by Braza et al. [18]: that is, transgranular cracks grow through the grains to form microchips. This wear mode causes substantial weight loss of the PSZ rollers.

A grain pullout wear mode was observed on the Al_2O_3-TiC composite rollers, as shown in the scanning electron micrographs in Figures 4a and 4b. Figure 4a is a worn area after the first 10 hours of testing and Figure 4b shows the same area after 10 more hours of testing. Grain boundary fracture is the predominant mode of failure in this composite ceramic material. It is known that the glassy grain boundary phases in alumina constitute the primary cause of fracture [19,20].

The lower wear rate of the hot-pressed and the pressureless sintered silicon

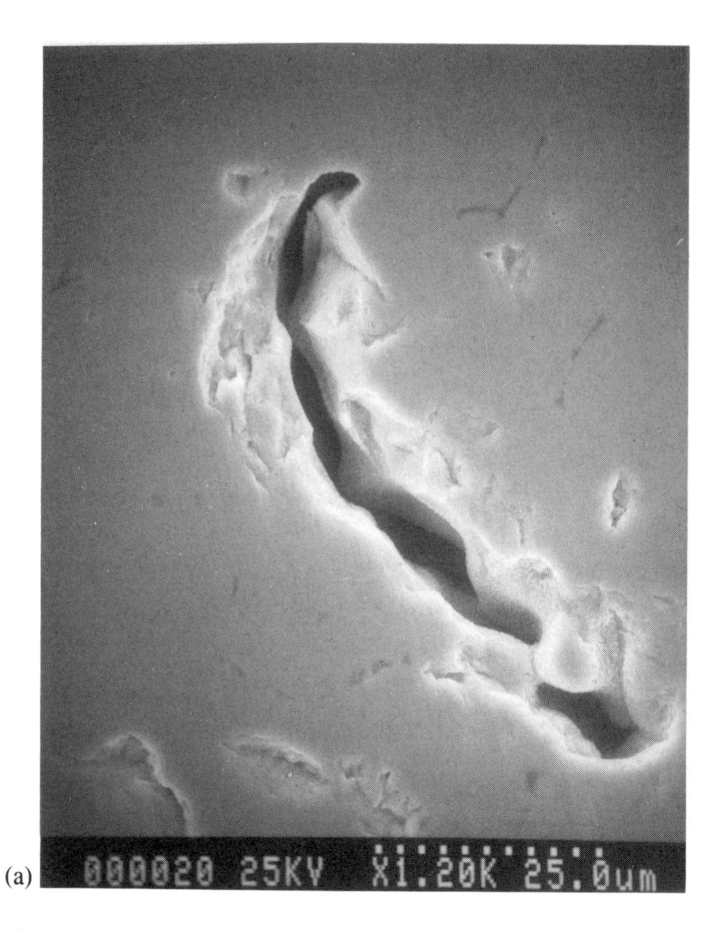

(a)

Figure 3 Scanning electron micrographs of partially stabilized zirconia: (a) unworn surface and (b) the same region after 100 hours of testing.

nitride can be ascertained by the micrographs of the worn surfaces. Figures 5 and 6 show scanning electron micrographs of the unworn and worn surface of hot-pressed and pressureless sintered silicon nitride, respectively. These micrographs show that wear has occurred by a submicrometer chipping mode in both materials. Even the large surface pores in the pressureless sintered silicon nitride (Fig. 6a) had no influence on its performance. On the other hand, the reacted silicon nitride exhibited a grain pullout wear mode, as shown in Figure 7: the difference in wear modes seen in the same worn region accounted for the variation in wear rates of the three silicon nitride materials. That is, grain pullout produced a greater volume loss than submicrometer chipping.

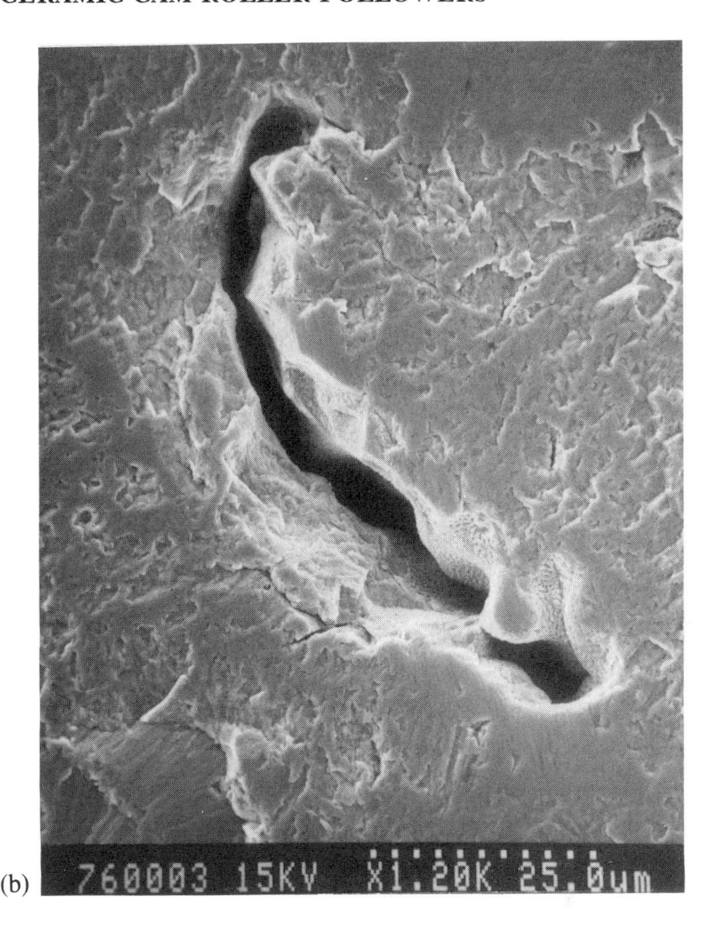

(b)

III. MOTORIZED VALVE TRAIN TESTS

A. Materials and Test Apparatus

The lubricated rolling contact tests provided guidance for the possible use of the ceramic materials as cam roller followers in automobile engines. This section presents results from motorized valve train tests to show how these ceramic materials can perform satisfactorily in long-term tests. The same ceramic materials were used as roller followers, with the exception of the PSZ, which was supplied by a different manufacturer and had a much smaller grain size (1–2 μm). For comparison purposes, production AISI 52100 steel roller followers were also tested. The pressureless sintered and reacted silicon nitride were chosen because they exhibited two different wear modes in rolling contact experiments described in Section II.

The motorized valve train apparatus used for the evaluation of long-term tribological performance of ceramic materials is shown in Figure 8; a detailed

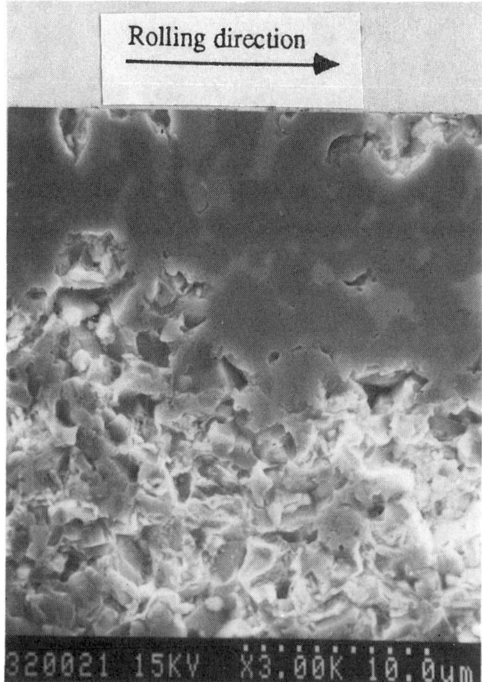

Figure 4 Scanning electron micrographs of the alumina–titanium carbide composite: (a) after 10 hours of testing and (b) the same region after an additional 10 hours of testing.

(a)

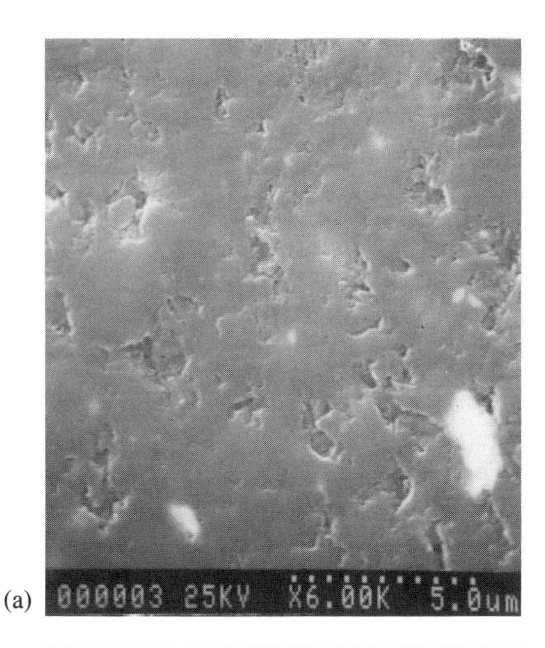

(b)

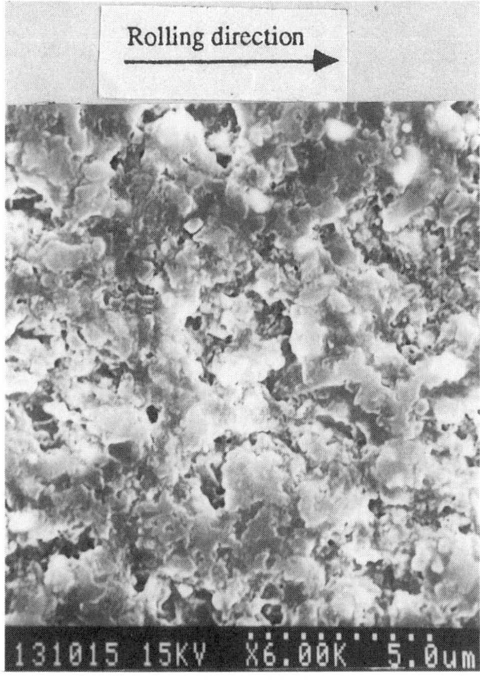

Figure 5 Scanning electron micrographs of (a) unworn and (b) worn surface of hot-pressed silicon nitride.

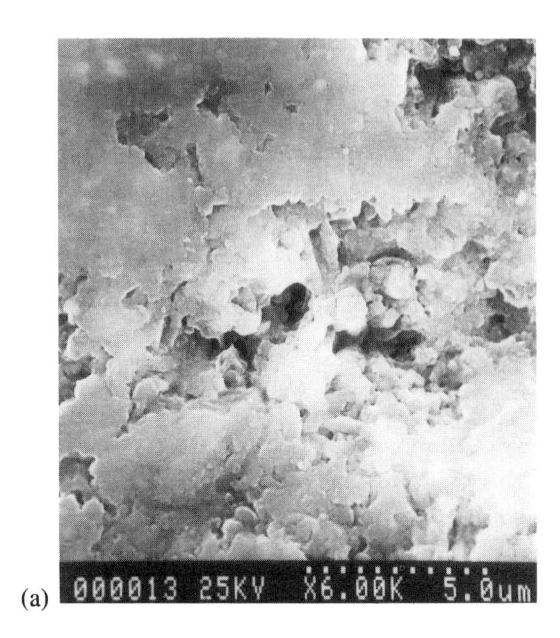

(a)

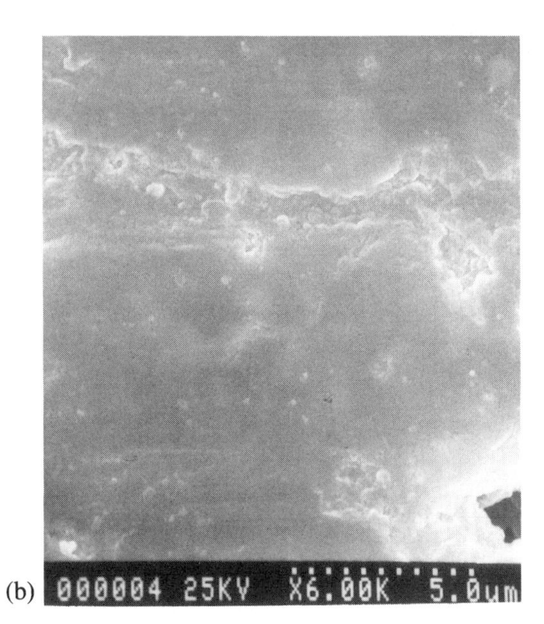

(b)

Figure 6 Scanning electron micrographs of (a) unworn and (b) worn surface of pressureless sintered silicon nitride.

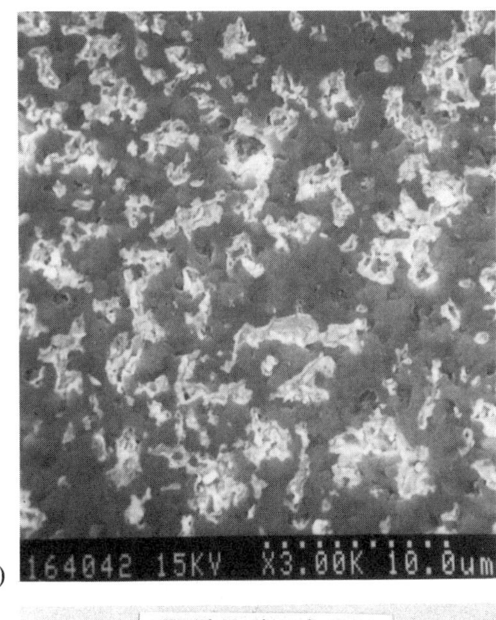

(a)

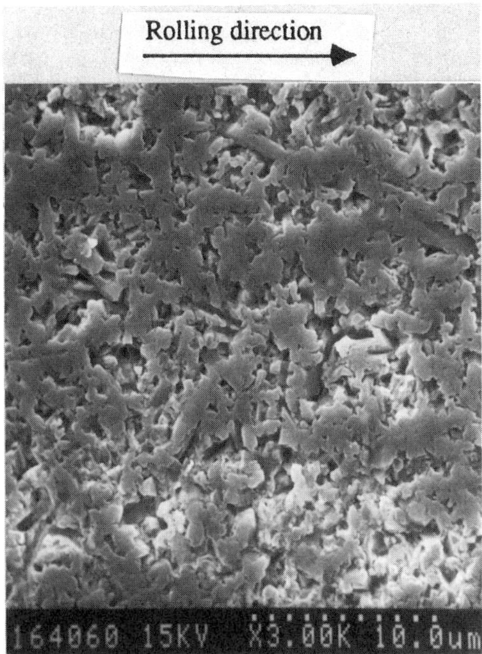

(b)

Figure 7 Scanning electron micrographs of (a) unworn and (b) worn surface of reacted silicon nitride.

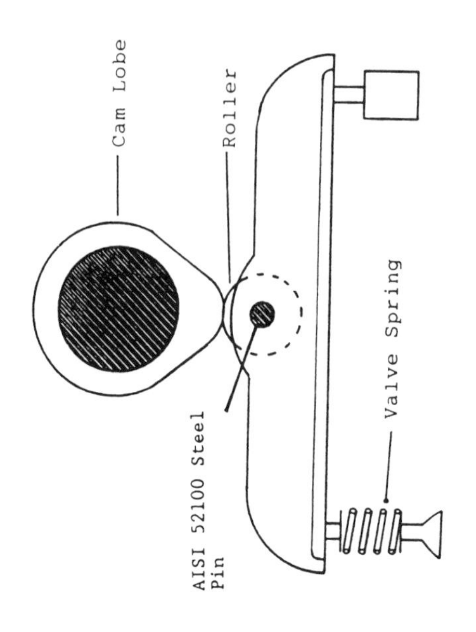

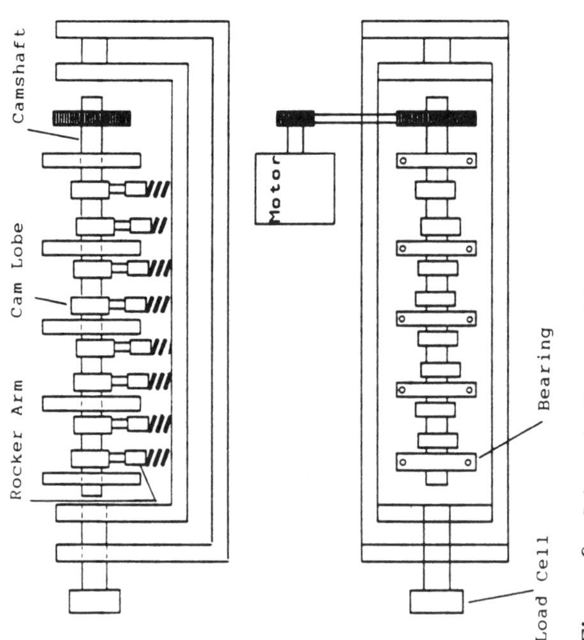

Figure 8 Schematic diagram of the motorized valve train apparatus.

description of the apparatus can be found elsewhere [8]. It essentially consists of a nodular cast iron camshaft containing eight cam lobes, which is driven by a 15 kW variable speed motor. The surfaces of the cam lobes were induction hardened to 55 Rockwell C. The cam lobes were driven against eight steel or ceramic rollers. The roller followers had the same dimensions and geometry as shown in Figure 1, where the journal shaft, made of AISI 52100 steel, was inserted through the inner diameter of the ceramic rollers and the ends of the pin were held fixed within the rocker arm assembly. The steel rollers had 18 roller pins of 1.5875 mm (0.0625 in.) diameter between the steel pin and the inner diameter of the outer roller. The static load at the contact area is primarily governed by the stiffness of the valve spring and the cam lobe profile. However, there is also a dynamic load, which was not measured. The contact areas were lubricated by a jet of mineral oil at a pressure of 0.2 MPa. The lubricant contained a small amount of an oxidation inhibitor and a pour point depressant, but no detergents and viscosity index improvers or antiscuff additives were added. The temperature of the lubricant was maintained at 88°C. The frictional torque was measured by a load cell.

B. Experimental Details

The maximum spring load applied on each roller was 1634 N such that the Hertzian contact stress varied between 1.26 and 1.48 GPa depending on the elastic modulus of the roller materials. The experiments were conducted at camshaft speeds of 250 and 3000 rpm, which correspond to engine speeds of 500 and 6000 rpm. The selection of these speeds enabled evaluation of their wear life under extreme operating conditions. The duration of each test was 100 hours; the camshaft speed was changed every 2 hours from 250 to 3000 rpm. The total duration of the tests was 900 hours. The tests were conducted with two pressureless sintered silicon nitride, two reacted silicon nitride, two alumina–titanium carbide composite, and two AISI 52100 steel rollers. The wear of the cam lobes was measured by the reduction in the largest diameter, and the wear of the rollers was measured by the change in the diameter and weight loss. The weight loss was measured to an accuracy of 0.02 mg. Also surface roughness measurements were carried out on the rollers at the end of each 100-hour test. Replicas were taken from the nose of the cam lobes and the rollers after each 100-hour test interval for further examination under the scanning electron microscope to record the progression of wear. In the case of partially stabilized zirconia, all eight rollers were tested for 800 hours at a constant camshaft speed of 1500 rpm. These rollers and the cam lobes were not measured for wear; however, the worn surfaces were examined under the scanning electron microscope at the end of 800 hours of testing.

C. Results and Discussions

The reduction of the largest diameter of the cam lobes with different roller follower materials is shown in Figure 9 as a function of test length. The values

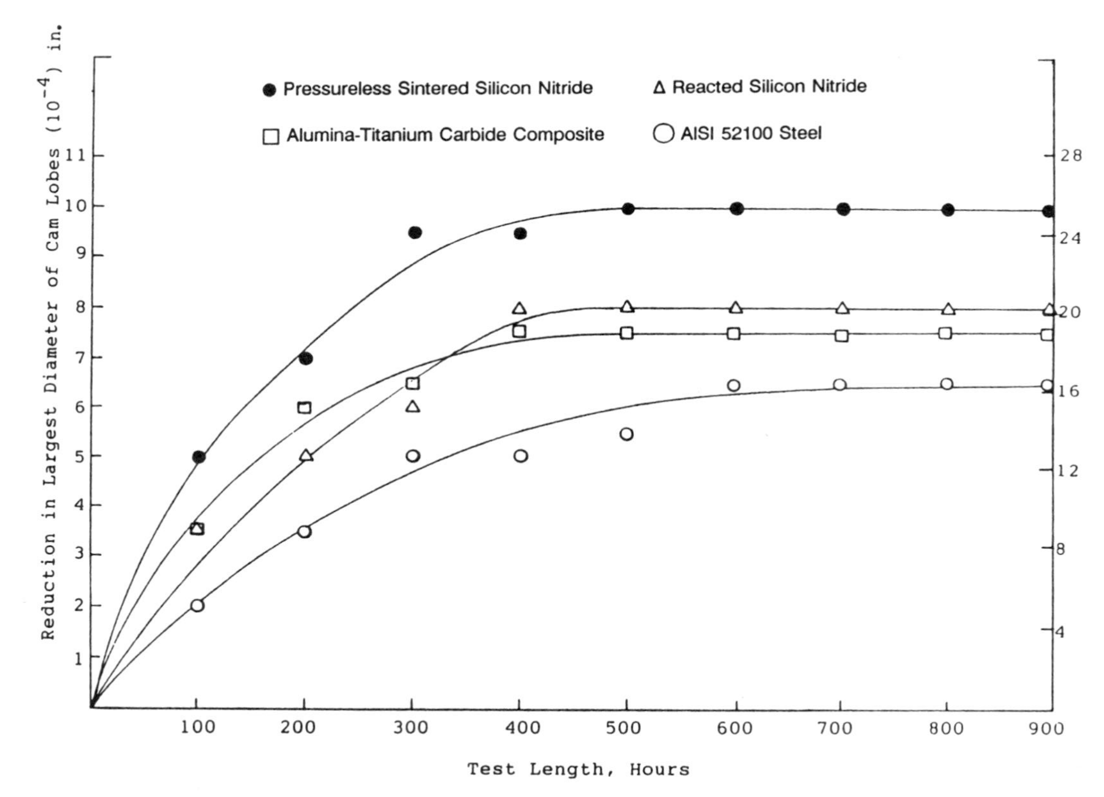

Figure 9 Wear of cam lobes tested against different roller follower materials.

are the average reduction of the two cam lobes tested against the same material pairs. It can be seen that the initial wear of the cam lobes was high as a result of run-in; after 400 hours no appreciable reduction in diameter of the cam lobes was observed. The cam lobes tested against the ceramic rollers exhibited slightly larger reduction in diameter than those tested against the steel rollers, possibly because of the lower elastic modulus of steel compared to ceramics, which produces a lower Hertzian contact stress. Abrasive wear by the harder ceramic is also considered to be a contributing factor. However, the wear of the cam lobes against ceramic rollers was in the acceptable range.

Initially, a linear caliper was used to measure the change in the diameter of ceramic roller followers. It was found that the ceramic roller followers had an insignificant change in the diameter. After 400 hours of testing, the reduction in diameter was lower than 2.5×10^{-3} mm, which is within the accuracy of the instrument. In subsequent test intervals, the change in weight of the rollers was also recorded using a single pan balance. Interestingly, it was observed that the rollers showed a slight gain in weight (≈ 0.03 mg) at the end of some of the 100-hour test intervals. This result may have been due to the formation of a transfer film on the surface. In the case of the steel rollers, the reduction in diameter was 5×10^{-3} mm after 400 hours and 7.6×10^{-3} mm after 900 hours.

In the laboratory rolling tests described in Section II, differences in wear constants were observed among the three types of ceramic and also among the three types of silicon nitride. However, in the motorized valve train apparatus, the presence of both static and dynamic loads prevented the observation of significant difference in wear among the different ceramic materials at the end of 900 hours of testing.

1. Scanning Electron Microscopy of the Ceramic Rollers

Scanning electron micrographs of the replica of the worn surface of pressureless sintered silicon nitride are shown in Figure 10 at the end of 100, 400, and 900 hours of testing. The original grinding furrows are still visible after 100 hours of testing (Fig. 10a). The white areas represent the pores or flaws on the surface, which may have been generated during grinding. After 400 hours, many of the grinding furrows disappeared (Fig. 10b). However, profilometer traces indicated that no change in surface roughness occurred on the worn surface between 100 and 400 hours, even though submicrometer chipping was observed at the end of 400 hours of testing. Both microchipping and removal of original grinding furrows accounted for the slight reduction in diameter of the rollers. At the end of 900 hours of testing, the worn surface still showed some microchipping (Fig. 10c), but still no change in the surface roughness was observed.

A scanning electron micrograph of the replicated worn surface of the reacted silicon nitride after 100 hours of testing is shown in Figure 11a. The grinding furrows are not very evident. The morphology of the worn surface appeared

(a)

Figure 10 Scanning electron micrographs of the worn surface of pressureless sintered silicon nitride after testing for (a) 100, (b) 400, and (c) 900 hours.

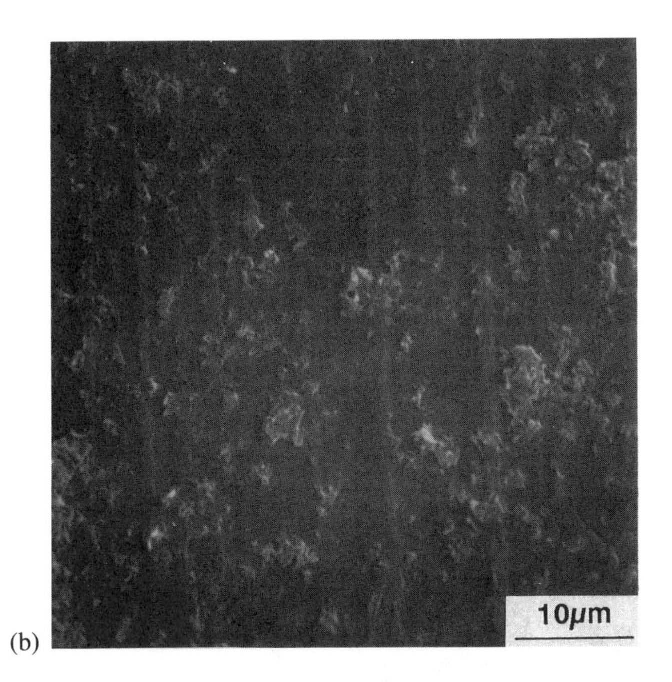

(b)

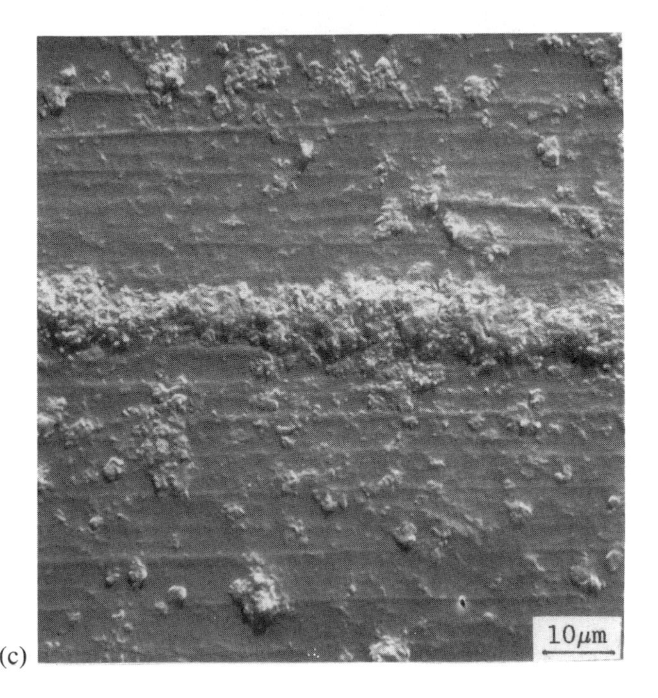

(c)

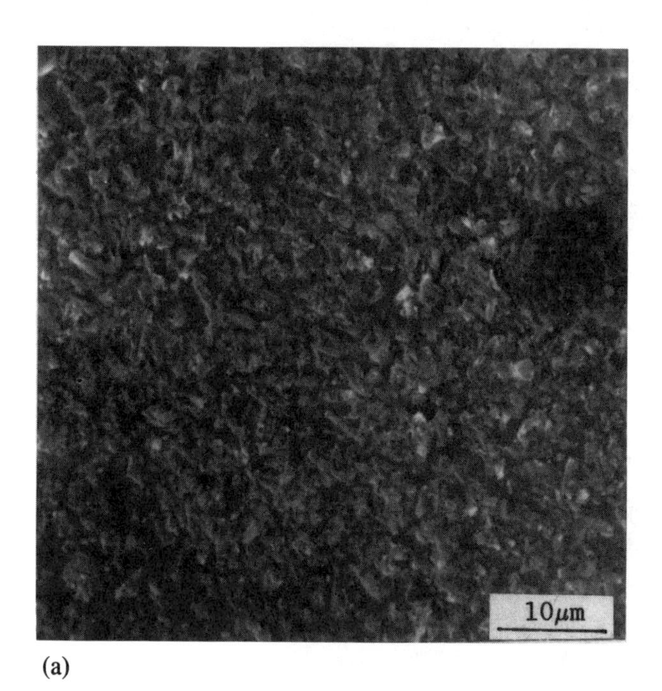

(a)

Figure 11 Scanning electron micrographs of the worn surface of reacted silicon nitride after testing for (a) 100, (b) 400, and (c) 900 hours.

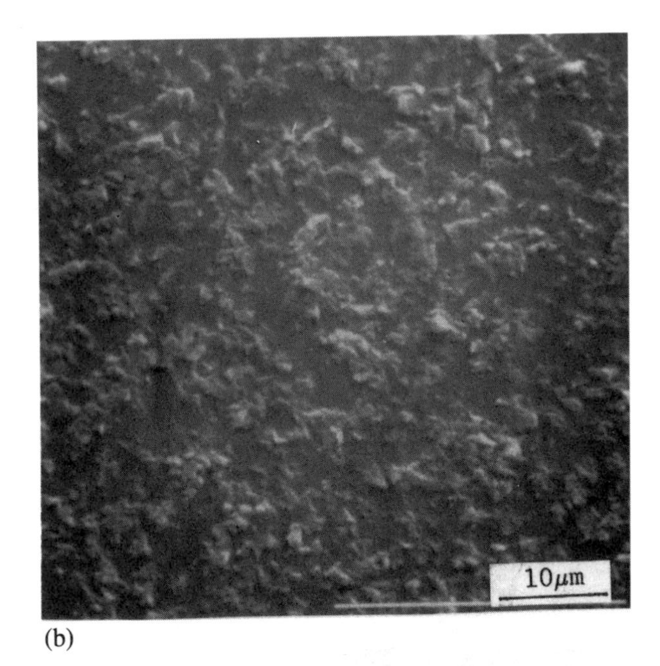

(b)

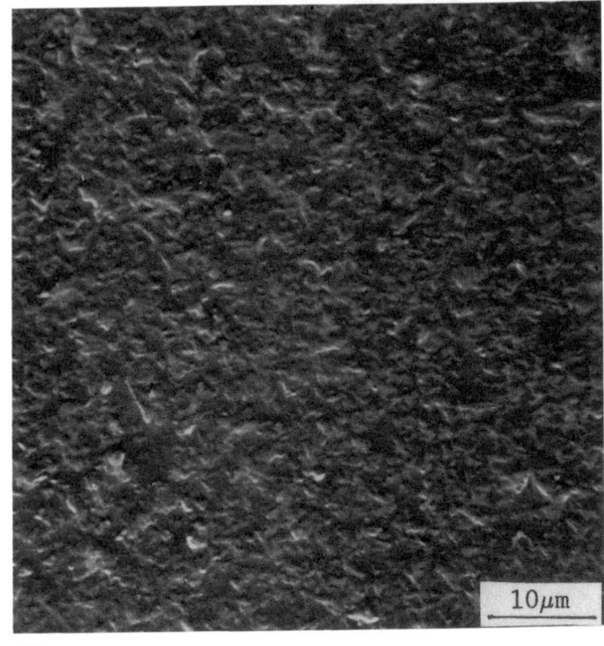

(c)

different from that of the pressureless sintered material, perhaps because of the microstructure of the material. After 400 hours, the worn surface exhibited smooth regions, as shown in Figure 11b. The surface roughness changed from 0.12 μm after 100 hours to 0.07 μm after 400 hours of testing. No further polishing was observed at the end of 900 hours of testing (Fig. 11c).

The replicated worn surface of the alumina–titanium carbide composite roller after 100 hours of testing is shown in Figure 12a. The worn surface contained

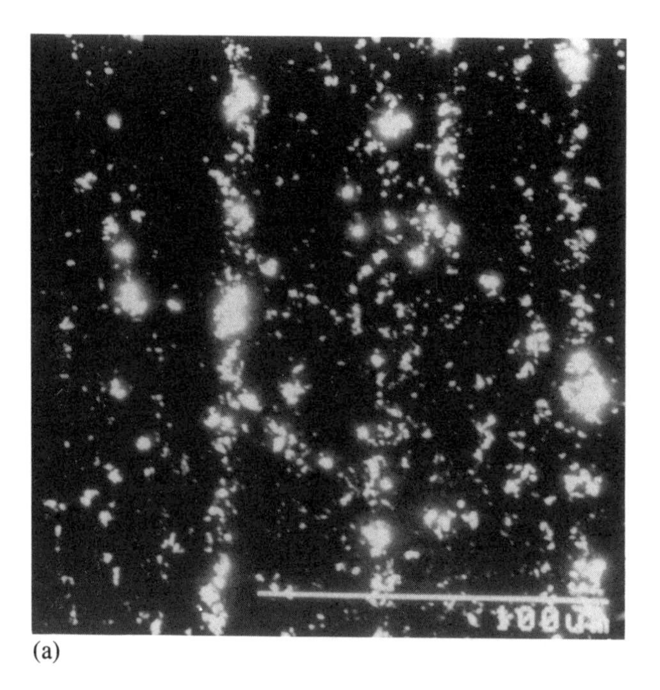

(a)

Figure 12 Scanning electron micrographs of alumina–titanium carbide composite after testing for (a) 100, (b) 500, and (c) 900 hours.

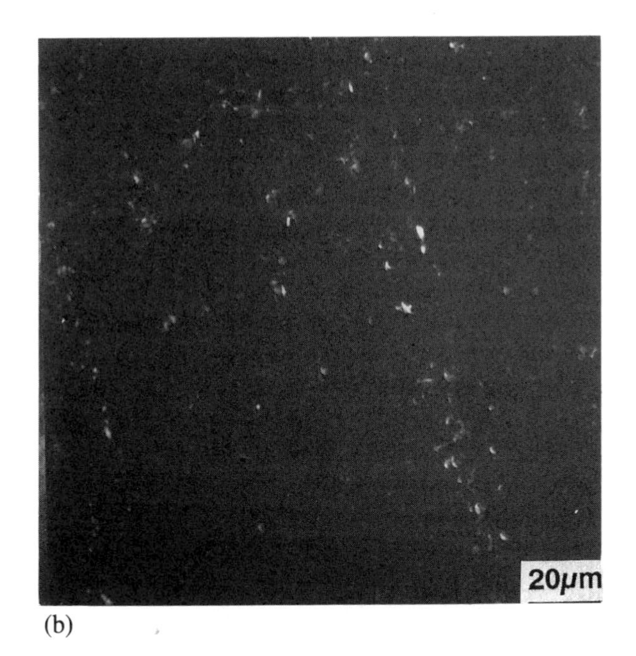

(b)

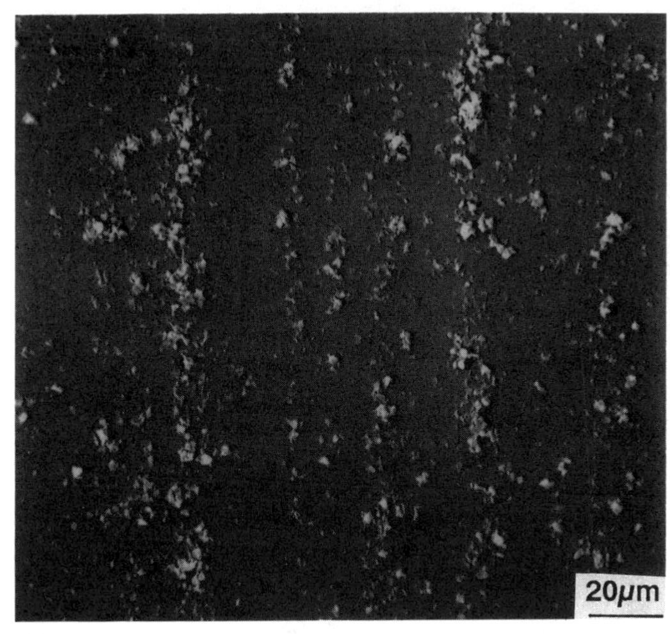

(c)

several small flaws and pores, which were found along the circumferential direction, suggesting that they were generated during the finishing operation. Many of these surface flaws are removed after 500 hours of testing as shown in Figure 12b. The surface appeared polished, although some fractured regions remained. Examination of the fractured regions at higher magnification showed evidence of intergranular fracture leading to pullout. However, this could have been caused during the finishing operation. The appearance of the worn surface after 900 hours did not change appreciably, as shown in Figure 12c.

The worn surface of the partially stabilized zirconia, observed at the end of 800 hours of testing, was found to contain several networks of microcracks over the entire contact area (Fig. 13). Examination at higher magnification clearly showed intergranular fracture. The same wear mode was also observed in rolling contact experiments, which caused greater wear.

The surface of steel rollers exhibited abrasive wear at the end of 900 hours of testing. No surface cracks could be observed. The steel pins inside the ceramic rollers were found to be covered with a patchy surface film. Occasionally abrasive wear marks were also observed.

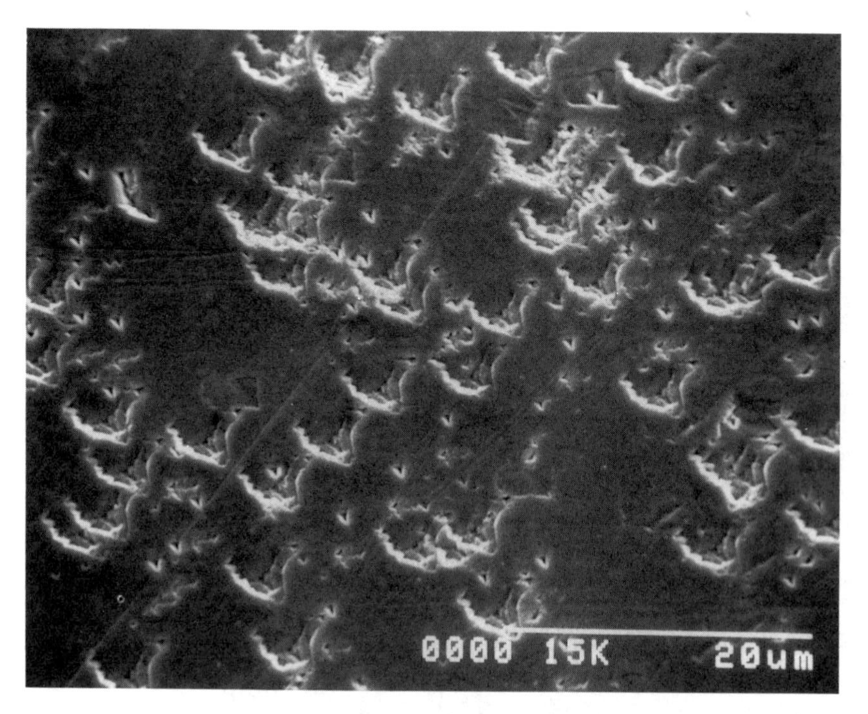

Figure 13 Scanning electron micrograph of worn surface of partially stabilized zirconia.

2. Scanning Electron Microscopy of the Cam Lobes

Replicas were taken from the nose of the cam lobes at 100-hour intervals and examined under the scanning electron microscope. Figure 14 shows micrographs of the cam lobes after 800 hours of testing. In general, the cam lobes tested against the ceramic rollers appeared smooth and polished. Some worn regions were found to be covered by surface films. Light abrasive wear marks were also occasionally observed, which may have been left from the grinding of the cam lobes. The small white regions in Figure 14c are the graphite nodules partially removed as a result of wear. Figure 14d shows the worn surface of the cam lobe tested against the partially stabilized zirconia roller, featuring a band of fractured regions believed to have been produced by wear. Sections were cut through the worn regions of several cam lobes after 900 hours of testing to observe any subsurface microcracking. These sections were carefully polished with 1 μm diamond paste followed by etching in 2% nital solution to reveal the microstructure. Figure 15 shows cross sections of the regions that were tested against the reacted silicon nitride and the alumina–titanium carbide composite; there is no evidence of subsurface crack formation or propagation.

IV. SUMMARY

In lubricated rolling contact tests, the friction coefficients of the alumina–titanium carbide composite, partially stabilized zirconia, and three types of silicon nitride (hot-pressed, pressureless sintered, and reacted) were virtually the same when tested in contact with nodular cast iron rollers. However, the wear constants varied. The partially stabilized zirconia exhibited the largest wear constant, which was due to the microchipping wear mode. Wear in the alumina–titanium carbide composite and the reacted silicon nitride occurred by grain pullout. This wear mode produced lower wear constants than microchipping, which was observed in partially stabilized zirconia, but higher than the submicrometer chipping of the pressureless sintered and the hot-pressed silicon nitride.

The wear of pressureless sintered silicon nitride, reacted silicon nitride, and alumina–titanium carbide composite roller followers after 900 hours of testing was found to be negligible, and these materials performed successfully in motorized valve train tests. The partially stabilized zirconia roller followers developed several networks of microcracks after 800 hours of testing and thus were not considered to be suitable for cam follower applications.

The wear of the cam lobes tested against steel and ceramic roller followers increased for the first 400 hours as a result of run-in, and no appreciable wear was observed after that. The wear of the cam lobes was predominantly due to polishing, and no subsurface cracking was observed.

There did not exist any strong correlations between the wear rates of the

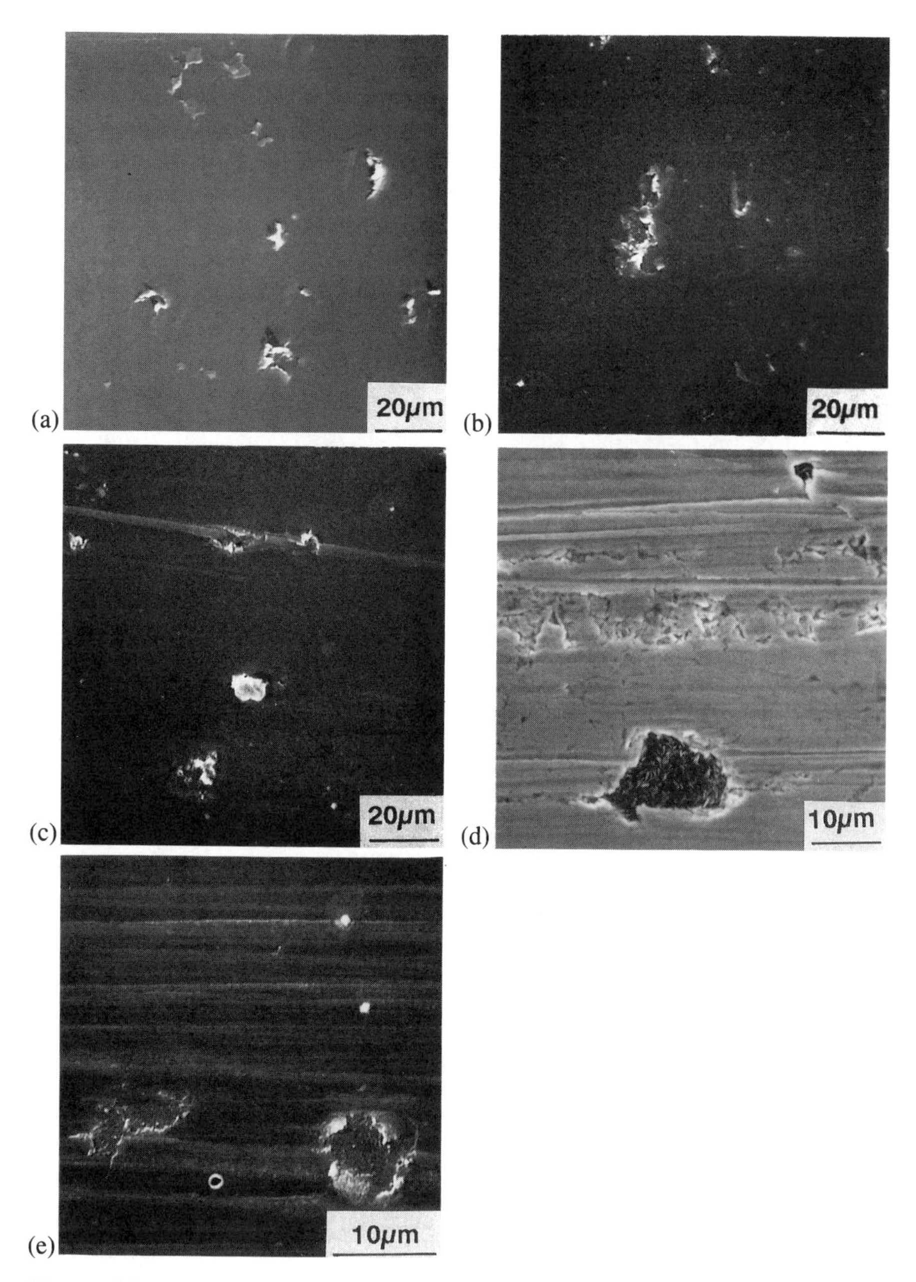

Figure 14 Scanning electron micrographs of cam lobes tested against (a) pressureless sintered silicon nitride, (b) reacted silicon nitride, (c) alumina–titanium carbide composite, (d) partially stabilized zirconia, and (e) steel rollers.

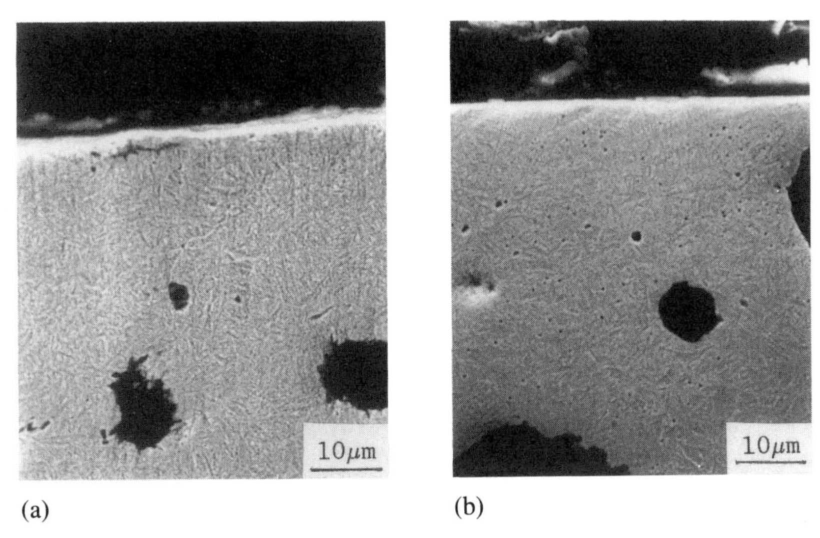

Figure 15 Sections through the worn cam lobes of (a) reacted silicon nitride and (b) alumina–titanium carbide composite.

ceramic materials in laboratory rolling tests and the reduction in diameter of the ceramic materials in motorized valve train tests. Two conjectures are offered to explain this lack of correlation. First, the weight loss measured in rolling tests was not large enough to change the diameter of the ceramic roller. Second, in the single-element tests, the ceramic rollers did not experience a traction force on the surface, as they do in motorized valve train tests. This traction force polishes the cam lobes, and cast iron material was transferred to the harder ceramic roller. This transfer film covered some of the grinding furrows or surface defects that were generated during the finishing operation.

These results indicate that hot-pressed, pressureless sintered, reacted silicon nitride, and alumina–titanium carbide composite rollers investigated in the present research should be considered to be potential candidates in engine valve train applications. However, successful use of these materials will depend on production of the components with the desired dimensional tolerance at an acceptable cost.

REFERENCES

1. J. T. Neil, *Mech Eng. 106*(3), 36 (1984).
2. R. A. Harman and C. W. Bearsley, *Mech. Eng. 106*(5), 22 (1984).
3. D. W. Richerson, *Modern Ceramic Engineering*, Dekker: New York, 1982.
4. J. W. Lucek and J. G. Hannoosh, in *Engineered Materials for Advanced Friction*

and Wear Applications, F. A. Smith and P. J. Blau (Eds.), ASM International: Metals Park, OH, 1988, p. 205.

5. R. T. Cundill, Society of Mechanical Engineers, (London), Publication No. 33 (1990).
6. H. M. Dalal, Y. P. Chiu, and E. Rabinowicz, *ASLE Trans. 18*(3), 211 (1975).
7. B. Bhushan and L. B. Sibley, *ASLE Trans. 25*(4), 417 (1982).
8. A. K. Gangopadhyay, H. S. Cheng, S. T. Harman, and J. M. Corwin, SAE International Congress and Exposition, Detroit, Paper No. 900401 (1990).
9. D. Zhu and H. S. Cheng, *Proceedings of the 17th Leeds–Lyon Conference in Tribology*, 1990.
10. H. Ishigaki, I. Kawaguchi, M. Iwasa, and Y. Toibana, *J. Tribol. 108*, 514 (1986).
11. T. E. Fischer and H. Tomizawa, *Wear, 105*, 29 (1985).
12. H. Tomizawa and T. E. Fischer, *ASLE Trans. 30*, 41 (1987).
13. J. Breznak, E. Breval, and N. H. Macmillan, *J. Mater. Sci. 20*, 4657 (1985).
14. J. F. Braza and P. A. Braza, in *Advances in Industrial Tribology—1990*, Y. P. Chung, (Ed.), Society of Tribologists and Lubrication Engineers: Park Ridge, IL, 1991.
15. J. F. Braza, H. S. Cheng, and M. E. Fine, *Scripta Met. 21*, 1705 (1987).
16. J. F. Braza, H. S. Cheng, and M. E. Fine, *Tribol. Trans. 32*(4), 439 (1989).
17. T. M. Sullivan, U. S. Patent No. 4,961,913 (1990).
18. J. F. Braza, H. S. Cheng, M. E. Fine, A. K. Gangopadhyay, L. M. Keer, and R. E. Worden, *Tribol. Trans. 32*(1), 1 (1989).
19. M. V. Swain, *Wear, 35*, 185 (1975).
20. A. K. Gangopadhyay, M. E. Fine, and H. S. Cheng, *Lubr. Eng., 44*(4), 330 (1988).

15

Ceramics in Mechanical Face Seal Applications

Ramesh Divakar

The Carborundum Co.
Niagara Falls, New York

ABSTRACT

The continually increasing demands on maximum attainable operating limits in mechanical sealing systems have necessitated the search for and development of new seal face materials. Ceramic seal materials have successfully substituted metals and alloys in seal applications by virtue of their superior physical, chemical, and tribological properties. Among ceramics of different types, the carbides have made excellent seal material candidates and are primarily responsible for significant improvements in seal performance. Seal evaluation criteria have not been standardized; however, the *PV* test is frequently used to assess operating limits. Seal failures are related to material characteristics and operating conditions. Several wear/failure mechanisms may operate simultaneously, thus adding to the severity of a particular application.

I. INTRODUCTION

Sealing devices of various types and configurations are available for mechanical systems. These may be contact or noncontact seals, dynamic or static seals. The shafts on which the seals are housed may be subjected to rotary or reciprocating motion. Mechanical face seals or end face seals are dynamic seals in which sealing takes place between surfaces either narrowly separated or in sliding contact. Such contact seals depend on the sealing fluid for lubrication of the sliding parts.

One of the most critical functions of a mechanical seal is to provide safe and reliable sealing of the working fluid. In performing this function, the seal also carries the hydraulic load. Overall seal performance is determined not only by the intrinsic characteristics of materials used in the seal assembly but also by a combination of factors including seal design and the operating conditions. However, it is reasonable to conclude that material properties impose an upper limit on the severity of the application that a mechanical seal can withstand. In general, a decrease in energy dissipated by the mechanical seal as a result of frictional losses, thermal losses, and viscous fluid losses is desired.

Ever since the development of mechanical face seals, there has been a constant thrust to improve their capabilities to meet harsher and more demanding operating conditions. Ceramic materials, in general, exhibit improved mechanical and thermal properties in addition to their excellent chemical inertness. These characteristics have made them very attractive candidates for sliding and sealing elements. The operating limit of a mechanical seal is usually defined in terms of the product PV, where P is the face pressure at the sealing interface and V is the mean velocity. In the late 1930s, carburized steel running against leaded bronze [with a PV limit of approximately 30 bar m/s (87,500 psi ft/min)] was considered to be acceptable [1]. Today, ceramic seal faces mated with a carbon–graphite counterfaces allow for a PV limit in excess of 175 bar m/s (500,000 psi ft/min). Mechanical face seals are good examples of the substitution of ceramics for metals being necessitated by higher performance requirements.

Mechanical face seals are used in a wide range of industrial pumping applications—from extreme high pressure, high speed operations in corrosive environments to milder service requirements. Table 1 shows some common examples of seal applications. Normally the fluid being sealed also acts as the lubricant. Fluid media encountered in these applications are generally poor lubricants. Therefore, the lubrication at the contact interface ranges from almost negligible to a situation where a hydrodynamic regime exists.

Figure 1 shows a schematic of a simple mechanical seal assembly. In its basic

Table 1 Common Seal Applications

Industry	Media	Issues
Chemical processing	Extremely corrosive	Toxicity: environmental safety; corrosion
Nuclear power	Radioactive	Safety; rapid maintenance
Paper	Wood pulp, cellulose	Corrision
Shipbuilding	Seawater, water, oil	Corrosion; abrasion
Sugar	Solids from raw juice, clarice, sludge	Contamination
Oil	Crude oil	Abrasion
Automotive	Engine coolant	Reliability

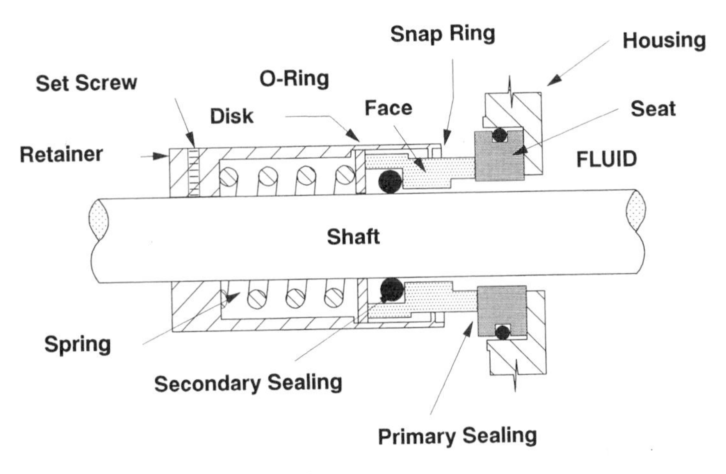

Figure 1 Schematic of a mechanical seal assembly.

form, the design of such a seal consists of radial planar surfaces, normal to the shaft axis, acting as a thrust bearing [2]. The two main components of interest are the stationary seal seat attached to the pump housing and the rotating seal face attached to the shaft. Contact takes place between the two while the shaft, by design, allows for positive closing forces to allow for wear (usually by means of spring-loading). This basic seal concept accounts for a large percentage of the seals found in pumps, compressors, and agitators.

II. DESIRABLE SEAL CHARACTERISTICS

A. General Requirements

There is no doubt that the life of a seal is determined to a large extent by the state of lubrication at the sealing interface. It is also known that besides seal design and operating conditions, several material parameters either directly or indirectly influence the retention a fluid film at the interface. These include thermal and physical properties of the seal material and surface topography. In many cases, the thin layer of fluid assumed to be present between the sliding faces does not prevent contact between the two surfaces [3,4]. Tribological properties of the materials are extremely important whenever contact occurs as a result of insufficient separation between the two seal faces at the interface. While it is not possible to expect all the desirable characteristics in any single class of materials, ceramics in general have several attractive features that make them excellent seal candidates. Some of these characteristics are discussed in the subsections that follow.

1. Physical Properties

Hardness is an important parameter because it imparts abrasion resistance [5]. The higher the hardness, the better the abrasion resistance of the material . High hardness could also lead to plowing of the softer counterface material, with the production of a "transfer" layer at the sliding interface. The role of a transfer layer in the sliding process is significant and is discussed in a later section. A high elastic modulus improves dimensional stability at high pressures. In addition, it plays an important role in determining the thermal shock resistance of ceramics, as discussed below.

2. Thermal Properties and Thermal Shock Resistance

The thermal properties of seal materials are extremely important because they not only influence interfacial temperatures but also control thermal distortion and have a major impact on thermal stress fracture, more commonly referred to as thermal shock. In seals used in pumps operating at high pressures and speeds, frequent dry running situations can occur in which the frictional heat generates high temperatures instantaneously at asperity contacts. This causes temperature gradients between adjacent volume elements of the material near the interface. When the cold lubricant reenters the interface region, these regions in the seals can experience sudden reversals in temperature. Such excursions in temperature set up transient stress distributions. When combined with the mechanical component of the applied stress (compression plus shear), these stresses can exceed the characteristic strength of the material and thus cause failure. Since thermal shock is a fairly significant problem for many mechanical seal applications, especially in ceramics, the following paragraphs treat this subject in some detail.

Thermal stress failures are common in many situations involving either the manufacture or use of not only technical ceramics but also glasses, whiteware, and refractories [6]. In general, thermal stresses can arise in a variety of ways. The existence of expansion coefficient mismatch or anisotropy in a material, the imposition of a nonuniform temperature on a body, and incomplete relaxation of strains upon cooling all can lead to thermal stresses. It is well known that in most ceramics, except some glasses and vitrified ceramics, stress relaxation due to plastic deformation, such as creep, occurs only at temperatures exceeding 1000°C. Given the brittle nature of ceramics in general, it is therefore understandable that ceramics are more prone to thermal stress failures than metals. In light of arguments presented for the case of seals (see above), the discussion of thermal shock due to transient temperature and stress distributions (as opposed to quasi-static conditions) is appropriate. The stress field at the sealing interface is usually estimated by a superposition of the mechanical and thermal components of stress. The mechanical component arises from normal and shear tractions at the contact interface, and the thermal component is produced by temperature gradients due to frictional heating. There have been several treatments of this type of problem either via integral transform techniques [7,8] or via finite element analysis [9,10]. When boundary conditions can be simplified, analytical solutions have also been developed.

There are two approaches to evaluating the thermal stress resistance of ceramics. The first approach, based on the theory of elasticity [11], attempts to predict the thermal shock resistance of ceramics by calculating the maximum thermal stresses required for crack nucleation. In general, the thermal stress resistance parameter (designated in the literature as R, R', R'', etc.) can be defined as the maximum temperature drop that a material can withstand prior to failure. Kingery [12] has considered three extreme cases for which analytical solutions exist. In the following three cases, a nondimensional quantity, the Biot modulus, is defined thus:

$$\beta = \frac{rh}{K} \tag{1}$$

where

r = dimension of specimen
h = heat transfer coefficient
K = thermal conductivity

Case a: Instantaneous temperature change ($\beta >> 1$)

$$R = \frac{\sigma(1 - \gamma)}{E\alpha} \tag{2}$$

Case b: Constant rate of heat transfer ($\beta < 1$)

$$R' = \frac{K\sigma(1 - \gamma)}{E\alpha} \tag{3}$$

Case c: Constant rate of heating or cooling

$$R'' = \frac{K\sigma(1 - \gamma)}{\rho c \ E\alpha} \tag{4}$$

where

E = elastic modulus (i.e., Young's modulus)
α = coefficient of thermal expansion
γ = Poisson's ratio
σ = fracture strength
ρ = density
c = specific heat
K = thermal conductivity

In reality, the situation in case a is not realized; Equation (2) allows for neither the time-dependent nature of transience nor finite rates of heat transfer (i.e., $\beta < \infty$) [6]. Case b represents the simplest approximation to the transient stress analysis, whereas case c defines the maximum allowable rate of surface heating (or cooling) prior to crack initiation.

The second approach, proposed by Hasselman and others [13–15], for the determination of thermal shock resistance, assumes that flaw initiation cannot be avoided and that the flaws already exist in the ceramic. When such a material is subject to thermal stresses, catastrophic failure would result if the total available elastic energy exceeded the total fracture energy required to propagate a crack over an area equivalent to the cross-sectional area of the material [14]. On this basis, two additional thermal shock parameters can be defined, (viz., R''' and R'''') as follows:

$$R''' = \frac{E}{\rho^2(1 - \gamma)} \tag{5}$$

for materials with similar fracture surface energies, and

$$R'''' = \frac{E\gamma}{\rho^2(1 - \gamma)} \tag{6}$$

for materials with different fracture surface energies (λ).

For crack initiation and instantaneous thermal stress fracture to be avoided, Equations (2)–(4) require that the material possess high values of fracture strength and thermal conductivity and low values of specific heat, Young's modulus, coefficient of thermal expansion, density, and Poisson's ratio. On the other hand, for high thermal shock resistance, Equations (5) and (6) require materials to exhibit high values of surface energy and Young's modulus, and low values of fracture strength. The latter two requirements are in direct contradiction to those needed for Equations (2)–(4). Hence, if attempts to prevent thermal fracture initiation for a material are unsuccessful, increasing Young's modulus and fracture strength can be counterproductive, per Equations (4) and (5) [16]. Parameters R''' and R'''' apply to severe thermal environments, where crack initiation cannot be avoided, or where the material contains preexisting flaws from processing. There are other extensions to the two approaches outlined here, including the Weibull treatment of flaws in ceramics, but these are beyond the scope of this chapter.

In summary, the use of these thermal shock parameters should be based on a knowledge of the severity of the thermal environment and validated by careful experimental observations. Such observations indicate that for mechanical seals, the parameters R' and R'' often are more applicable when materials are mated with carbon–graphite [1].

In addition to the characteristics described in the preceding paragraphs, there are other properties that determine seal performance. Since seals are used in many corrosive applications, their corrosion resistance is important. Degradation of the interface due to corrosive attack could enhance the wear rate of the material and lead to a loss of sealing. A material that is otherwise acceptable may be unable to withstand corrosive environments. Ceramics, however, are generally chemically inert and exhibit good corrosion resistances.

B. Role of Sliding Interface

Material combinations that exhibit low friction and low wear rates are obviously desirable. In this sense, seal face materials must be compatible. The sliding interface plays a key role in shaping the tribological behavior of the seal and thus its overall performance. Fluid film stability is dependent on the nature of the interface and the dynamics of the system. Indeed, in practice, attempts are made to tailor the initial surface topography of the seal surface to allow for pockets of lubricant to be trapped between the peaks and valleys of the surface. Since contact does occur between the sealing faces, their surface topography is expected to change with time. Therefore, it is unclear whether the goal of retaining lubricant pockets is assured by an initial surface topography. It is also unclear whether a steady state topography and/or wear rate is achieved in seal applications as in dry sliding.

It is well known that a transfer layer is formed on the surfaces of one or both sliding components. There have been numerous observations of material transfer in metals and nonmetals in both lubricated and dry sliding modes (e.g., see Refs. 17–20). This layer essentially consists of an intimate mixture of very fine crystallites from both sliding surfaces and the products of reaction with the environment. The very nature of the sliding interface is changed when these layers form. It is the properties of these surface films that determine the magnitude of the dynamic friction coefficient [21]. Additionally, it has been shown that fragments of wear debris are structurally and chemically similar to the transfer layer and, therefore, are derived from the transfer layer [17,22]. In boundary-lubricated carbon–metal contacts, transfer layers have been found to be essential for low friction and low wear [21]. However, the interaction between such fine-scale, highly surface-active crystallites and the fluid (lubricant) medium is largely unknown.

It seems clear from the discussion above that it is desirable both to promote and to retain transfer layers at the sliding interface. The choice of sliding materials must be made with this aspect in mind. Indeed, one of the main reasons for choosing carbons in sliding applications is their tendency to form transfer films on the seal surfaces. However, not all transfer layers are desirable—only those that promote solid film lubrication.

III. CERAMIC SEAL MATERIALS

Seal materials may be generally classified into three main classes, namely metals, ceramics, and polymers. Each of these material systems has its advantages and disadvantages and, therefore, the systems are selectively used depending on the application. As mentioned earlier, the state of the art today involves the use of advanced ceramics in sealing applications. This discussion is restricted to ceramic seal materials and their distinguishing features, although the term "ceramic" is used in a much broader sense than is conventional. Therefore, metal-bonded "cermets" such as tungsten carbide and carbon–graphites are briefly reviewed as well. Although oxide ceramic materials are included, more emphasis is placed on carbon and silicon carbide materials because of their widespread use. Silicon nitrides have seen only limited use because of their inferior thermal properties compared to the carbides. Table 2 outlines a general scheme of classification and gives some typical examples in each category.

Table 3 shows some important physical properties of selected ceramic materials.

A. Oxide Ceramics

The most often used oxide ceramic seal material is alumina (Al_2O_3). It is available commercially in several grades of purity, with impurity contents typically ranging

Table 2 A General Classification of Seal Materials

Metals
Steel; alloys of Cr, Ni, Mo; cast materials; bronze
Spray or diffusion-coated with oxides or carbides of Cr, Ni
Sintered WC with 6–12% of binder (Ni, Ni/Cr, Co)
Ceramics
Oxides
alumina (several grades of purity)
zirconia
Nonoxides
Carbon: natural graphite, carbon–graphite, electrographite
Silicon carbide: pressureless sintered SiC, CVD SiC, siliconized graphite, reaction sintered SiC
Polymers
Polytetrafluoroethylene

Table 3 Physical Properties of Selected Seal Materials

Material	Density (kg/m^3)	Hardness[a] (GPa)	Elastic modulus (GPa)	Compressive strength (MPa)	Coefficient of thermal expansion (10^{-6}/K)	Thermal conductivity ($Wm^{-1}K^{-1}$)
Al_2O_3 (99.7%)	3900	1700–2300	350	2000–3500	6.6–7.0	21–29
ZrO_2 (PSZ)	5740	1250	210	2000	10.2	2.5
WC (6–12% Co)	14,300–14,900	1230–1640	550–630	3200–6200	4.6–7.0	80–115
Carbon–graphite (Sb impregnated)	2100–2300	84[b]	26–27	280–350	4.0–4.7	7–13
Siliconized graphite	2650	2500[b]	135	600	3.0	125
Reaction-sintered SiC	3090	1900[c]	380	1000	5.0	130
Pressureless sintered SiC	3100	2800[c]	410	3900	4.0	126

[a]Hardness conversions from one unit to another are material dependent and are available only for selected materials (e.g., steel).
[b]Scleroscope.
[c]Knoop hardness.

from 0.1 to 10% by weight or higher. It is made by conventional ceramic processing techniques. The purity of Al_2O_3 determines its mechanical properties and chemical resistance to a large extent. Its comparatively low thermal conductivity (see Table 2) limits its thermal stress resistance; hence its usage generally is confined to medium pressure ranges ($p < 25$ bar) [23]. Figure 2 shows a typical microstructure of 99.8% Al_2O_3 with equiaxial grains approximately 2 μm in size.

Partially stabilized zirconia (ZrO_2) containing Y_2O_3 or MgO as a stabilizing agent is an interesting material because it offers enhanced fracture toughness compared to other ceramic materials. However, it is used less than Al_2O_3 primarily because of its poor thermal properties. In milder seal applications, corrosion resistance limits its use. Among commonly encountered reagents in the chemical processing industries, ZrO_2 is corroded rapidly by HCl and HNO_3 [24].

B. Nonoxide Ceramics

Carbon is a generic term used to describe a wide range of materials from naturally occurring graphite to electrographite to impregnated carbon–graphite. Natural

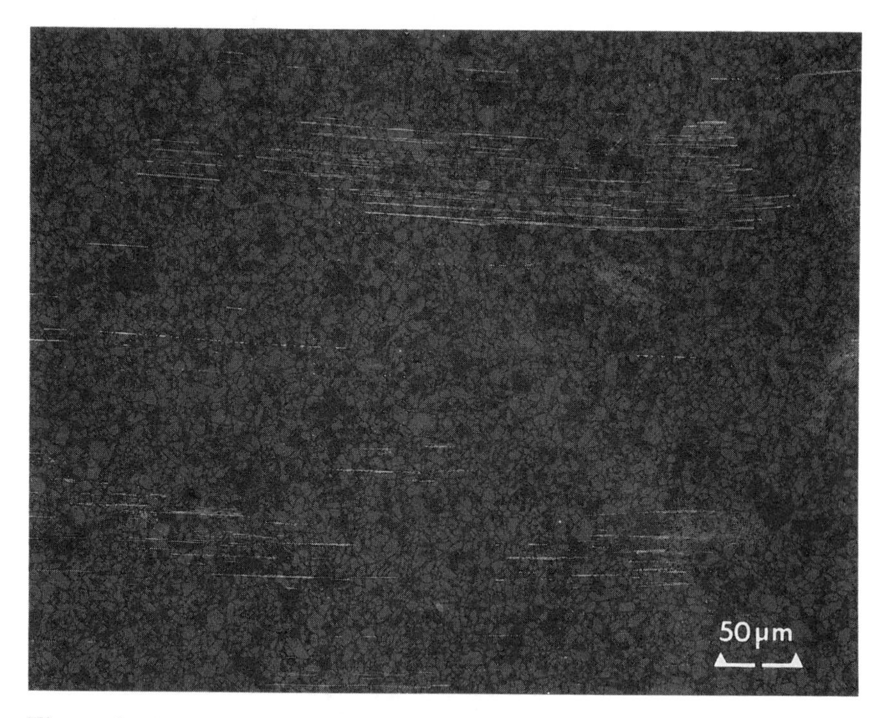

Figure 2 A typical microstructure of 99.8% alumina.

graphite has a hexagonal crystalline structure in which sheets of hexagonally arranged carbon atoms (basal planes) are stacked successively on top of each other. Within the sheets, bonding is covalent and atoms are closer together. The intratomic spacing between the basal planes is much greater, and bonding is due to weak van der Waals forces and π-electron interactions. During sliding, shear occurs primarily between the basal planes, which are easy to shear, thus leading to a low value of friction coefficient [21].

Most manufactured carbons are multiphase materials containing different amounts of the graphitic component. Their structure and properties vary widely, depending on the raw materials involved and the type of processing. Various carbonaceous materials such as coke derived from coal, residues from oil distillation, and carbon blacks are mixed together with a suitable binder such as pitch or tar, and pressed to desired shapes. They are then fired to approximately 1000 °C in a protective atmosphere to produce carbon–graphite [25]. Some grades are heat treated to above 2000 °C, where graphitization occurs to an extent that is controlled by the types of raw material used. This product, called electrographite, has improved thermal conductivity. The term "electrographite" is often used to identify materials that have been graphitized by heating via the passage of electric current through the structure. Carbon and graphite materials produced in such fashion are usually porous. Their properties are improved by impregnating them with resins, metals, inorganic salts, and so on, to yield lower permeability and other desirable properties.

If the definition of ceramics could be extended to cermets, tungsten carbide (WC) would be appropriate for discussion here. Several grades of WC differ in their composition because of the different binders that are used in their fabrication. The binder used is usually Ni, Co, or Ni-Cr at levels of 6–12 wt%. While WC materials are much tougher than other ceramics, their corrosion resistance is impaired by the occurrence of selective attack of the binder phase.

It is fairly clear that it was primarily the emergence of the silicon carbide group of materials that contributed to a substantial increase in the capabilities and service life of mechanical seals [23]. Silicon carbide (SiC) comes in several commercial varieties: chemical vapor deposited (CVD) SiC, siliconized graphite (SiC), reaction-bonded SiC, and pressureless sintered SiC. CVD SiC or pyrolytic SiC is a crystalline coating deposited on a graphite substrate at high temperatures using a gas such as methyl trichlorosilane. The gases react and pyrolize onto the graphite substrate but do not react with the substrate itself, thus forming a mechanical bond between the coating and the substrate. The usefulness of CVD SiC is limited by the thermal and mechanical differences between the coating and substrate and the weak bonding between them [26]. Siliconized graphite is also a coated form of SiC on a graphite base, made usually via a chemical vapor reaction (CVR) process. Here, depending on the permeability of the graphite substrate, a thin layer of the surface of the substrate is chemically converted to silicon carbide. Some form of impregnation is usually necessary to eliminate residual porosity.

Reaction-bonded SiC and pressureless sintered SiC are manufactured by more traditional means of processing ceramics. The major difference between these two forms is the sintering process and the resultant microstructure. Reaction sintering (or reaction bonding) is conducted in a silicon environment, where silicon reacts with graphite and the carbon residue of the binder to form silicon carbide. Residual silicon metal fills the pores in the material, resulting in a microstructure consisting of approximately 8–12 wt % free Si, depending on SiC grain size, graphite, and binder content, plus a matrix of SiC. Pressureless sintered SiC is produced by sintering SiC in the presence of a small amount of sintering aids (usually < 2 wt %) such as boron, carbon, or aluminum. A typical microstructure of pressureless sintered SiC (Fig. 3) consists of a single-phase, equiaxial microstructure of SiC grains, which are roughly 6–8 μm in size. While the microstructure of pressureless sintered SiC contains no free silicon, the dark regions have been associated with porosity, B_4C, or C particles [27–29]. Figure 4 shows a typical microstructure of reaction-sintered SiC, with bright regions indicating the free silicon phase in a matrix of SiC. Both reaction-sintered SiC and pressureless sintered SiC possess excellent thermal shock resistance. In addition to its superior hardness, the latter material exhibits perhaps the best corrosion resistance among ceramic seal materials.

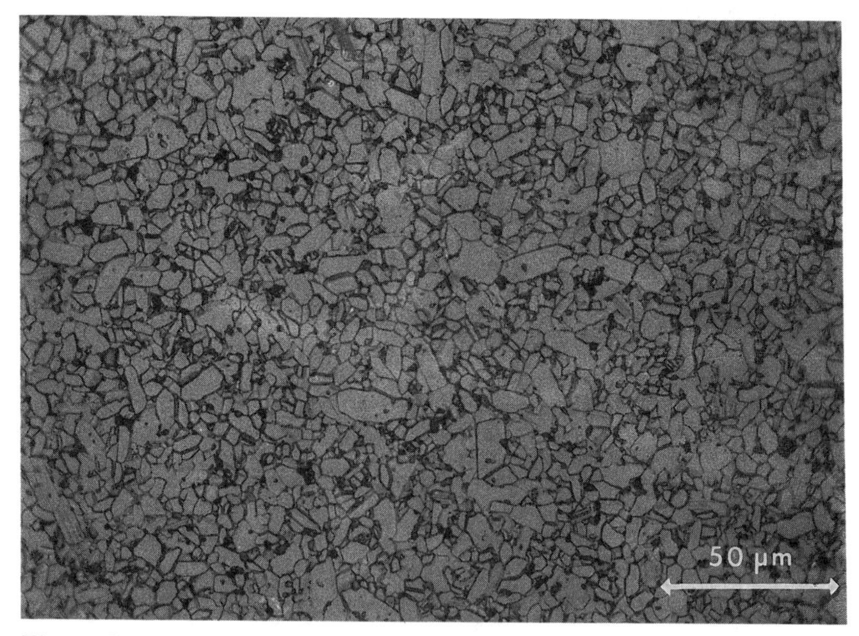

Figure 3 A typical microstructure of pressureless sintered SiC.

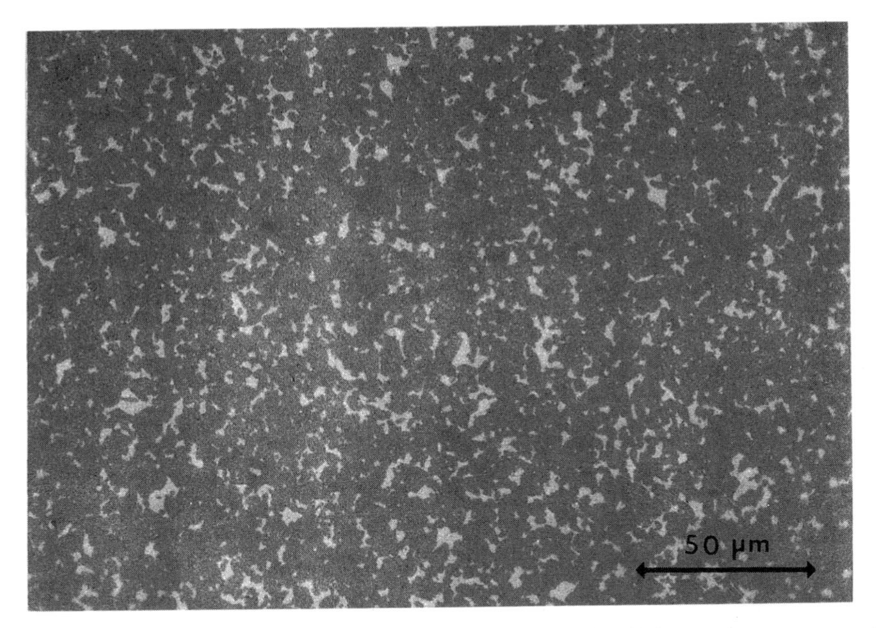

Figure 4 A typical microstructure of reaction-sintered SiC; lighter regions correspond to Si metal.

IV. SEAL EVALUATION CRITERIA

As noted earlier, it is not easy to evaluate a mechanical seal with the aim of separating the contributions to performance of the seal material itself from the design of the seal and the operating parameters. Unfortunately, despite this complexity, none of the tests used to measure seal performance is standardized throughout the industry.

A. Performance Limits

The most common approach to seal evaluation is to determine a performance limit or *PV* limit of the seal for a given design and medium that is consistent with the intended application. The term *PV* is defined by the following expression:

$$PV = [\Delta p(b - k) + P_F]V_m \tag{7}$$

where

Δp = differential pressure
b = balance ratio
k = pressure gradient factor
P_F = spring pressure
V_m = mean seal face velocity

The total pressure at the seal interface is due to two components, namely differential hydraulic pressure acting on the seal and the spring pressure. In Equation (7), the hydraulic balance ratio b is simply the ratio of the hydraulically loaded area to the nominal area of contact at the seal face, and it is determined by seal design. The pressure gradient factor k may take values from 0 to 1; for water it is usually taken to be 0.5 [30]. The power consumption P_w of a seal is related to the PV value by the expression

$$P_w = (PV)Af \tag{8}$$

where
A = nominal area of contact
f = coefficient of friction

The value of the coefficient of friction f depends on the operating lubrication regime. It may range from 0.02 to 0.1; usually an intermediate value is used. Alternatively, it can be calculated if P_w is known.

The performance limits for a seal material combination are usually established by conducting a series of PV tests. Since the test has not been standardized, the exact methodology may vary. Usually, a number of tests are run for a sufficient length of time (50–100 hs) at each level of PV. The temperature of the medium is held constant during the test. The surface topography of the seal materials is measured both before and after each test. Several parameters obtained via profilometry may be used to measure surface roughness (R_a, R_t, R_z , etc.)* The wear of each seal is determined, usually by the change in the seal height as measured from a suitable reference point. In addition, the power consumption (hence, coefficient of friction) and leakage rate are continuously monitored. Seal surfaces are examined by optical and electron microscopy to determine the extent of wear. Figure 5 is a schematic representation of wear rate of a seal material (either the rotating or the stationary seal) versus PV. Usually, when carbon is being used as the expendable seal face, its wear rate is plotted against PV. The dashed line could represent a maximum allowable wear rate to ensure a service life of a known duration. In Figure 5 it is obvious that material C performs the best, since it can withstand a much higher PV level than materials A and B; materials A and B have reached their PV limits, as indicated by the large increase in wear rates. Some representative data for PV limits for different seal material combinations are cited by Johnson and Schoenherr [30].

*R_a: is the arithmetic average of the distance of the filtered or unfiltered surface roughness profile from its mean line, R_t is the maximum peak-to-valley distance in the assessment length, and R_z is the mean of all peak-to-valley heights in the assessment length. The assessment length is an integer multiple of sampling length.

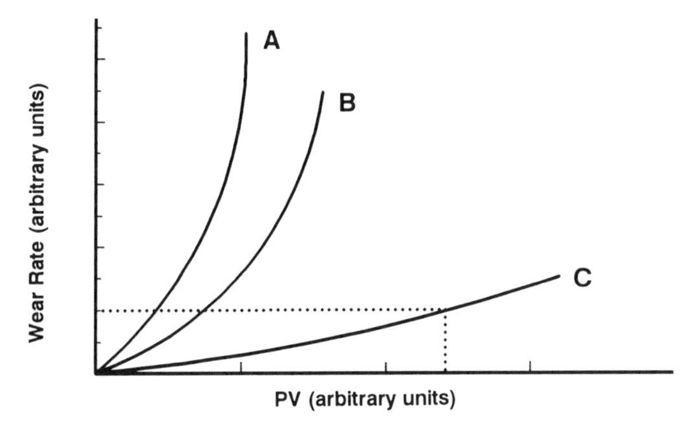

Figure 5 Representation of typical data from a *PV* test.

There are several variations of the *PV* test. The hydraulic loads maybe stepped up by known increments at fixed intervals of time (say, 30 min). Temperature measurements can be carried out on the stationary seal close to the sliding interface. The liquid temperature differential across the seals can also be measured. The temperature at the seal face can be estimated using finite element analysis with known boundary conditions [31].

B. Temperature Margins

The seal chamber or seal housing is the area enclosing the seal assembly (and the medium) in a pump (see Fig. 1). The dissipation of frictional heat occurs mainly by conduction through the body of the seal material. The primary heat sink is the fluid medium in the seal chamber (i.e., fluid far away from the seal interface). For the chamber fluid to operate as a heat sink, there must be a temperature differential between the interfacial fluid (i.e., fluid in contact with the seal interface) and the chamber fluid. When the maximum temperature of the chamber fluid is such that the interfacial fluid begins to boil, vaporization of the interfacial fluid takes place, leading to excessive wear at the interface. This unstable condition is usually marked by "puffs" of vapor leakage or fluctuating frictional torque [32]. This limiting chamber temperature can be plotted against the sealed pressure to produce an envelope of operating conditions of a seal for a given speed and fluid.

The difference between the boiling point of the fluid at the seal chamber pressure and the maximum chamber temperature is defined as the required temperature margin ΔT_r. The available temperature margin ΔT_a is defined as the difference between the boiling point of the fluid at the seal chamber pressure and the temperature of the fluid in the seal chamber under service conditions. To avoid unstable seal operation, ΔT_r must be greater than ΔT_a. Figure 6 schematically represents the operating envelope in water for a seal (where SiC and C constitute the sliding pair), while Figure 7 shows the dependence of ΔT_r on the operating speed and pressure [33].

C. Other Criteria

In addition to the above-mentioned tests, there are special tests for small seals. Small seals may be classified as those fitted on shafts with diameters less than 20 mm (e.g., dishwasher seals, water pump seals). They normally operate at relatively low differential pressures compared to larger seals but may run at higher speeds. Generally, *PV* levels encountered in, for example, water pump seal service are relatively low. In practice, such seals see a range of operating speeds and frequent stops and starts [34]. Typical areas of concern with small seals are grooving/abrasion of the seal surface by hard deposits that crystallize out from coolants, noise from vibrations, thermal shock, and vapor leakage.

Tests used to characterize or evaluate small seals attempt to incorporate the above-mentioned operating characteristics. A good example is an automotive water pump seal. A typical test of such a seal may use a mixture of antifreeze and water at an operating temperature of 95°C. In addition to providing a number

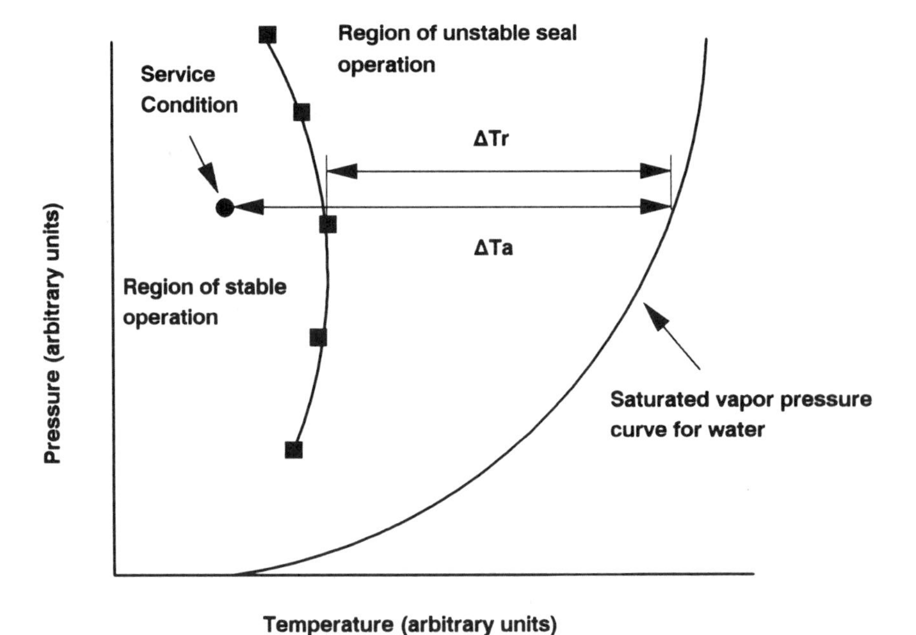

Figure 6 Schematic representation of a seal operating envelope in water; solid squares are typical experimental data. (Adapted from Ref. 33.)

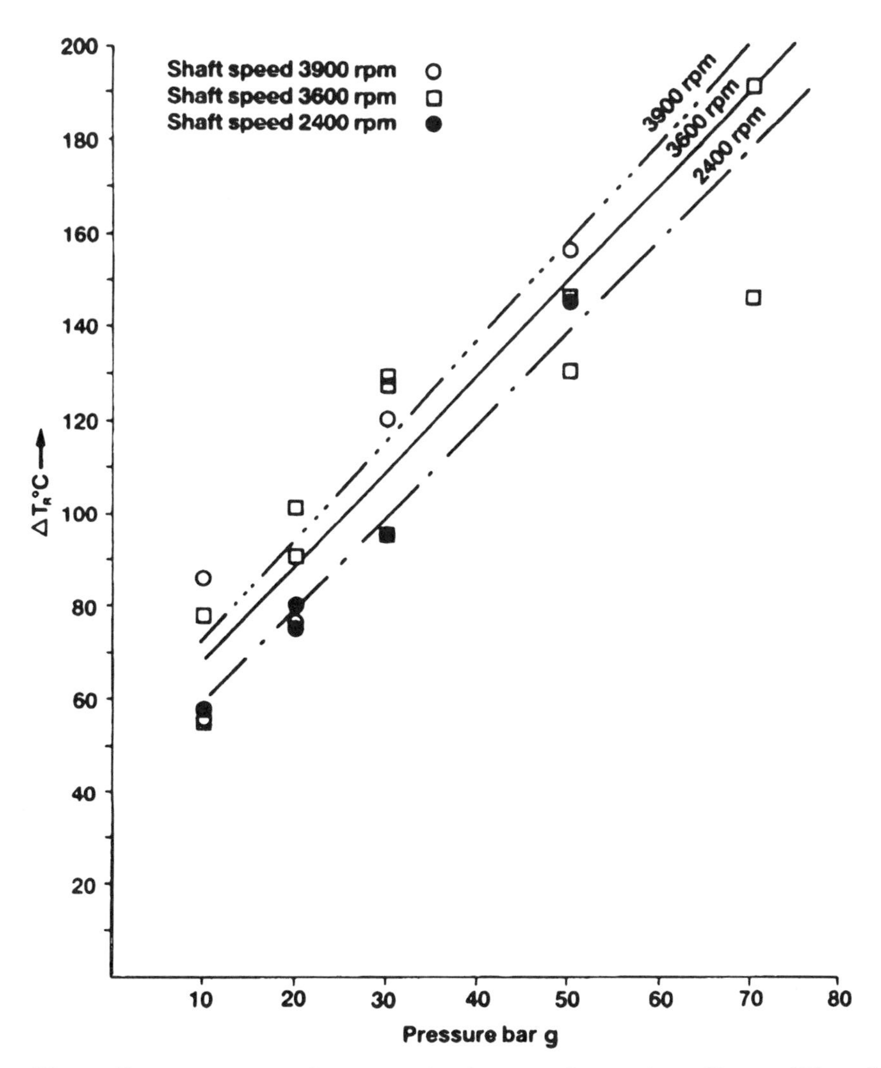

Figure 7 Dependence of ΔT_r on speed and pressure for a carbon–silicon carbide seal running in water. (From Ref. 33.)

of stops and starts, sliding speeds may vary from 1000 to 7000 rpm, with periods of stable running at constant speed. Measured parameters and monitored variables include seal speed, frictional torque, seal leakage, temperature, and surface topography via profilometry before and after each test. Specialized test rigs can be set up to conduct dry and wet (lubricated) sliding tests. Associated variations include cyclically changing from dry to wet running at increasing speeds to assess thermal shock characteristics of the sealing materials, short (100 hours) and long-term tests (300–500 hours).

Finally, it is necessary to mention tests that are used for the determination of corrosion resistances of materials in the fluid medium in which they are expected to function. This is especially relevant to seals used in chemical processing industries. The conventional method of the determining corrosion resistance of seal materials is by immersion corrosion tests. Steady state corrosion rates are found by noting weight changes after definite intervals of time. Testing is usually carried out for an extended duration—say, 500 hours—to ensure attainment of a steady state corrosion rate. Needless to say, such testing, although simple, is time-consuming and often not accurate enough when corrosion rates are low. Electrochemical corrosion techniques, on the other hand, are much faster, more elegant and accurate. They are a well-established family of techniques and have been used extensively in metals and alloys. Recently, they have been used to investigate corrosion rates in ceramics [35]. These techniques exploit the electrochemical nature of corrosion reactions and the charge transfer step (i.e., electron transfer) that such reactions entail. By imposing an external signal and monitoring the polarization behavior of the sample (or the working electrode) in comparison with a reference electrode, kinetic information can be derived on the corrosion reaction in the reagent/electrolyte of interest.

Table 4 shows corrosion rates of selected ceramics obtained from immersion corrosion tests of up to 300 hours in aqueous media typical of chemical processing

Table 4 Corrosion Rates of Ceramic Seal Materials

Reagent	Temperature (°C)	Corrosive weight loss ($mg/cm^2/yr$)[a]			
		α-SiC	RSSiC[b] (12% Si)	WC (6% Co)	99% Al_2O_3
98% H_2SO_4	100	1.8	55.0	>1000	65.0
50% NaOH	100	2.5	>1000	5.0	75.0
52% HF	25	<0.2	7.9	8.0	20.0
85% H_3PO_4	100	<0.2	8.8	55.0	>1000
70% HNO_3	100	<0.2	0.5	>1000	7.0
45% KOH	100	<0.2	>1000	3.0	60.0
25% HCl	70	<0.2	0.9	85.0	72.0
10% HF + 57% HNO_3	25	<0.2	>1000	>1000	16.0

[a]*Corrosion weight loss guide*:

>1000 $mg/cm^2/yr$	Completely destroyed within days
100 to 999 $mg/cm^2/yr$	Not recommended for service greater than a month
50 to 100 $mg/cm^2/yr$	Not recommended for service greater than a year
10 to 49 $mg/cm^2/yr$	Caution recommended, based on the specific application
<0.2 to 9.9 $mg/cm^2/yr$	Recommended for long-term service

[b]RSSiC, reaction-sintered SiC.

industries [36]. As noted earlier, sintered silicon carbide (α-SiC) exhibits outstanding chemical inertness.

V. SEAL FAILURES

Seal failures can be caused by any number of factors, including improper handling of brittle materials such as ceramics. It is difficult to isolate the effects of variables such as design, fluid, operating parameters, or materials on seal failure. It has been reported that in centrifugal pumps, 70% of the failures were directly attributable to improper pump maintenance and only a small fraction of failures resulted from excessive wear [37]. However, as pointed out in earlier sections, thermal and physical properties may indirectly influence the degree and extent of damage during operation. Excessive thermal and mechanical loads can lead to thermal distortion or "coning." This results in heavy contact near the inner or outer diameters of the seal with little or no contact elsewhere. Fluid starvation at these heavy contact locations can then cause excessive wear. Another common cause of seal failure is "heat checking" due to thermal shock. This is usually evidenced by radial cracks on the seal surfaces accompanied by leakage. Still another cause of excessive leakage, especially prevalent in small automotive water pump seals, is the buildup of deposits from the fluid on the seal surfaces. These hard deposits are also responsible for abrasion of the surfaces as noted below.

When the seal material is responsible for failures, it is conceivable that several mechanisms could occur in parallel. It is also highly likely that one of these mechanisms acts as a catalyst for the initiation or acceleration of another, thus synergistically increasing the probability of material damage. For example, freshly created surfaces at the interfaces due to adhesive or abrasive wear can be chemically attacked by the fluid. The corroded material is weaker and can be easily worn away by a subsequent adhesive or abrasive wear event.

A. Adhesive Wear

Adhesive wear is perhaps the most frequent wear mechanism encountered in seals, especially in the boundary or mixed lubrication regimes of operation. It involves contact and interaction between asperities on the two sliding surfaces. This situation is also encountered in dry sliding at low loads and speeds. Figure 8 shows a wear track of sintered SiC with evidence of plowing and smearing on the surface. Transfer layers play a dominant role in this type of wear process. Shear occurs along the weakest shear plane (such as a basal plane in graphite) parallel to the direction of sliding. When shear occurs either at or very close to the sliding interface, wear is minimal. Additives to the carbon that promote the formation of transfer films should be beneficial in reducing wear [21].

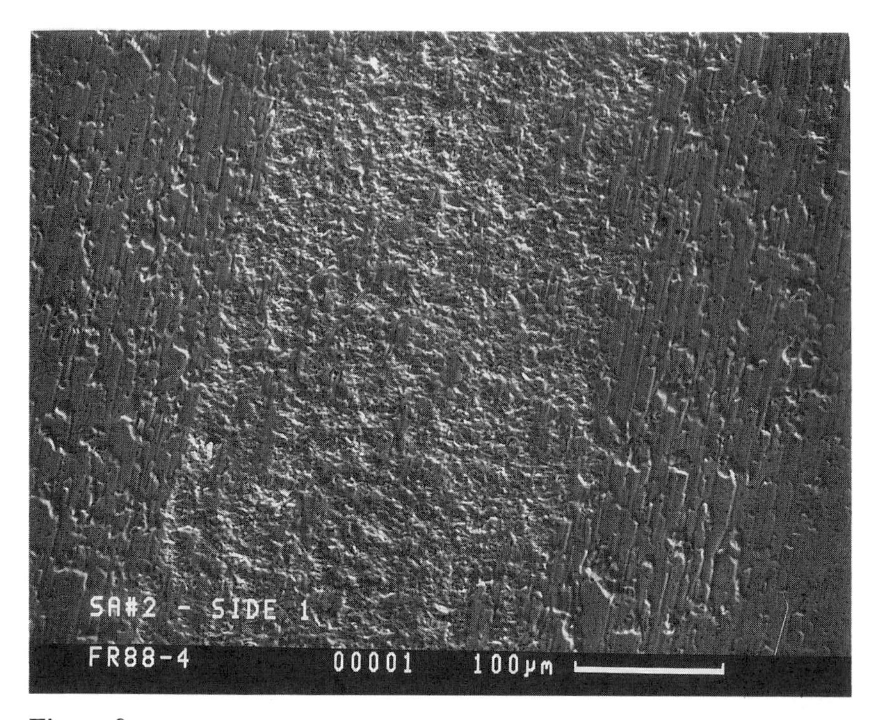

Figure 8 Scanning electron micrograph of a wear track of self-mated sintered SiC after unlubricated sliding at 10 N load and a speed of 0.1 m/s.

B. Abrasive Wear

Abrasive wear usually results in damage to seals when wear of the mating faces produces debris that abrades either or both surfaces. Often debris and sharp asperities on one of the surfaces can cut and remove material from the other surface. Zum Gahr [5] gives a detailed treatment of abrasive wear. Abrasive wear of seal materials such as carbon–graphite is also likely when impurities or hard nongraphitic constituents contained in them are released during the sliding process. Additionally, abrasive particles produced from downstream mechanical components and corrosion products are carried into the sealing gap by the fluid medium. Effective filtration or other means of preventing these particles from reaching the sealing gap should decrease abrasive wear.

Failures in water pump seals are frequently due to abrasion of the sliding surfaces by hard deposits formed from the coolant. This results in increased wear of the surfaces and/or increased leakage because the surfaces are physically separated by the occurrence of these deposits. These deposits may be silicates or zeolites depending on the chemistry of the coolants involved and the impurities dissolved in them. For example, it was shown in water pump end face seal tests [38] that adhesive deposits would form only when the sealed liquid contained Fe dissolved from the metal parts contacting the liquid, P from additives in the coolant, and Ca from the water used in the formulation of these coolants. As is common with chemical reactions, an increase in fluid temperature accelerates the deposition process and the resulting wear and leakage. Since frictional heat generated at the interface as a result of contact can be either transmitted to the fluid or dissipated away from the interface via conduction through the seal itself, it is clear that the thermal conductivity of the sliding material is an important factor in determining heat loss into the fluid.

C. Corrosive Wear

Corrosion is an electrochemical process and is a result of chemical incompatibility between the seal material and the process fluid or lubricant that is being sealed. The corrosion phenomenon can range from uniform corrosion to a more insidious type of corrosion (i.e., pitting). As with other chemical reactions, the kinetics of corrosion generally increase with temperature. Frictional heating at asperities or the increase in ambient temperature of the medium accelerates corrosive attack. As mentioned in the preceding section, thermal properties of the seal material play an important role in determining how fast interfacial heat is dissipated away and, therefore, indirectly determine corrosive damage. In many instances these corrosive products may promote abrasion.

D. Blistering

Blistering is a term commonly used to describe failure in carbon seal materials. Most carbon or carbon–graphite seals are commonly impregnated with resins or metallic elements to improve their properties. Although not well understood, it is believed that blistering is caused by differential expansion of the materials used in impregnation and the carbon matrix, subsequent evaporation of liquid lubricant within subsurface pores, and lifting of the surface [30]. Thermal stress cracks may also serve as sites for liquid entrapment. Figure 9 shows an example of extensive blistering on the wear track of a phenolic resin impregnated carbon–graphite seal that was mated with a reaction-sintered SiC. It has been suggested that the surface roughness of the mating seal is important in determining the extent of the blistering phenomenon. Matt lapping of tungsten carbide has reduced the blistering tendency in standard carbon grades [25]. Another approach to improving resistance to blistering in carbon is the use of alternate impregnants such as antimony, or a combination of impregnants.

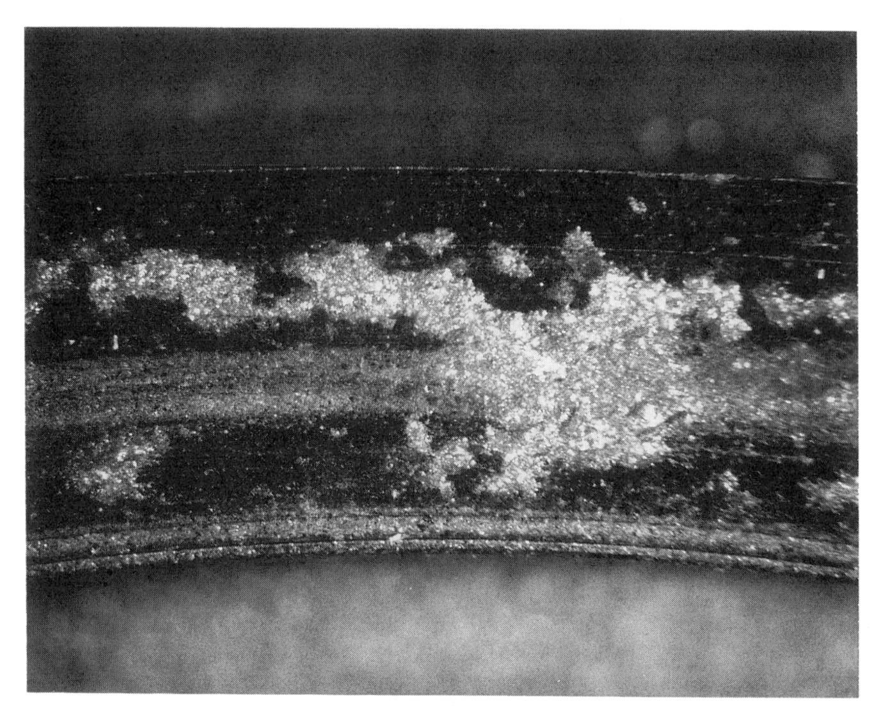

Figure 9 Wear track of a phenolic resin impregnated carbon–graphite seal showing extensive blistering after testing at 195 bar m/s against reaction-sintered SiC.

VI. SUMMARY

Improvements in properties, processing, manufacture, and knowledge related to ceramic-based design have enabled advanced ceramics to successfully replace metals as sealing elements in mechanical face seals. Ceramics are used today in mechanical face seals in a variety of mild and severe service applications. They exhibit many of the desirable characteristics that allow for greatly improved seal capability and service life. In material trends, Al_2O_3 grades are being gradually replaced in high volume mechanical seal applications by the SiC family of materials because of higher performance requirements. WC materials are still used where other ceramic materials are not available in all sizes and shapes or where brittleness is an issue. Carbon–carbon composites may become important as a replacement for carbon–graphite as a result of their potentially superior mechanical properties. However, such substitutions are not attractive economically at present [39].

From a materials standpoint, overall seal performance is determined by a combination of physical, chemical, and tribological characteristics of the materials being used. Material transfer processes on the seal surfaces play a key role in controlling the tribological behavior. Seal evaluation criteria vary widely because of a lack of standardization. However, the *PV* test is commonly used to establish performance limits. Seal failures are characterized by one or more wear modes that may occur in synergy.

REFERENCES

1. J. W. Abar, Rubbing contact materials for face type mechanical seals, *Lubr. Eng. 20*, 381 (1964).
2. E. W. Fisher, in *Standard Handbook for Mechanical Engineers*, T. Baumeister and L. S. Marks (Eds.), McGraw-Hill: New York, 1967, pp. 8–188.
3. F. E. Kennedy and J. N. Grim, Observation of contact conditions in mechanical face seals, *ASLE Trans. 27*, 122 (1984).
4. F. E. Kennedy, J. N. Grim, and C. K. Chuah, An experimental/theoretical study of contact phenomena in mechanical face seals, in *Developments in Numerical and Experimental Methods Applied to Tribology*, Butterworths: London, 1984, pp. 285.
5. K. H. Zum Gahr, Microstructure and Wear of Materials, Elsevier: Amsterdam, 1987.
6. R. Morrell, Thermal stress and thermal shock in ceramics—A survey of industrial problems and a review of test methods and thermal stress history, *NPL Report No. Chem. 66* (April 1977).
7. V. C. Mow and H. S. Cheng, Thermal stresses in an elastic half-space associated with an arbitrarily distributed heat source, *Z. Angew. Math. Phys. 18*, 500 (1967).
8. J. H. Huang and F. D. Ju, Thermomechanical cracking due to a moving friction load, *Wear, 102*, 81 (1985).
9. F. E. Kennedy and S. A. Karpe, Thermocracking of a mechanical face seal, *Wear, 79*, 21 (1982).
10. A. Mishra and T. Prasad, Residual stresses due to a moving heat source, *Int. J. Mech. Sci. 27*, 571 (1985).
11. B. A. Boley and J. H. Weiner, *Theory of Thermal Stresses*, Wiley: New York, 1960.
12. W. D. Kingery, Factors affecting thermal stress resistance of ceramic materials, *J. Am. Ceram. Soc. 38*(10), 3 (1955).
13. D. P. H. Hasselman, Elastic energy at fracture and surface energy as design criteria for thermal shock, *J. Am. Ceram. Soc. 46*(11), 535 (1963).
14. D. P. H. Hasselman, Unified theory of thermal shock fracture initiation and crack propagation in brittle ceramics, *J. Am. Ceram. Soc. 52*(11), 600 (1969).
15. R. W. Davidge and G. Tappin, Thermal shock and fracture in ceramics, *Trans. Br. Ceram. Soc. 66*(8), 405 (1967).
16. D. P. H. Hasselman, Thermal stress resistance parameters for brittle refractory ceramics: A compendium, *Ceram. Bull. 49*(12), 1033 (1970).
17. P. Heilmann, J. Don, T. C. Sun, D. A. Rigney, and W. A. Glaeser, Sliding wear and transfer, *Wear, 91*, 171 (1983).

18. L. H. Chen and D. A. Rigney, Transfer during unlubricated sliding wear of selected metal systems, *Wear*, *105*, 47 (1985).
19. M. Kerridge and J. K. Lancaster, The stages in a process of severe metallic wear, *Proc. R. Soc. London, Ser. A*, *236*, 250 (1956).
20. S. Norose and T. Sasada, Mutual transfer of rubbing materials and the mixing structures formed in lubricating oil, *J. Jpn. Soc. Lubr. Eng., Int. Ed. 1*, 5 (1980).
21. J. K. Lancaster, The wear of carbons and graphites, in *Treatise on Materials Science and Technology*, Vol. 13, D. Scott (Ed.), Academic Press: Orlando, FL, 1979, p. 141.
22. D. A. Rigney, L. H. Chen, M. G. S. Naylor, and A. R. Rosenfield, Wear processes in sliding systems, *Wear*, *100*, 195 (1984).
23. W. Schopplein and D. Zeus, Sliding materials—The state of the art and development trends, technical literature, Feodor Burgmann Dichtungswerke GmbH, Wolfratshausen, Germany.
24. H. H. Sturhhahn, W. Dawish, and G. Thamerus, Applications and properties of sintered zirconia. II. Wear behavior and application examples, *Ber. Dtsch. Keram. Ges.* *52*(4), 84–86 (1975) (in German).
25. K. Wakely, Mechanical seals: Some developments in face materials, *Tribol. Int.* *19*(4), 198 (1986).
26. R. Lashway, Silicon carbide for mechanical shaft seals, presented at the Ninth International Conference on Fluid Sealing, Paper No. A4 (April 1–3, 1981).
27. R. F. Davis et al., Microanalytical and microstructural analyses of boron and aluminum regions in sintered alpha silicon carbide, *Scanning Electron Microsc., Part 3*, 1161–1167 (1984).
28. R. Sherman, Auger analysis of hot-pressed and sintered silicon carbide, *J. Am. Ceram. Soc.* *68*(1), C7–C10 (1985).
29. R. Browning, J. L. Smialek, and N. S. Jacobson, Multielement mapping of α-SiC by scanning auger microscopy, *Adv. Ceram. Mater.* *2*(4), 773 (1987).
30. R. L. Johnson and Karl Schoenherr, Seal wear, in *Wear Control Handbook*, American Society of Mechanical Engineers: New York, 1980, p. 727.
31. N. D. Barnes, R. K. Flitney, and B. S. Nau, Mechanical seal performance—Face material test procedure, Technical Note No. 3012, BHRA, Cranfield, U.K., 1989.
32. R. K. Flitney and B. S. Nau, Mechanical seal performance—Procedure for the determination of temperature margin, Technical Note No. 3010, BHRA, Cranfield, U.K., 1989.
33. P. J. Dolan, D. Harrison, and R. Watkins, Mechanical seal selection and testing, *Proceedings of the 11th International Conference on Fluid Sealing*, Cannes, France, 1987, pp. 1–17.
34. A. J. Ryde-Weller, M. T. Thew, and R. Wallis, The performance of small commercial mechanical seals using various face materials when subjected to intermittent and variable speed running, *Proceedings of the 12th International Conference on Fluid Sealing*, Brighton, U.K., 1989, pp. 342–366.
35. R. Divakar, S. G. Seshadri, and M. Srinivasan, Electrochemical corrosion techniques for corrosion rate determination in ceramics, *J. Am. Ceram. Soc.* *72*(5), 780 (1989).
36. The Carborundum Company, technical literature, 1990.

37. R. K. Flitney and B. S. Nau, Reliability of mechanical seals in centrifugal process pumps, *Proceedings of the 11th International Conference on Fluid Sealing*, Cannes, France, 1987, pp. 17–45.
38. K. Kiryu, K. Tsuchiya, Y. Yonehara, T. Shimomura, and T. Koga, An investigation of deposit formation on sealing surfaces of water pump end face seals, 43rd Annual Meeting, Society of Tribologists and Lubrication Engineers, Cleveland, Preprint No. 88-AM-5E-1, (1988).
39. G. A. Green and F. J. Tribe, Carbon–carbon composite materials for submarine propeller shaft seals, *Proceedings of the 11th International Conference on Fluid Sealing*, Cannes, France, 1987, pp. 611–630.

16

Ceramics for Magnetic Recording Applications

S. Chandrasekar and T. N. Farris

Purdue University
West Lafayette, Indiana

B. Bhushan

Ohio State University
Columbus, Ohio

ABSTRACT

The friction and wear behavior of ceramics under lightly-loaded, sliding-contact conditions are discussed in the context of their application to magnetic recording systems. A recording-head, slider material should have low adhesion to the mating material, high hardness and Young's modulus, and a high thermal conductivity among its principal characteristics. Based on the requirements and our current understanding of the tribology of ceramics, potential slider and overcoat materials are identified for further study of a next generation of magnetic recording systems. Potentially attractive slider materials include single-crystal diamond, cubic boron nitride (CBN), and sapphire.

I. INTRODUCTION

A. Overview of Magnetic Recording Systems

Conventional magnetic recording is accomplished by the relative motion of a magnetic disk or tape against a stationary or moving read/write magnetic head. Figure 1 shows a schematic of typical thin-film rigid disk consisting of a substrate, an undercoat, magnetic layer, a hard ceramic overcoat, and a thin layer (1–4 nm) of fluorocarbon lubricant (not shown). Sometimes there is an additional layer between the magnetic medium and the undercoat to control the magnetic properties of the recording layer. Typical overcoat materials are diamondlike carbon, SiO_2,

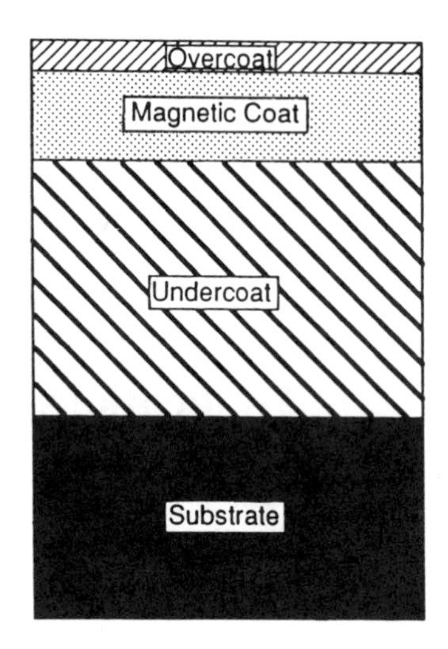

Figure 1 Schematic of thin-film disk.

and ZrO_2-Y_2O_3. For more details on the description and composition of the layers, see Mee and Daniel [1] and Bhushan [2]. The magnetic head sliders are generally made of ceramics such as Mn-Zn and Ni-Zn ferrites, calcium titanate, aluminum oxide/titanium carbide and yttria-stabilized zirconia/aluminum oxide–titanium carbide composites. A head–disk interface tribology resides in lightly loaded (mean pressure ~ 10 kPa), marginally lubricated sliding conditions. The storage density of a Winchester-type magnetic disk is a strong function of the head-to-disk spacing. For high-density, high-resolution recording, close proximity between the head and disk is essential. Under steady operating conditions, a load carrying air film (~0.1–0.4 μm) is formed, preventing direct contact between the head and the disk. Physical contact between head and disk, however, does occur during start/stop and at isolated points during flying [2].

The drive toward high-density and high-resolution read/write recording systems has necessitated research into systems with ultralow flying gaps (<0.1 μm) between the head and the disk. The friction between the head and the disk at the contact zones and wear of the magnetic head and disk media is therefore of great interest. A low coefficient of friction between head-slider materials and the disk surface is highly desirable for a number of reasons. A high coefficient of friction generally results in excessive local heating and contact temperatures, high wear of both head and disk materials, and excessive power requirements. Sometimes, high friction leads to severe stick–slip behavior resulting in errors during the read-write process.

B. The Role of Ceramics

In the quest for head–slider and thin-film rigid-disk overcoat materials with good tribological properties in magnetic recording systems, ceramics prove to be a natural choice because of their unique properties. Ceramics in general have high hardness, low adhesion to contacting surfaces, and are chemically inert in a wide range of environments [3–5]. Furthermore, they retain these properties over a much wider temperature range than metals [6]. In view of these unique properties, ceramics are found to have low friction and wear under a wide range of sliding contact conditions [4]. Since there are a number of ceramics and ceramic composites available, a natural question that is posed is: What is the best ceramic or class of ceramics for use in a given tribological system? The answer is by no means obvious. The purpose of this chapter is to unravel some of the underlying factors that control the friction and wear of ceramics based on a thorough review of results in the literature. Furthermore, potential ceramic slider materials and disk overcoats are identified for future study.

The chapter begins with a review of relevant mechanical and physical properties and crystal structures of several ceramics. These include ceramics that have already found wide applicability in head–disk and head–tape systems and ceramics that are potential contenders. Next, the friction and wear behavior of these ceramics in ceramic–metal and ceramic–ceramic sliding contact is presented using a wealth of experimental data available in the literature. Based on these results, some of the fundamental properties of ceramics that influence their friction and wear behavior are identified. The ensuing discussion concentrates on how this interrelationship between properties and tribological behavior can be used to select a suitable ceramic or class of ceramics for magnetic recording applications, where certain friction and wear mechanisms dominate. A theme that runs throughout this chapter is: "The macroscopic friction and wear behavior of ceramics is a consequence of the fundamental mechanical and physical properties existing at the microscopic level."

II. STRUCTURE AND PROPERTIES OF CERAMICS FOR TRIBOLOGICAL APPLICATIONS

In this section, the structure and properties of several ceramics of interest to tribological applications are reviewed.

A. Diamond

Diamond is the hardest material known. It is available in both single crystal and polycrystalline form (bonded). Diamond crystallizes in the modified face-centered cubic (fcc) structure. The carbon–carbon bonding is completely covalent, which accounts for its high hardness.

Single-crystal diamond can be classified into four categories: types Ia and Ib and types IIa and IIb. The classification is made primarily on the basis of the amount of nitrogen impurity in the material [7–9]. Type II natural diamond contains the least amount of nitrogen and, therefore, has a higher thermal conductivity than type I [7]. In type I diamond, the dissolved nitrogen content can range up to 2500 ppm for natural Ia diamonds and up to 500 ppm for synthesized Ib diamonds. Most of the natural diamonds (~98%) are type Ia; only a few are type Ib. No type Ia diamonds have ever been synthesized in the laboratory [10]. Synthetic diamonds are of type Ib and types IIa and IIb. Table 1 lists the various physical properties of diamond and Table 2 lists the various mechanical properties.

The hardness of diamond (~8000–10,000 kg/mm^2) is about twice that of cubic boron nitride (CBN), the next hardest material known. At near room temperature diamond has the highest thermal conductivity among any material; the thermal conductivity is almost five times as large as that of copper. This, coupled with its low thermal expansion coefficient, gives it high resistance to thermal shock in spite of its brittle nature [11]. In sliding contacts, the high thermal conductivity of diamond enables faster heat dispersion from the interface and, hence, significantly lower interface temperatures. Many of the mechanical properties of diamond single crystal, such as hardness and strength, are a function of crystal orientation. On an average, the hardest face is the (110) plane, while the (111) and the (100) planes have comparable hardness [9,12]. The (110) plane also has a high atomic density [8]. Some of these hardness values are listed in Table 2. Diamond has a very high Young's modulus (~1000 GPa), higher than that of any other ceramic (Table 2). The Young's modulus of diamond exhibits little anisotropy, while the Poisson ratio varies between 0.1 and 0.29 [8]. While diamond is an almost ideal brittle solid at room temperatures, plastic deformation and significant dislocation mobility are observed at elevated temperatures around 1800°C [8,13]. Also, significant plastic deformation can occur if fracture is inhibited by hydrostatic stress [8,14].

Diamond is thermodynamically stable only at high pressure (see the phase diagram in Fig. 2), but is chemically inert at room temperature. However, various gases such as hydrogen are known to be physically adsorbed very rapidly on a clean diamond surface. This has been illustrated using surface analytical techniques by Pepper [15,16] and Lurie and Wilson [17]. Spear [13] discusses C–H bond formation (with atmospheric hydrogen) to satisfy the dangling sp^3 bonds of the surface carbons on the various planes of diamond. The C–H bonding on the (100) face does not satisfy all the dangling bonds due to steric hindrances. Graphitization of diamond occurs in air sometimes at temperatures as low as 1000°C and in vacuum at around 1400–1700°C [10].

Polycrystalline diamond is synthetic and typically of two different types. One form consists of diamond particles embedded in a tough matrix (usually metals such as nickel) forming a hard composite mass. Another is a polycrystalline mass formed by direct diamond-to-diamond sintering without use of a binder. (Often metallic impurities such as cobalt segregate at the grain boundaries.) The advantages of polycrystalline diamond compared to single-crystal diamond are its greater toughness and its more uniform wear independent of direction [9]. This is because the easy cleavage planes of the constituent crystals are randomly oriented with respect to one another, making crack propagation through the mass difficult. Polycrystalline diamond bonded through a metal matrix, however, has

Table 1 Physical Properties of Selected Bulk Ceramics

Material	Density (g/cc)	Melting or decomposition temp./max. operating temp. (°C)	Specific heat (J/g°K)	Thermal conductivity ($W \cdot m^{-1} k^{-1}$)	Electrical conductivity $(ohm \cdot m)^{-1}$	Coeff. therm. expansion $\times 10^6$ (°C)
Single-crystal diamond						
Type Ia	3.51			900	$<10^{-18}$	
Type Ia	3.51	4000/1000	0.525		$<10^{-18}$	
Type IIa	3.51			2000	$<10^{-18}$	1.0
Type IIb	3.51				$<10^{-5}$	
Polycrystalline diamond	3.50	4000/1000	0.525	800	$<10^{-15}$	1.0
Cubic boron nitride	3.45	3250/1250	0.51	300–600	10^{-6}–10^{-0}	5–9
Silicon carbide	3.21	2800/1700	0.76	90–120	10^{3}	4.3
Silicon nitride (hot pressed)	3.18	1900/1500	0.72	33	$<10^{-8}$	3.0
β'-Sialon	3.10	1900/1500	0.70	20–25	10^{-10}	2.7
Sapphire	3.90	2000/700	0.95	35	10^{-12}	7.1
Polycrystalline-aluminum oxide (hot-pressed)	3.90	2000/700	0.92	30	10^{-9}	7.1
Y_2O_3-partially stabilized zirconia (PSZ)	6.10	2760/1500	0.63	2.2	10^{-8}	11
Aluminum oxide/titanium carbide	4.20	—	—	—	—	—
Ni–Zn ferrite	5.30	—	0.71	8.7	$<10^{-3}$	—
Mn–Zn ferrite	5.06	—	0.68	4.6	—	—
Calcium titanate	3.95	—	—	—	—	—
Barium titanate	4.32	—	—	—	—	—
Borosilicate glass	2.20	1400/300	0.085	1	$\sim 10^{-14}$	3.5–5.0
Glass ceramic (Macor)	2.30	/1700	—	3.5	10^{-14}–10^{-11}	0.4–2.0
Hafnium carbide	12.7	3930/700	0.19	13	2×10^{6}	6.3
Titanium carbide	4.9	3060/1000	0.52	27	1.6×10^{6}	7.2
Chromium oxide (Cr_2O_3)	5.21	2300/1000	0.73	7.5	—	5.4
Silicon dioxide (fused silica)	2.2	1800/900	0.67	1.2	10^{-12}	0.9
Tungsten carbide	15.8	2780/500	0.20	29	5×10^{6}	5.2
Tungsten carbide–cobalt (94%–6%)	15.1	—	0.21	90	—	5.4
Boron carbide	2.52	2420/600	0.92	26	200	4.3

Table 2 Mechanical Properties of Selected Bulk Ceramics

Material	Knoop hardness (kg/mm^2)	Bending fracture strength (MPa)	Fracture toughness $K_{IC}(MPa \cdot m^{1/2})$	Young's modulus (GPa)	Poisson's ratio ν
Single-crystal diamond					
(110)<110>	9500		—		
(111)<110>	9000	1050	—	1000	0.2
(100)<110>	8600		—		
Polycrystalline diamond	8000	1000	—	900	0.2
Cubic boron nitride	3500–4750	—	—	660	0.15
Silicon carbide[a]	2300–2900	725	—	440	0.17
Silicon nitride (HP)	1600–1800	900	3.5–4.0	300	0.26
Sialon (HP)	1800	850	5.5	300	0.28
Sapphire (single crystal)[b]	2000	—	—	400	0.23
Aluminium oxide (HP)	2000	500–600	4.9	400	0.23
Partially stabilized zirconia (PSZ)	1100–1300	700–1000	8.5–9.5	210	0.24
Aluminum oxide/titanium carbide	2300	590	—	—	—
Ni–Zn ferrite	750	200	~1.0	170–190	—
Mn–Zn ferrite	650	190	~1.0	170–190	—
Calcium titanate	950	—	—	—	—
Barium titanate	1050	—	—	—	—
Borosilicate glass	530	30–50	—	60	0.2–0.23
Glass ceramic (Macor)	650	150–170	—	100	0.25
Hafnium carbide	2700	240	—	410	0.18
Titanium carbide	2800	860	—	450	0.19
Chromium oxide (Cr_2O_3)	1300	95	—	—	—
Silicon dioxide (fused silica)	550–750	50–120	—	75	0.17
Tungsten carbide	2100	700	—	690	0.19
Tungsten carbide–cobalt (94%–6%)	1500	700–2000	—	640	0.28
Boron carbide	3200	300–350	—	400	0.19

[a]Hardness of 2300 kg/mm^2 for the $(10\bar{1}0)$<0001> system and 2900 kg/mm^2 for the (0001)<1120> system.
[b]The hardness of 2000 kg/mm^2 is for the (0001)<1120> system.

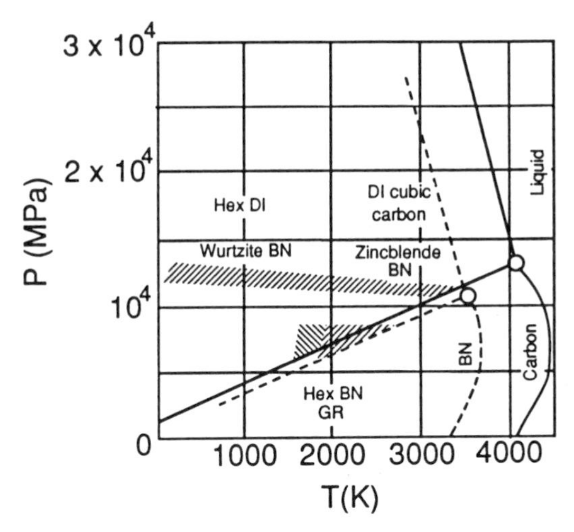

Figure 2 Pressure-temperature phase diagram for carbon and boron nitride (BN). (DI, diamond; GR, graphite) [9].

lower hot strength and thermal conductivity compared to the single crystal because of the inferior properties of the matrix material. Its hardness is also about 1000–1500 kg/mm^2 lower than that of single-crystal diamond. Polycrystalline diamond is used in grinding wheels, cutting tools, and dies.

Diamond and diamondlike carbon films can now be synthesized [13,18,19]. Processes that utilize a plasma decomposition of various hydrocarbons (referred to as plasma-enhanced chemical-vapor deposition, or PECVD) and physical vapor deposition (sputtering) of carbon have been used to form films with hardness as high as 1000–3000 kg/mm^2. These films are referred to as diamondlike carbon (DLC) and their structure is predominantly amorphous when observed by various surface analytical techniques. Some areas within the films are composed of small randomly oriented graphite crystallites ($\leq$2 nm), with a small percentage of sp^3 bonded carbon atoms ($\leq$5%). DLC films deposited by CVD processes are generally harder than similar films deposited by sputtering [18]. The diamondlike films typically contain hydrogen as an impurity. Hydrogen loss and graphitization of these films have been observed sometimes at temperatures as low as 400°C [20]. Diamondlike carbon films have been used as an overcoat for thin-film magnetic disks [2], where they serve as hard, protective, wear-resistant and corrosion-resistant coatings for the magnetic media.

The CVD processes are also used to deposit crystalline diamond films by utilizing temperature and pressure conditions under which graphite is clearly the stable form of carbon. However, various kinetic factors (not all of which are yet fully understood) allow crystalline diamond to be produced by the chemical reaction

$$CH_4(g) \xrightarrow[\textit{activation}]{} C(\text{diamond}) + 2H_2(g).$$

Besides methane, several other carbon-containing gases can also be used. Table 3 gives some of the growth conditions for vapor-deposited diamonds. The coating of diamond is extensively reviewed in a recent article by Spear [13] and diamondlike carbon coatings are reviewed by Angus et al. [18]. It is now possible to deposit a uniform thickness of high-quality diamond films on silicon and other substrates.

B. Cubic Boron Nitride

Cubic boron nitride (CBN) has properties quite analogous to diamond including high hardness ~ 4500 kg/mm^2, good chemical inertness, and high thermal conductivity. In fact, its thermal conductivity is much higher than all other except diamond (see Table 1). The similarity in the properties of diamond and CBN can be inferred from the similarity in crystal structure. The phase diagram for CBN in Figure 2 closely parallels that of diamond. Like carbon, boron nitride exists as soft graphitelike, layer–lattice form, i.e., hexagonal boron nitride (HBN), and as a hard cubic form (CBN). The crystal structure of CBN is similar to that of diamond. But an important crystallographic difference between diamond and CBN lies in the latter's lack of a center of symmetry in its crystalline structure. Also, the weak bonding in CBN is along six (011) planes, while the natural cleavage surfaces of diamond are the four (111) planes. The cleavage planes are an important expression of the structural differences between diamond and CBN. In an abrasive grain, the cleavage mode of CBN may be an advantage in self-sharpening, whereas in a single-point tool or gem, the presence of six

Table 3 Growth Conditions of Vapor-Deposited Diamonds

	Typical conditions	Best crystals
Substrate temperature (°C)	700–1000	950–1050
Total pressure (kPa)	0.013–12	5.3–12
CH_4 (in H_2) (mol%)	0.1–5.0	0.1–1.0
Total flow (sccm)[a]	20–200	50–100
Filament temperature (°C)	1800–2500	2000–2500

[a]Standard cubic centimeter per minute.

cleavage planes can be a disadvantage. CBN oxidizes in air or in oxygen at around 450°C forming a hard B_2O_3 protective layer that prevents further oxidation up to ~ 1300 °C [21]. Conversion to the hexagonal form occurs above 1500°C. The surface energy of the (011) plane is somewhat lower than that of diamond. Sometimes CBN crystals fracture along the (011) family of planes when heated to 900–1000°C. This thermally activated fracture is thought to occur due to the release of residual strain in the crystals [21]. As a general rule, nitride- and boride-forming metals will react with CBN, aluminum being particularly noteworthy in this respect [21]. A number of borates and borosilicate glasses are compatible with boron nitride, which is the reason for the use of glass binders in grinding wheels containing CBN abrasive. Significant dislocation mobility exists at temperatures above 1300°C.

In polycrystalline form, CBN is stable in air to somewhat higher temperatures than diamond. One of the dominant advantages of CBN over diamond is that it has a very low affinity for iron. Hence, it is used as an abrasive in the grinding of steels. More details on the structure and properties of CBN may be found in References 21 and 22. The similarity with graphite and diamond is also demonstrated when thin films of boron nitride are grown by physical vapor deposition (PVD) and plasma-enhanced chemical-vapor deposition (PECVD) processes. The structure of these films is predominantly amorphous with small crystallites of the CBN phase, very similar to that of diamondlike carbon films. Also, PECVD-grown films do not exhibit as high a hardness as those generated by PVD techniques [19]; such films are, therefore, used mainly as corrosion-resistant coatings. The hardness of PVD films is in the range 1000–2000 kg/mm^2.

C. Carbide and Nitride Ceramics

Carbides and nitrides of several metals such as titanium and tungsten, and nonmetals such as boron and silicon fall under this class of hard materials. The hardest among these are *silicon carbide*, *titanium carbide*, and *boron carbide*. Hardness and other important physical/mechanical properties of these materials are listed in Table 1 and 2. All of the hard transition metal carbides tend to have a close-packed structure of metal atoms with the much smaller carbon atom occupying the interstitial position. The metal–carbon bonding in these structures is generally intermediate between covalent and metallic. However, in the case of *single-crystal silicon carbide* (SiC), because of the similar electronegativities of the silicon and carbon atoms, the bonding is completely covalent. This is responsible for the higher hardness of SiC relative to the other carbides [6].

Shaffer [23] determined the Knoop hardnesses of single-crystal β-SiC with relation to crystal orientation. The hardness was highest ~ 2900 kg/mm^2 on the basal close-packed (0001) plane when the indenter was oriented parallel to the $<10\bar{1}0>$ or $<11\bar{2}0>$ directions. In comparison, the hardness was lower on the

$(10\bar{1}0)$ or $(11\bar{2}0)$ faces ~ 2300–2700 kg/mm^2. The hardness of SiC is considerably lower than that of diamond or CBN. Silicon carbide oxidizes in air to form SiO_2; the rate of oxidation increases with increasing temperature. The presence of water vapor in the oxidizing atmosphere greatly speeds up this reaction. SiO_2 has a much lower hardness (~600–800 kg/mm^2) than SiC. Highly oxidizing atmospheres can, therefore, lead to rapid attrition wear of SiC. Polycrystalline SiC has a lower hardness (~ 2400 kg/mm^2) than the single crystal. The thermal properties of β-SiC are generally inferior to those of diamond and CBN (Table 1).

Boron carbide has a higher hardness (~2800 kg/mm^2) than SiC but its role is an abrasive material is severely limited due to its chemical reactivity with a number of metals [24]. Hence, it is a poor tribological material for applications and a poor abrasive for grinding.

Tungsten carbide (hardness ~ 2100 kg/mm^2) and titanium carbide (hardness ~ 2800 kg/mm^2) are widely used in machining applications. TiC is generally used in the form of coatings on WC substrates in metal cutting tools. Hafnium carbide and chromium carbide (Cr_3C_2) are other carbides with potential for tribological applications. The properties of refractory carbides are reviewed well by Storms [25].

Nitride structures are similar to carbides, with the bonding again being mixed covalent-metallic. The hardness of these materials generally decreases with decreasing covalent contribution to the bonding. Hot-pressed silicon nitride, in particular, is an excellent material for many bearing applications because of its high fracture toughness, low density, good stability, and good hot-hardness [26,27]. Titanium nitride has a very low affinity for metals and is widely used in cutting tools. TiN has a golden appearance and is wear-resistant; it is commonly used for decorative purposes, e.g., watches, jewelry, etc. Hafnium nitride is much harder than TiN and TiC at elevated temperatures and is a candidate for high-temperature wear-resistance applications. Most of the carbide and nitride materials can be synthesized in thin-film form by a variety of PVD and CVD processes. The development of epitaxial films of these materials could increase their application potential.

D. Oxide Ceramics

Most important of the oxide ceramics are single-crystal aluminum oxide (sapphire), polycrystalline Al_2O_3, ZrO_2, Cr_2O_3, and SiO_2. Sapphire has a hexagonal close-packed structure with a mixture of covalent and ionic bonds (Table 4). The thermal conductivity of sapphire is around 30 $W \cdot m^{-1}K^{-1}$ (Table 1), which is an order of magnitude greater than that of zirconia. Its hardness and Young's modulus are also significantly higher than that of zirconia, chromium oxide, and silica (Table 2). Buckley [5] has pointed out that the friction and wear behavior of sapphire is quite analogous to hexagonal metals, a consequence

Table 4 Crystal Structure of Selected Ceramics

Material	Structure	Bonding	Lattice Constant (Å)
Diamond	diamaond cubic (8 C atoms/unit cell)	100% covalent	3.567
Cubic boron nitride	cubic (zinc blende type, 4B and 4N atoms per unit cell)	~25% ionic	3.615
Silicon carbide	β-cubic α-hexagonal >2000°C	covalent	a = 3.08 c = 15.11 (α)
Silicon nitride	hexagonal	mixed[a]	c = 2.92 a = 7.61 (β)
Sialon	hexagonal	mixed	c = 2.91–3.0 a = 7.61–7.72
Sapphire (Al_2O_3)	hexagonal	—	—
Y_2O_3-partially stabilized zirconia	tetragonal (partially stabilized)	—	—
Ni–Zn ferrite	spinel cubic	—	—
Hafnium carbide	face-centered cubic	mixed	4.64
Titanium carbide	face-centered cubic	—	4.328
Silicon dioxide (fused silica)	amorphous	—	—
Tungsten carbide	hexagonal	—	a = 2.90 c = 2.84
Chromium oxide (Cr_2O_3)	hexagonal	—	—
Boron carbide	rhombohedral	—	—

[a]"Mixed" refers to a combination of covalent and ionic bonding.

of the similarity in their structure. Plastic deformation is found to occur at contacting interfaces of sapphire crystals [5] and in indentation hardness tests [28]. The preferred slip system is the (0001) basal plane and $[11\bar{2}0]$ direction because of close packing. Polycrystalline aluminum oxide ceramics have good stability and hardness at high temperatures (Tables 1 and 2); hence, they are widely used as abrasives in grinding wheels.

Among the recently developed oxide ceramics, partially stabilized zirconia (PSZ) offers an excellent potential for tribological applications [29,30]. Zirconia has three different crystal structures: cubic above 2600 K, monoclinic below 1440 K, and tetragonal between these temperatures. These can also be seen in the ZrO_2–Y_2O_3 phase diagram in Figure 3a. The transition from tetragonal to monoclinic at 1440 K is martensitic and cannot be suppressed by quenching. Furthermore, there is a volume increase of around 4% and an angular shear of around 9° during this transition. It has been known for many years that addition of greater than 17 wt% of oxides such as yttria (Y_2O_3) and magnesia (MgO) can stabilize the cubic form even at room temperature producing fully stabilized zirconia. However, much more important than this is partially stabilized zirconia (PSZ) which has a higher fracture toughness. An addition of about 5 to 12% by wt of Y_2O_3 to ZrO_2 partially stabilizes the tetragonal phase at room temperature. Figure 3a shows the zirconia-rich end of the ZrO_2–Y_2O_3 phase diagram [31]. A

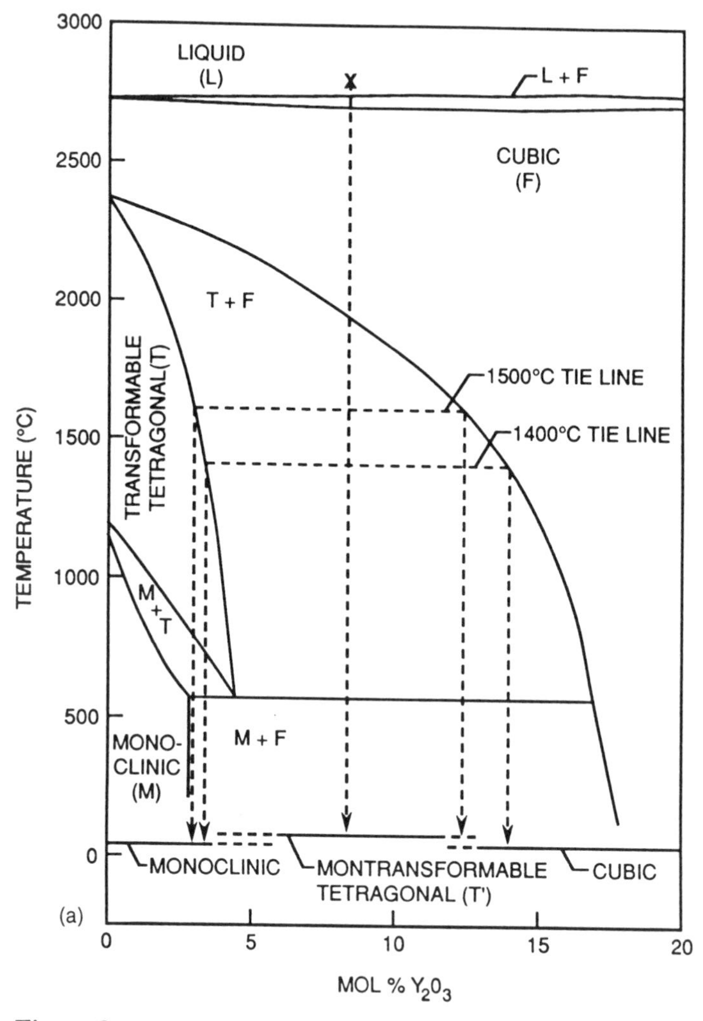

Figure 3 Phase diagrams. (a) Y_2O_3-partially stabilized zirconia [31]; (b) Sialon [36].

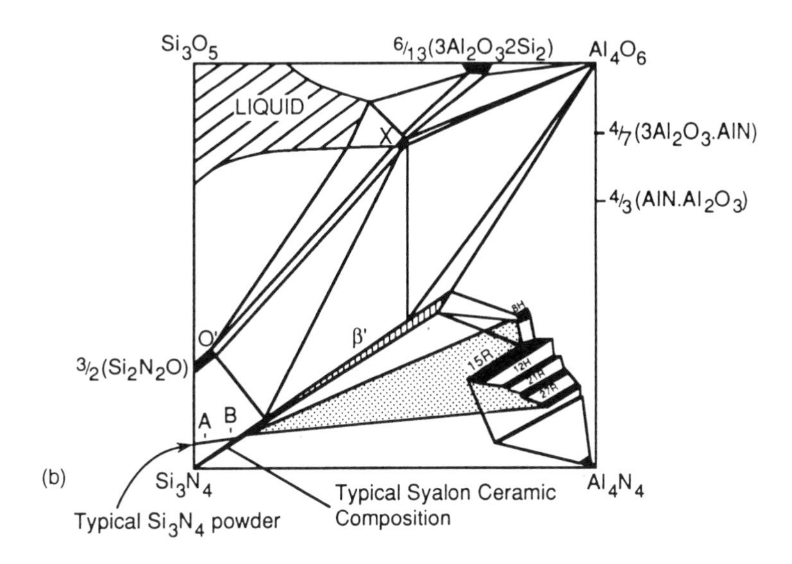

typical Y_2O_3–PSZ contains around 9 mole % Y_2O_3. When a solid solution of this mixture is cooled from about 2000°C to room temperature, it contains tetragonal precipitates in a monoclinic matrix. The precipitates are coherent with the matrix. If a propagating crack enters a grain containing tetragonal precipitates, the coherency stresses in the vicinity of the crack are reduced and the precipitates transform to the monoclinic form with its associated expansion and shear. The large compressive stresses resulting from this volume expansion retard crack initiation and propagation.

PSZ has a high fracture toughness, around 10 $MPa \cdot m^{1/2}$ (Table 2), which is almost twice as great as that of other ceramics, and an excellent tensile strength. Chemical stability of PSZ is also very good with little or no oxidation and with resistance to attack by most acids and alkalis. PSZ, however, has a low thermal conductivity of about 2 $W \cdot m^{-1} K^{-1}$ (Table 1), which is an order of magnitude less than that of alumina and almost two orders less than that of diamond. Heat dissipation from sliding interfaces of friction couples involving zirconia can, therefore, be a problem. More details about other types of PSZ may be found in Porter and Heuer [32] and Garvie [33].

Sputtered coatings of PSZ and SiO_2 are used for disk overcoats [2]. Plasma sprayed coatings of zirconia, aluminia–titania composite, and chrome oxide are commonly used for magnetic tape heads. Zirconia/alumina–titanium carbide composites are also used as slider materials. Al_2O_3 particles have also been used in the magnetic coating in particulate magnetic media. Cr_2O_3 has found applications as an abrasive and as a wear-resistant coating in high-performance gas bearings [34,35]. Lapping tapes with Cr_2O_3 as an abrasive are commonly used in the computer industry; it is also used as an additive in the magnetic disks. Sputtered Cr_2O_3 is used as a wear-resistant overcoat.

E. Sialon

Sialons are synthesized by reacting a mixture of silicon nitride, silica (SiO_2), aluminum oxide, and aluminum nitride, as shown in the phase diagram of Fig. 3b. Its formula can be represented as $Si_{6-z}Al_zO_zN_{8-z}$, where z denotes the number of nitrogen atoms substituted by oxygen atoms (z can take on values in the range 1–5). The partial substitution of silicon and nitrogen by aluminum and oxygen in the silicon nitride lattice results in a compound having greater strength and chemical stability than silicon nitride [36]. The resulting microstructure also gives sialon a higher toughness relative to many other ceramics (Table 2). Sialon ceramics are finding increasing applications in heat engine components, wire-drawing dies, and cutting tools. A complete characterization of sialon ceramics may be found in Spencer [37]. The important physical and mechanical properties of sialon can be found in Tables 1 and 2.

F. Ferrites

Manganese–zinc (Mn–Zn) and nickel–zinc (Ni–Zn) ferrites are commonly used as recording head materials in magnetic recording systems. They provide a good compromise between magnetic properties and friction and wear for a magnetic ceramic material. Both of these ferrites have the spinel cubic structure. The magnetic properties of ferrites arise from the location of the Mn, Ni, and Zn ions in the spinel lattice. Ni–Zn ferrite (700–750 kg/mm^2) is slightly harder than Mn–Zn ferrite (~600–650 kg/mm^2) [38]. In single-crystal Ni–Zn and Mn–Zn ferrites, the (110) plane on an average has the highest hardness [39]. Both the microhardness and the bending strength of polycrystalline Ni–Zn ferrite increase with decreasing grain size; the strength dependence is found to obey a Hall–Petch type relationship, with the strength being proportional to the $(\text{grain size})^{-1/2}$ [38]. A recent study [40] has shown that the bending strength of lapped Ni–Zn ferrite also increases after annealing. In relation to the other ceramics that have been discussed, Mn–Zn and Ni–Zn ferrites have significantly lower hardness, bending strength, and fracture toughness; see also Tables 1, 2, and 4. Mn–Zn ferrites have a high saturation magnetization and relatively low resistivity ~ 10^5–10^9 ohm-cm) allowing applications at high frequencies [6]. The thermal conductivity of Ni–Zn ferrite is about twice as high as Mn–Zn ferrite (Table 1), but inferior to that of many other ceramics.

G. Oxide Ceramic Composites

An Al_2O_3/TiC composite (70% Al_2O_3-30% TiC) has been used as a slider material mainly in hard-disk applications. It has much higher hardness and bending strength than Mn–Zn or Ni–Zn ferrite [41] (see Table 2). Another ceramic in the same class is Al_2O_3/TiC (87% Al_2O_3–13% TiC). However, there is not much information available in the literature on the various physical properties of these ceramics. A similar class of composite ceramics, an example of which is graphite whisker-reinforced Si_3N_4 ceramic, are finding use in tribological applications such as in metal cutting tools and dies. These are much harder and significantly tougher than the Al_2O_3/TiC composite.

H. Titanates

Polycrystalline calcium and barium titanates have also been used occasionally as slider materials in magnetic recording systems [2]. These are somewhat harder than polycrystalline Mn–Zn ferrite, as can be seen from Table 2. Barium titanate is the hardest of the three titanates listed in Table 2. All three of these titanates have the cubic–perovskite structure and constitute an important class of ferroelectric ceramics. Calcium titanates have found limited applications in composite sliders. Wear resistance of titanates is poor, probably due to their low fracture toughness.

I. Glasses and Glass Ceramics

There are a number of glasses and glass ceramics. Fotoceram is a ceramic obtained by converting photosensitive glass (Fotoform) to a crystalline state through heat treatment and exposure to ultraviolet radiation. It has higher hardness than its glassy counterpart. Its use as a rigid-disk slider material has been limited because of its low hardness and low fracture toughness and consequent poor wear resistance. Metallic glasses are formed by rapid quenching of certain alloy compositions to get an amorphous phase. They have good high-temperature properties and, in general, low adhesion to ceramics and metals. Macor is a glass ceramic noted for its easy machinability. it is relatively soft ($\sim$650kg/mm^2). For a review of glass and glass ceramic properties, see Corning's handbook of glass and glass ceramic properties [42].

III. FRICTION AND WEAR OF CERAMIC MATERIALS

In this section, the frictional behavior of various ceramic materials in sliding contact with other ceramics and metals under lightly loaded conditions is reviewed. Particular attention is given to experimental results that enable an understanding of the mechanism of ceramic friction and wear.

A. Tribological Material Systems

1. Single Crystal Diamond

Some of the earliest results on the deformation and friction of diamond were reported by Bowden and Tabor [43]. Since then, there have been a number of investigations of the friction and wear of diamond in contact with various materials [44–54]. Their results provide considerable insight into the tribological behavior of diamond.

Effect of Environment

The coefficient of friction of diamond on diamond in air is relatively low ($\mu \cong$ 0.05–0.1). These values were obtained in single-crystal diamond sliding experiments for a diamond stylus sliding against a diamond flat [(001) face] in air at light normal loads [49,55,56]. In vacuum under similar conditions, the friction is high ($\mu \cong 1.0$) but on admitting air into the vacuum chamber the friction coefficient falls to low values of around 0.05–0.1 [56]. Figure 4a gives the

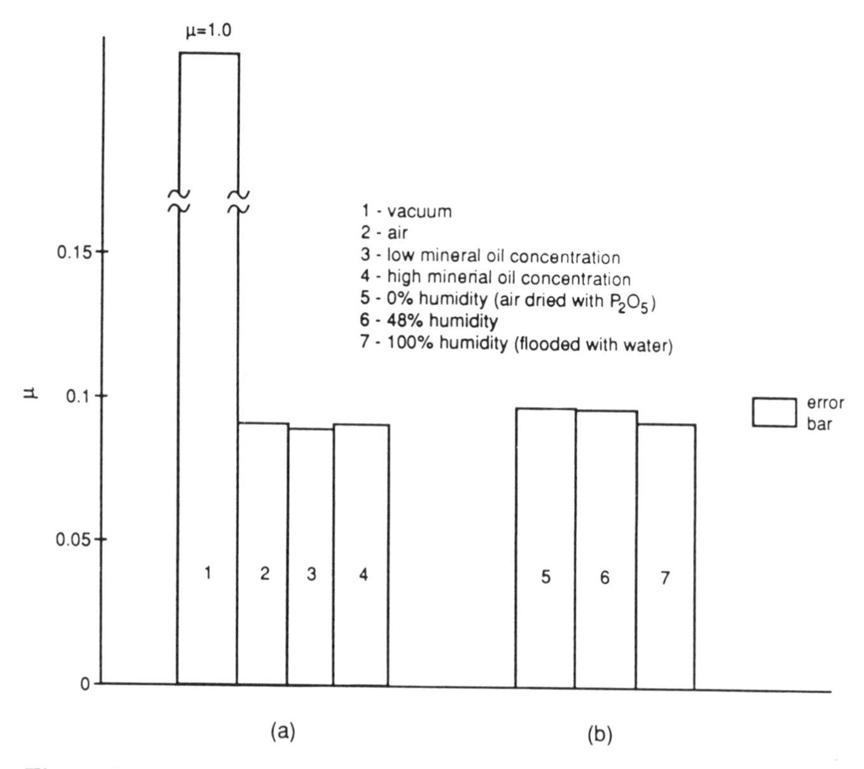

Figure 4 The friction of diamond sliding against diamond in various environments under lightly loaded conditions. (a) air and oil; (b) different humidities [46,49].

friction coefficients for diamond sliding against diamond in several environments. From this it is seen that the friction coefficient in air is almost unchanged in the presence of lubricants such as oil and water. The friction coefficient of diamond against diamond is also unaffected by changes in the relative humidity (Fig. 4b). This was observed in the experiments of Seal [44–46] who investigated a range of conditions from dry air (dried with P_2O_5) to almost 100% humidity (by flooding the surfaces with water). This observation is particularly important for magnetic recording systems that may be required to perform under varying humidity conditions.

Hydrogen adsorbs readily on the surface of diamond [13]. The adsorbed hydrogen plays an important role in lowering the friction coefficient of diamond in contact with itself or metal [15,16,54]. Figures 5a and 5b show ionization loss spectra of the diamond (110) surface before and after exposure to hydrogen in a vacuum chamber. The disappearance of the K_O peak (Fig. 5a) in Figure 5b after exposure of the diamond surface to hydrogen suggests rapid growth of adsorbed hydrogen on the diamond surface. A decrease in the friction coefficient is observed at the same time (Fig. 5c) for the diamond–metal sliding system. Hayward [54] has recently also observed a significant reduction in the friction coefficient of single-crystal diamond sliding against diamond, when small quantities of hydrogen (10^{-5}–10^{-6} Torr) were bled into the vacuum chamber. The wear of diamond surfaces sliding against each other in vacuum was high but was negligible in a hydrogen environment with only a few atoms removed per pass [52].

Miyoshi and Buckley [48] observed metal transfer to the diamond surface during sliding friction experiments of (111) diamond against transition metals. The metal transfer was higher for those transition metals which exhibited a higher coefficient of friction. In general, for most of the diamond sliding systems studied, the wear was higher when the coefficient of friction was higher.

The significant reduction in the coefficient of friction in the presence of air, hydrogen, and other gaseous environments suggests that adhesion plays an important role in the friction and wear of diamond; the role of hydrogen in reducing the friction appears to be reasonably well understood. Adhesion is also apparent from the metal transfer observed in diamond–metal sliding contacts.

Frictional Anisotropy: The Effect of Crystallographic Orientation

In sliding friction experiments of a diamond stylus on a diamond flat, the friction coefficient is found to depend on the orientation of the flat and the stylus [44]. Figure 6a shows the variation in the friction coefficient for diamond on diamond on different crystallographic planes. On the (110) face of the flat, the friction coefficient is almost constant in all directions and has a value of 0.05. This is the "hard" face of diamond (with highest atomic density) (see Table 2). However, on the (100) cube face, the friction coefficient varies from 0.05 to 0.15 depending on

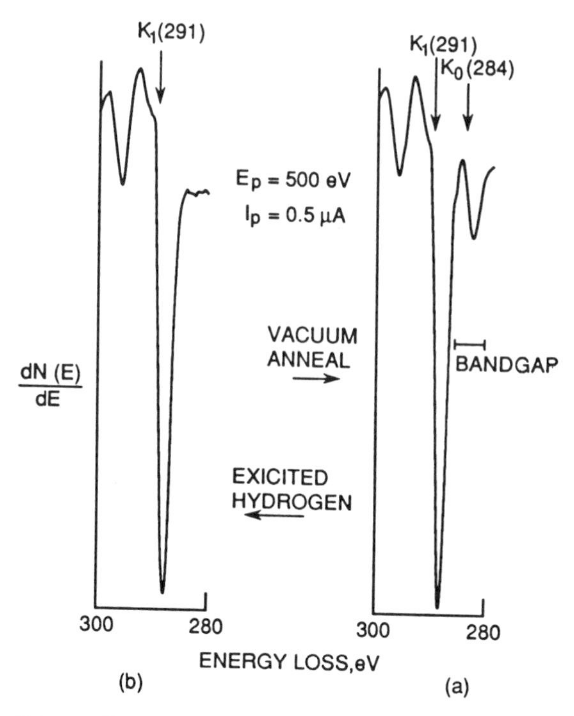

Figure 5 The effect of hydrogen on the diamond (110) surface. (a) Ionization loss spectra of annealed (~900°C) surface; (b) Ionization loss spectra of a fresh diamond surface exposed to hydrogen (note the disappearance of K_0 peak in (b) after exposure to hydrogen); (c) copper–diamond static friction coefficient as a function of diamond annealing temperature. ○, polished surface; □, clean diamond surface exposed to hydrogen; ◇, clean surface exposed to hydrogen and then annealed [16].

the direction of sliding [49]. Hayward [54] has observed also that the coefficient of friction of a diamond stylus is lower when sliding against the (111) and (110) faces of a diamond flat compared to the (100) face. It should be noted here that the (100) face is softer relative to the other faces (Table 2). Moreover, the (111) and (100) faces have a higher atomic density compared to the (100) plane for the diamond cubic structure. The anisotropy in the friction coefficient is also seen in Fig. 6b, where the friction coefficient for the diamond–diamond system is plotted for various directions along the (100) face of the diamond flat. The friction coefficient is significantly lower along the close-packed <110> direction compared to the <100> direction. The hardness along the <110> direction is also higher than that along the <100> direction on this face. As may be seen from the figure, the anisotropy becomes greater with increasing normal load. This has

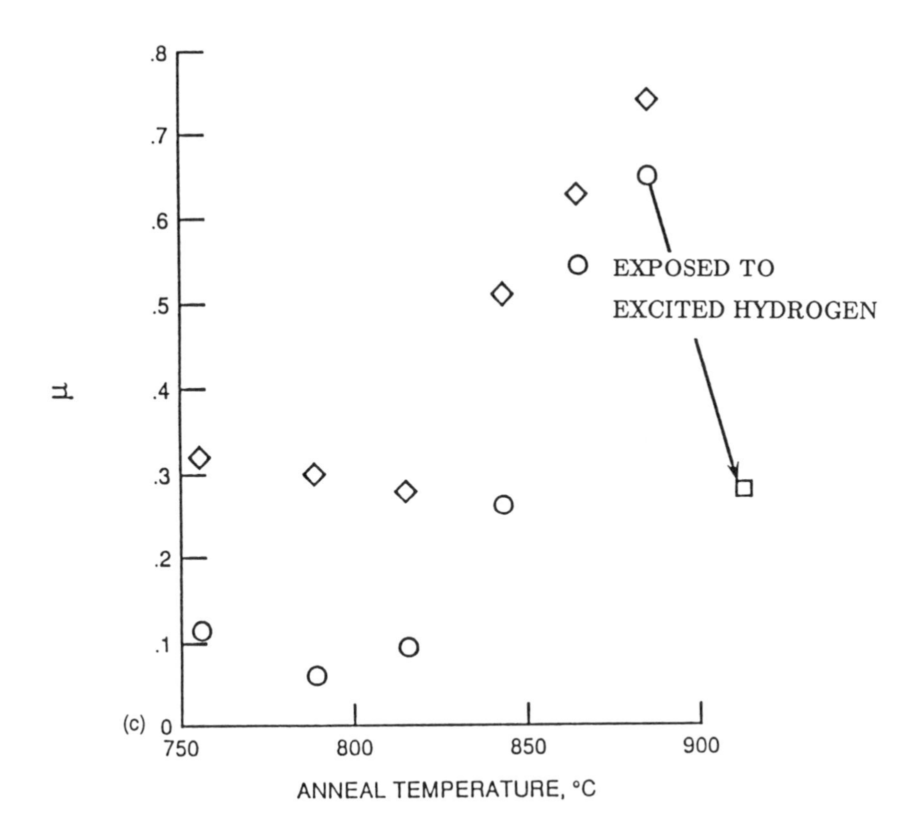

been attributed to the increase in the near surface damage, such as dislocation generation and cracking, occurring at higher normal loads [57,58]. The surface damage was characterized by cathodoluminescence experiments. The coefficient of friction increased with an increase in subsurface damage. A similar phenomenon has been observed in the friction of MgO crystals [59].

In all these instances, the frictional anisotropy exhibited symmetries that correlated with the crystallographic symmetries. Crystallographic planes and directions with highest atomic density generally have high hardness and low friction.

Effect of Normal Load and Sliding Velocity

Under lightly loaded conditions, the coefficient of friction of diamond is independent of the normal load, irrespective of the other sliding material [3,55]. However, when the normal load increases sufficiently to induce subsurface damage on one of the sliding surfaces, the coefficient of friction increases as noted earlier. The wear rate also increases with the onset of subsurface damage.

In a recent study [60], it was found that the coefficient of friction, for a diamond stylus sliding against a diamond flat, is relatively unaffected by the stylus tip geometry if the interface contact pressure is unchanged.

At low sliding velocities, the friction coefficient for diamond on diamond remains essentially constant as a function of sliding velocity (Fig. 6c). Interface temperatures at these velocities are bound to be low in view of the extremely high thermal conductivity of diamond (Table 1).

Effect of Surface Finish

Samuels and Wilks [60] have studied the effect of polishing direction on friction when a diamond stylus was slid against a diamond flat. The friction coefficients on both the (001) and (011) planes of the flat were greater in the two <100> directions (soft or easy polishing directions for diamond) which were parallel and perpendicular to the polishing lines on the diamond surface (in this case of the <100> direction) compared to the <110> (harder abrasion) direction; the latter is inclined at an angle of 45° to the polishing direction. At low contact pressures ~10 MPa or less between the stylus and the flat the friction coefficient on the (001) face was smaller for sliding in the direction of polishing compared to the friction coefficient of sliding in a direction perpendicular to the polishing direction. In all cases, however, the lowest friction coefficient was obtained for sliding in the <110> (hard to abrade) direction. Hayward [54] also obtained

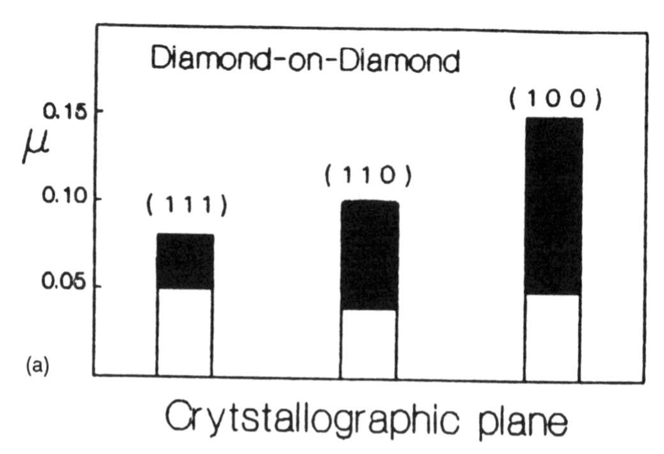

Figure 6 The anisotropy of diamond friction. (a) single-crystal diamond sliding against diamond on various crystallographic planes; typical values of μ (shaded regions denote a range); (b) single-crystal diamond sliding against diamond along various directions of the (100) face [58]. (c) The effect of sliding velocity on the friction of single-crystal diamond sliding against diamond [46].

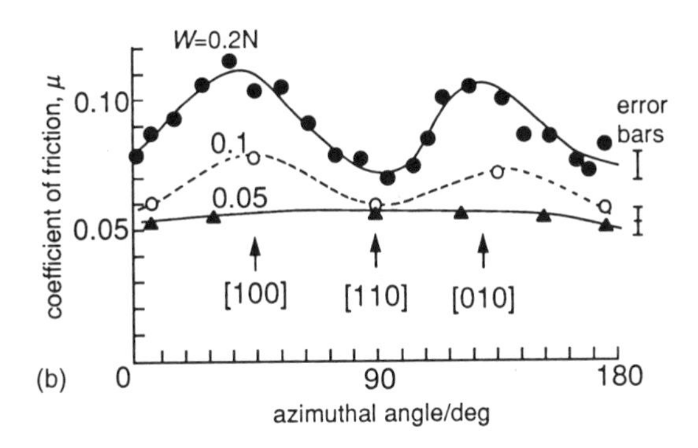
W=0.2N
0.1
0.05
coefficient of friction, μ
0.10
0.05
error bars
[100]
[110]
[010]
(b)
0
90
180
azimuthal angle/deg

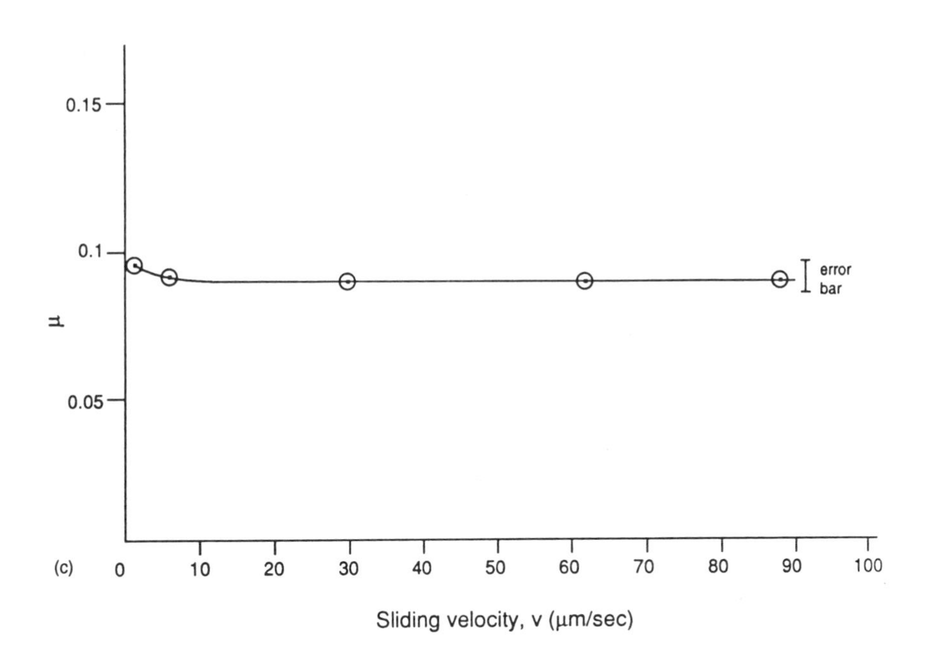
0.15
0.1
0.05
μ
error bar
(c)
0
10
20
30
40
50
60
70
80
90
100
Sliding velocity, v (μm/sec)

similar results. Repeated traversals of the stylus against the flat caused an increase in the coefficient of friction [60].

The importance of surface topography is also illustrated in several other experiments. A smooth diamond surface without any directional polishing lines, which was generated by a polishing process utilizing oxidizing agents on a cast iron scaife (lapping plate), gave coefficients of friction about a factor of 2 lower in all directions compared to diamond surfaces polished using a conventional process or an ion-beam technique [61]. These surface topography effects attributed to roughness on the diamond flat are not yet very well understood.

Effect of Repeated Traversals

During repeated traversals of a diamond stylus across a diamond flat the friction coefficient is observed to change [46,54]. On some faces of the flat, the friction coefficient increased with traversals; while on other planes, the friction coefficient decreased slightly. In general, increases in friction coefficient were also observed when one stylus track crossed another. The erratic behavior of the friction coefficient with repeated traversals on the different crystal planes is probably due to the different changes in the surface roughness of the flat as a result of wear.

Summary

The diamond single crystal has an extremely low coefficient of friction when slid against itself or metals in most environments. Hydrogen, in particular, appears to play a useful role in lowering the friction coefficient. The hardness is higher and the friction coefficient is lower on high atomic density crystallographic systems, for example, the (110) <100> system, compared to the less hard (100) face. Even on the (100) face there is frictional anisotropy, with μ values being lower in the close-packed directions such as <110>. At light loads, the friction coefficient of diamond against diamond is constant. But when the normal load reaches a value high enough to cause subsurface damage in one of the sliding members, the friction coefficient increases. The friction coefficient is practically independent of external lubrication, humidity, and sliding velocity (at low velocities).

2. Polycrystalline Diamond

Polycrystalline diamond (synthetic or man-made) is available in two forms:

1. Diamonds embedded in a metal matrix (typically 50–70% diamond in Ni or Si). This diamond is referred to as diamond composite.
2. Diamonds held together by direct diamond-to-diamond bonding. This is called diamond compact [9].

Mehan and coworkers [62,63] have studied the friction and wear of diamond compacts and composites sliding against diamond and various hardened alloys.

Tests were conducted in pin-on-disk and rotating ring-block configurations at sliding speeds of around 1 m/s. The coefficient of friction of diamond compact when sliding against itself and various metals is shown in Table 5. The friction coefficient of polycrystalline diamond compact (~0.1–0.15) is higher than that of diamond single crystal (~0.05–0.1). The use of a lubricant (mineral oil) slightly reduced the coefficient of friction. The wear rates for diamond compacts were nearly the same as for single-crystal diamond. Furthermore, diamond compacts show no anisotropy in the friction coefficient because of their polycrystalline nature. In the case of diamond composites at interface pressures exceeding about 200 MPa, the diamond particles have a tendency to dislodge from the matrix. Diamond composites have slightly higher μ values compared to compacts. Hayward and Field [52] observed in their studies of a diamond stylus sliding against a diamond composite with a cobalt matrix (in air), that the friction coefficient increased from 0.12 to 0.17 when the contact changed from diamond–diamond to diamond–cobalt.

Mehan et al. [62,63] report that the friction coefficient of both polycrystalline diamond composites and compacts are higher than single-crystal diamond but much lower compared to other hard ceramics such as Al_2O_3 and SiC ($\mu \sim 0.6$), when sliding against hardened alloys in a nitrogen atmosphere. Moreover, the wear rates of the metals sliding against diamond were lower than that for sliding against ceramics. The wear rates followed the same trend as the friction coefficient. In fact, the wear rate and the friction coefficient of a martensitic stainless steel sliding against diamond composite were much lower (almost by a factor of 20 for the wear rate) than those for sliding against itself, as shown in Figure 7. Increased roughness of the diamond flat resulted in greater coefficients of friction and wear rates of the metal. Polycrystalline diamonds, in general, have greater toughness than single crystals, and hence offer better wear and fracture resistance under extreme loading conditions such as in cutting tools and dies [9].

3. Cubic Boron Nitride

The frictional behavior of CBN is quite similar to diamond [64] except in the machining of steels. Here, the wear rate of CBN is markedly less due to the low

Table 5 Coefficient of Friction for Diamond Compact

Diamond Against	Coefficient of Friction in Air
Diamond	0.1–0.15
4620 steel	0.1–0.15
12% Cr steel	0.12
Synthetic fiber	0.08–0.2

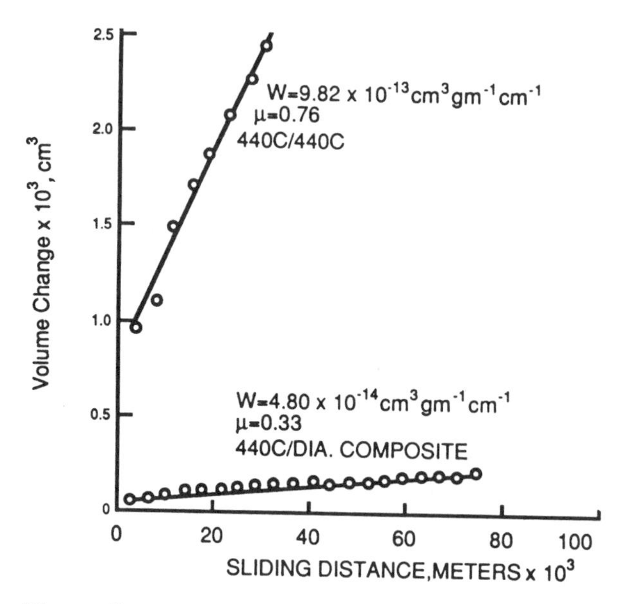

Figure 7 Wear rate and friction coefficient of a 440°C steel pin rubbed against itself and against a diamond composite: W, wear rate; μ coefficient of friction [62,63].

chemical affinity of boron nitride for steel, in contrast to diamond at high temperatures [65].

4. Silicon Carbide

The friction and wear of SiC under adhesive and abrasive wear conditions has been studied by Miyoshi and Buckley [66,67]. Figure 8a shows the variation in friction coefficient with temperature for single-crystal SiC sliding against a hemispherical iron rider (tip radius ~ 0.79 mm) in vacuum. Under these conditions of low normal load and low sliding velocity, adhesive wear dominates [66,67]. For single-crystal silicon carbide sliding against itself, the coefficient of friction was lowest on the preferred basal slip plane (0001) (Fig. 8b) (high hardness and high atomic density) and in the preferred slip direction $<11\bar{2}0>$. This is due to the low strength of the adhesive junctions at the sliding interface in the preferred slip system orientation [4]. The lowest wear rate of both the iron rider and the SiC crystal was also observed in the preferred slip system orientation. Under abrasive wear conditions, in which a diamond conical tip radius of ~ 5 μm was sliding against SiC single crystal, the lowest coefficient of friction and the lowest wear rate were obtained in the $<10\bar{1}0>$ direction on the (0001) plane, which corresponds to the direction with the highest hardness. The least amount of subsurface damage also occurred for this system in the abrasive wear experiments

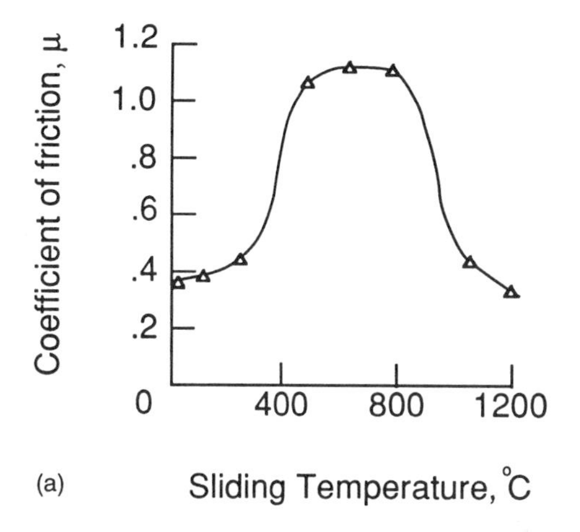

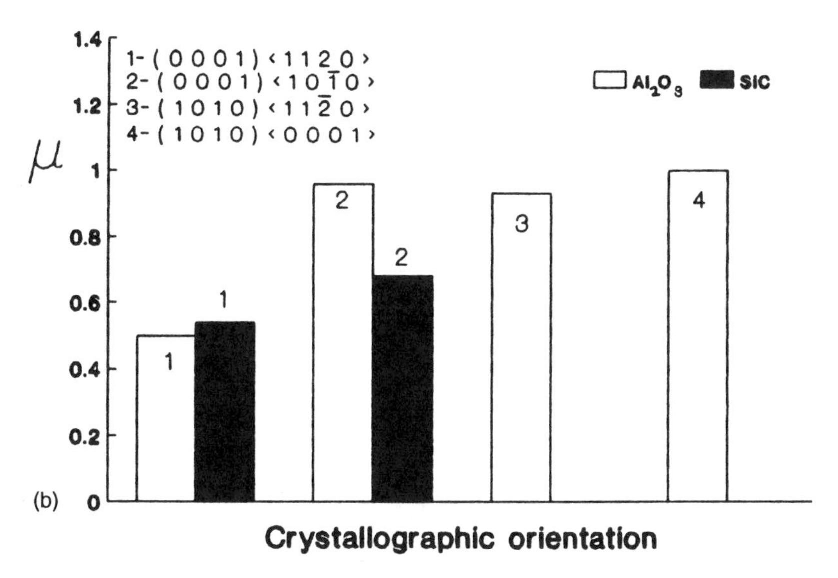

Figure 8 The variation of friction coefficient with temperature for a sintered polycrystalline silicon carbide surface sliding against a clean iron rider in vacuum [66,67] (a), and the effect of crystallographic orientation on the friction of sapphire–sapphire and single-crystal SiC–SiC in vacuum [5,66,69] (b).

[4]. The increase in the coefficient of friction in the temperature range 400–800°C (Fig. 8a) was due to the breakdown of adsorbed films on the SiC surface; while a large decrease in μ above 800°C was due to the graphitization of the SiC surface providing a lubricating action.

5. Silicon Nitride

In pin-on-disk experiments in air, Ishigaki et al. [27] have measured the friction and wear of silicon nitride in contact with itself and diamond sliders, at sliding speeds of up to 0.78 m/s. The friction and wear rate of Si_3N_4 sliding against itself increased with increasing sliding speeds. Higher sliding speeds, with attendant higher interface temperatures, result in a decreasing hardness of Si_3N_4. Among various types of Si_3N_4 hot-pressed ceramics sliding against themselves, the friction coefficient decreased with increasing fracture toughness and hardness. Ishigaki et al. [27] also found that the friction coefficient and wear rates were anisotropic (in hot-pressed Si_3N_4 against diamond) due to an orientation texture developed during hot-pressing. In general, both μ and wear rate were lower in the texture directions with a higher hardness. The friction and wear of Si_3N_4 also decreased slightly with increasing humidity.

6. Other Carbides and Nitrides

Cemented tungsten carbide bonded together with a cobalt matrix (typically ~ 94% WC and 6% Co) is widely used in cutting tool inserts for the machining of metallic materials. Titanium carbide and titanium nitride CVD coatings on WC substrates have also emerged as an important class of tool materials. These coated tool materials have lower friction against steel than WC; furthermore, wear due to diffusion is also lower with the coated tool materials [68].

Sialon has recently emerged as an important cutting tool material in the same class, especially for machining cast iron. Boron carbide has not found much application in cutting tools and grinding abrasives because of its chemical reactivity and brittleness [24].

7. Aluminum Oxide

The behavior of single-crystal Al_2O_3 (sapphire) in contact with various metals and against itself has been studied by Buckley [5] and Pepper [69]. In general, the coefficient of friction against metals in vacuum was lowest on the perferred slip plane (0001) in the preferred slip direction $[11\bar{2}0]$. This system also gave the lowest coefficient of friction for sapphire sliding against sapphire (Fig. 8b) similar to the case of SiC against SiC discussed earlier. In air, the friction coefficient of sapphire was low (~0.15). This is only a little higher than the friction coefficient of single-crystal diamond sliding against itself (μ ~ 0.05 to 0.1). For sapphire in contact with metals, friction coefficients were highest against those metals which were easily oxidized by Al_2O_3. There was good correlation between μ and the free energy of formation of the lowest metal oxide [4]. Pepper [69] observed that adhesion of metal–sapphire contacts was promoted by the presence of oxygen, but decreased in nitrogen atmospheres. He attributed this to the strong affinity of sapphire to chemisorbed oxygen on the metal surface. Wear, as measured by the transfer of metal to the sapphire surface, correlated well with the friction coefficient in these experiments [4]. In air, the coefficient of friction of polycrystalline alumina sliding against itself is around 0.2 (Fig. 9a) at room temperature, which is higher than that of sapphire on sapphire ($\mu \sim 0.15$) mentioned earlier.

8. Zirconia

The coefficient of friction of MgO partially stabilized zirconia (in air) sliding against itself is found to increase with increasing ambient temperature (Fig. 9a) up to around 700 K, similar to polycrystalline alumina sliding on alumina. Buckley [3] attributes such increases to the removal of adsorbents from the surface. However, the wear rate reaches a minimum around 625 K for these ceramics, as seen in Figure 9b. In fact, the maximum in the wear rate occurs close to the minimum friction condition. The presence of water increased the coefficient of friction of zirconia sliding on zirconia from 0.2 to 0.5 in air, while a long-chain fatty acid lubricant (0.5% stearic acid in n-tridecane) reduced it to 0.1. The coefficient of friction was also higher ($\mu \sim 0.4$–0.5) for PSZ in contact with soft metals (Cu, Al, Fe). In general, the friction and wear performance of PSZ is better than sialon and comparable to alumina. Phase transformation toughening plays an important role in influencing the tribological behavior. Under heavily loaded conditions, crack initiation and propagation in PSZ are reduced by the transformation toughening process [29,70,71].

9. Chromium Trioxide

Chromium oxide exhibits extreme chemical stability and good tribological properties. Cr_2O_3 grains are used as an abrasive in the lapping of hard surfaces. Chromium oxide as a solid and in the form of coatings is used for gas bearings and other sliding applications in extreme operating environments. Sputtered Cr_2O_3 coatings have been developed for a foil bearing that functions from room temperature to 650°C [72]. Among many material combinations tested, sputtered Cr_2O_3 sliding against detonation gun sprayed Cr_3C_2 coatings performed the best. Sputtered Cr_2O_3 coatings also performed superior to sputtered TiN coatings in ball-bearing applications [72].

10. Ferrites

Miyoshi and Buckley [39,73] conducted friction and wear experiments on Mn–Zn and Ni–Zn ferrites sliding against themselves and in contact with transition metals. The studies showed that:

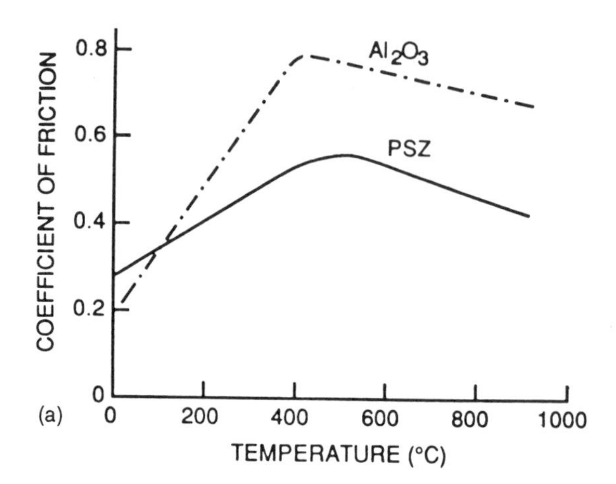

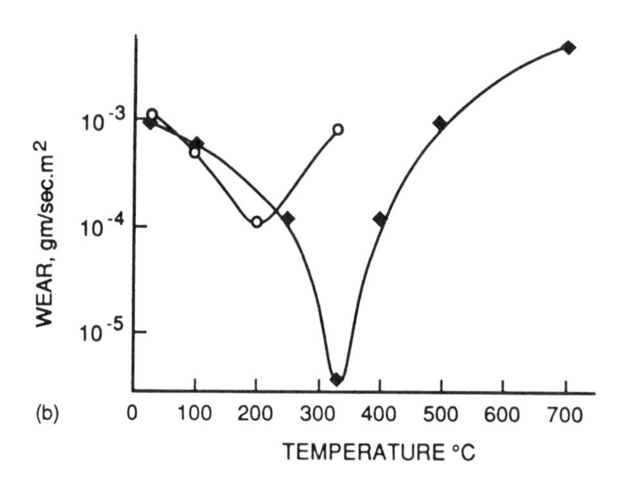

Figure 9 The friction and wear of MgO—partially stabilized zirconia and polycrystalline alumina in contact with themselves at various temperatures. (a) friction; (b) wear rate of PSZ–PSZ at 1: sliding speed 0.27m/s and 2: sliding speed 0.44 m/s [29,71].

1. Matched parallel high atomic density planes and crystallographic directions at the interface (for ferrite on ferrite) exhibited low coefficients of friction. Lowest μ was obtained along <110> directions on (110) planes.
2. For ferrites in contact with metals in vacuum, friction coefficients increased with increasing affinity of a metal for oxygen (in ferrite). μ correlated well with the free energy of formation of the lowest metal oxide (Fig. 10a). The higher the free energy, the greater the affinity of metal for oxygen and the higher the friction coefficient.
3. Presence of an oxygen atmosphere increased the coefficient of friction.
4. μ also decreased with increasing d-character of the metal bond (another measure of chemical activity of the metal) for ferrites in contact with transition metals. Figure 10b shows this in a variation of friction coefficient with percent d-character of the metal bond. The percent d-character of the metal bond has an important influence on the friction and wear of ceramics in contact with transition metals and this aspect is discussed further later.

Kehr et al. [38] observed an increase in the wear of Ni–Zn ferrites with decreasing grain size, when sliding against CrO_2 and γ-Fe_2O_3 tapes (Fig. 11).

11. Metallic Glasses

Miyoshi and Buckley [74] found that the friction of ferrous base metallic glasses in contact with aluminum oxide increased with temperature. Increase in friction was also obtained under sliding conditions that led to crystallization of the amorphous glass.

B. Properties that Affect Friction and Wear

The coefficient of friction and the wear process in ceramics under lightly loaded, marginally lubricated sliding contact conditions appears to be predominantly due to adhesion at the contacting interface. While there is no single property or parameter at present that can effectively characterize the adhesion process, it is still useful to qualitatively understand the factors influencing this interaction. In the light of the experimental results discussed so far, the factors influencing the adhesion and friction of ceramics under lightly loaded conditions are: crystallographic system; grain boundaries; environment and adsorbed layers; hardness; elastic modulus; thermal conductivity; onset of surface damage; fracture toughness; percent d-character of the bond; and tribocharging.

The effect of these factors on the friction and wear of ceramics is discussed in this section.

1. Crystallographic System

The coefficient of friction is generally lower when sliding occurs along the close-packed planes and directions in single-crystal ceramics. For example, the

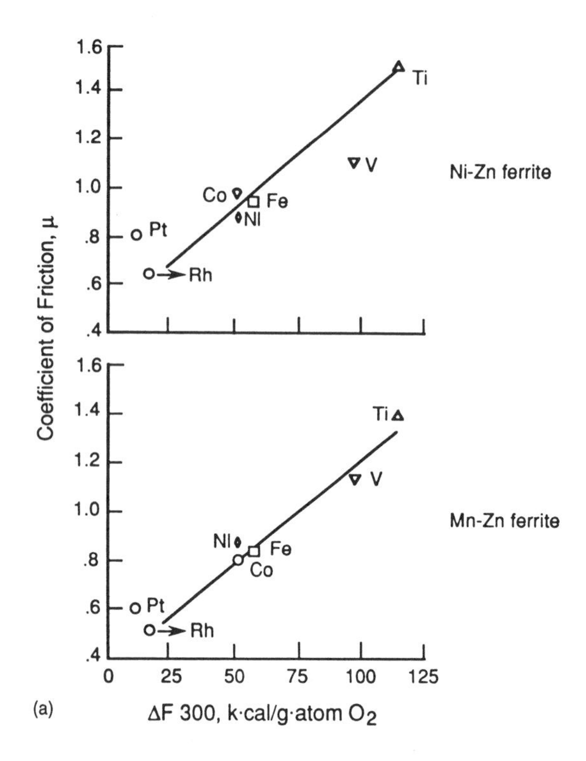

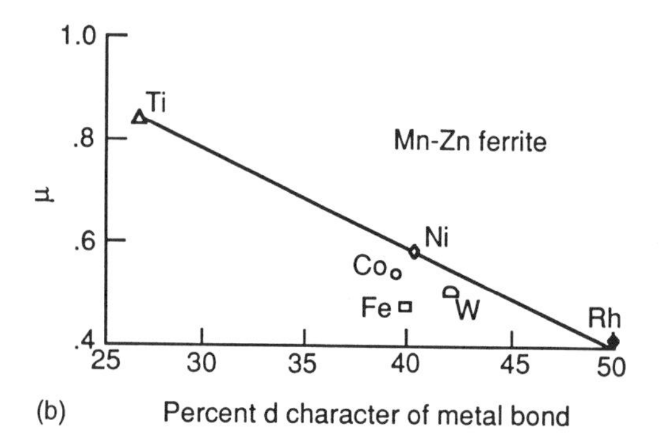

Figure 10 The coefficient of friction of various metals in contact with single-crystal Ni–Zn and Mn–Zn ferrite in vacuum. (a) as a function of the free energy of formation of the lowest oxide; (b) as a function of the percentage d-bond character of various metals [73].

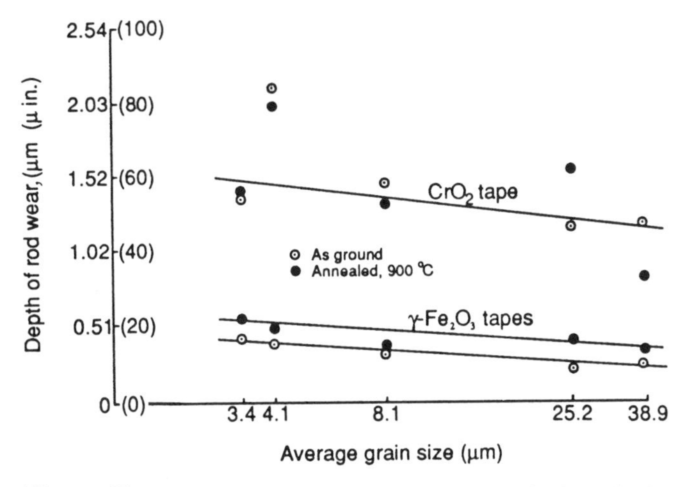

Figure 11 The effect of grain size on the wear of Ni–Zn ferrite rod in contact with Fe_2O_3 and CrO_2 tapes [38].

coefficient of friction of diamond on diamond is low on the close-packed (110) plane (Fig. 6a). Even on the less densely packed (100) plane, the friction coefficient is lower along the close-packed <110> direction (Fig. 6b). A similar observation can be made in the case of single-crystal sapphire on sapphire and SiC on SiC (Fig. 8b). Qualitatively, this may be explained in terms of surface energy. A close-packed system, in general, has a lower surface energy [75,76,77] and, therefore, lesser reactivity in comparison to a higher surface energy associated with less densely packed crystallographic face. The higher the surface energy, the greater the reactivity of the surface atoms. The denser packing also results in a larger hardness on the close-packed planes and in close-packed directions, relative to the less densely packed faces; compare, for example, the hardness of the diamond (110) <110> system ($\sim$9500 kg/cm^2) to the hardness of the diamond (100) <110> ($\sim$8600 kg/mm^2), which is a less closely packed system. The potential energy profiles along close-packed directions in close-packed planes is also smoother, which facilitates the motion of atoms on one surface across the other along these orientations [49,78,79] and possibly lower friction. A common example of this aspect is in the easy movement of dislocations in close-packed slip systems.

It appears, therefore, that adhesive forces between the contacting solids and the resistance to interfacial motion would be lower on close-packed crystallographic systems.

2. Grain Boundaries

Grain boundaries are high-energy sites, which are much more chemically active than the interior of the grain. Figure 12a shows that the friction coefficient increases dramatically whenever a single-crystal sapphire slider approaches the grain boundaries in a large grain tungsten disk. A similar dramatic increase occurs also for a copper single crystal approaching a grain boundary (Fig. 12b). Hayward and Field [51] have observed increases in the friction coefficient when a single-crystal diamond approaches the diamond–matrix interface in a diamond composite. These suggest that as long as slider surfaces do not suffer surface/subsurface damage, single-crystal materials are preferable from a friction point of view. For higher contact pressures, however, the improved toughness of polycrystalline materials offers natural advantages.

3. Environment and Adsorbed Layers

The adhesion at the sliding ceramic interface is dependent on the nature of the adsorbed layers present on the sliding surfaces and its interaction with the ceramic surface. The investigations of Bowden and Tabor [43] and Bowden and Hanwell [56] demonstrated that the coefficient of friction of diamond sliding against diamond decreased from about 1 in vacuum to a value of about 0.15 in air. Similar results have been reported elsewhere [46,54,80]. In particular, the adsorption of hydrogen on the diamond surface reduces the friction coefficient (Figs. 5a–c). The mechanism of hydrogen adsorption and the associated decrease

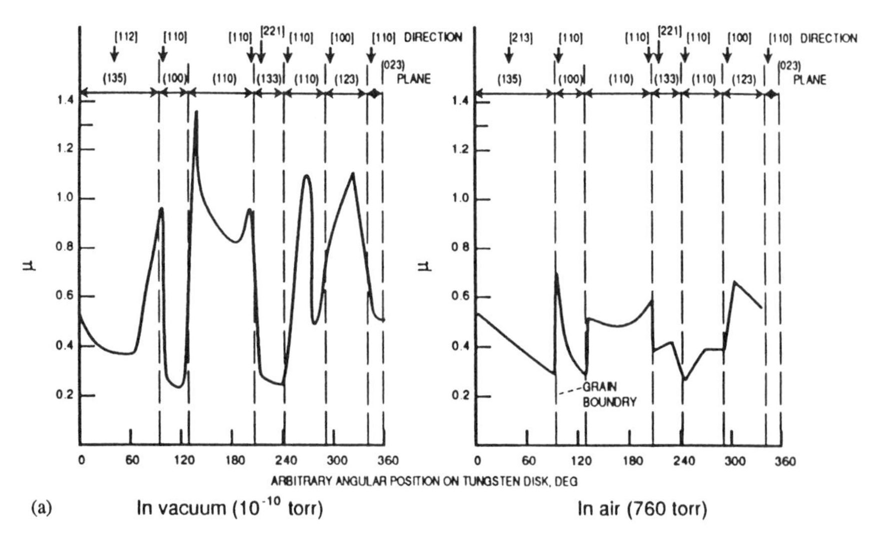

Figure 12 The effect of grain boundaries on friction. (a) the coefficient of friction of sapphire (10$\bar{1}$0) plane sliding against polycrystalline tungsten; (b) the friction of single-crystal copper sliding against polycrystalline copper (in both cases, note the increase in μ as the single crystal approaches a grain boundary of the mating polycrystalline material) [5,47].

in the friction coefficient appear to be related to the *dangling carbon bonds at the diamond surface*. It is known that the surface of diamond is covered by carbon atoms with free dangling bonds [13,81,82]. These bonds have a high affinity for hydrogen and it is well established that hydrogen adsorbs easily on the surface of diamond [15,16]. Pepper has shown that adsorption of hydrogen on the diamond surface makes the diamond devoid of any electronic states in the band gap, resulting in low adhesion and friction. On the other hand, the removal of hydrogen by heating or by exposure to vacuum results in a dramatic increase in the adhesion of diamond–metal interfaces [15,16]. It is also our hypothesis that the C–H bond at the interface is essentially similar to that in a hydrocarbon and the surface adsorbed layer acts as a natural lubricant. In recent studies, Chandrasekar and Bhushan [80,83] have reported low values for the friction coefficient of diamond when sliding against hard disks. Moreover, Hayward [54] and Seal [46] have discovered the presence of waxy debris during the sliding of diamond against diamond and this is believed to act like a lubricant. There is speculation, based on analysis of this debris, that it is made up of hydrocarbons, but this fact has not been conclusively established.

For sapphire sliding against metals such as Ni, the presence of oxygen

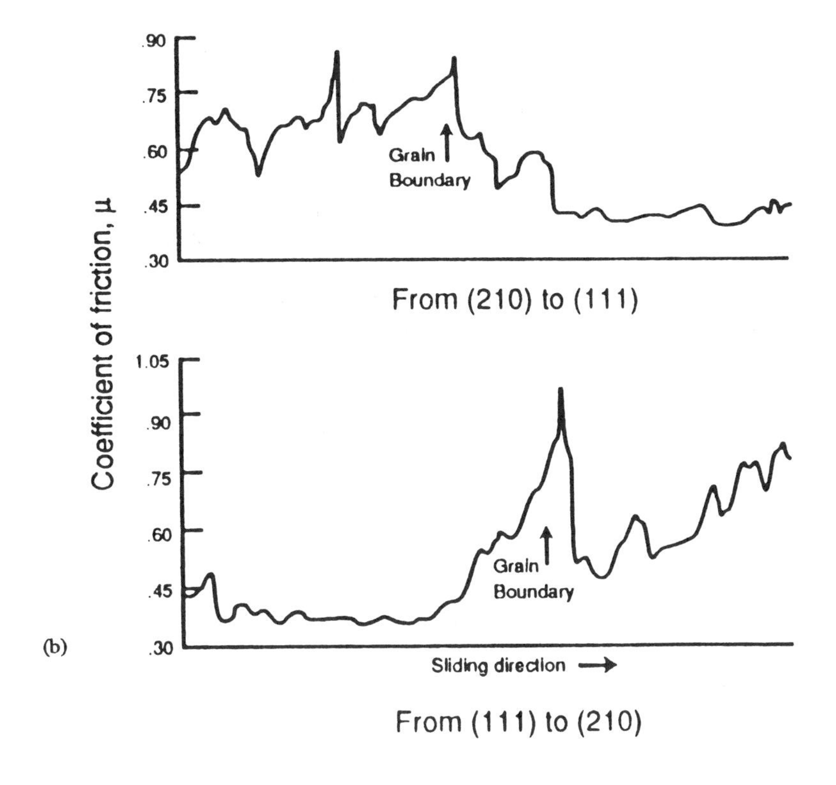

(b)

increases interfacial adhesion while chlorine has the opposite effect [69]. The mechanism of the interaction has not yet been fully understood in this case.

4. Hardness

When two surfaces are brought into contact, the real area of contact is inversely proportional to the hardness of the softer material. A smaller real area of contact results in lower adhesion and friction. This probably contributes in part to the low friction of diamond on diamond and sapphire on sapphire. We, however, suggest that under marginally lubricated conditions, the hardness (the effective hardness for composite structures) of the mating hard members should be matched as closely as possible.

5. Elastic Modulus

Tabor [49] has shown that the presence of atomic scale roughness on the sliding ceramic crystal surface causes an "elastic repulsive" force that forces the asperities apart. This, in turn, can cause a reduction in the friction coefficient. This "repulsive" force is, among other things, a function of the Young's modulus of the sliding material. Diamond has a very high elastic modulus and therefore this force will play a role in reducing the adhesion between near-atomically flat surfaces of diamond. This effect will also be important in other high elastic-moduli ceramics such as sapphire. This effect is in addition to the elastic spring back at contacting asperities, when the normal load is released. The elastic spring-back is larger if the Young's modulus of the sliding materials is higher. A consequence of this would be a lowering of the adhesion for diamond sliding against diamond.

6. Thermal Conductivity

The analysis of Jaeger as discussed in Bowden and Tabor [43] shows that sliding interface temperatures are, among other things, inversely proportional to the sum of the thermal conductivities of the two sliding objects. A similar dependence is also found for head–media contacts [2]. Lower interface temperatures generally result in less degradation of the interface and in lower adhesion. Diamond has the highest thermal conductivity among the ceramics in Table 1 followed by CBN. Single-crystal ceramics, in general, have a higher thermal conductivity compared to their polycrystalline counterparts.

7. Fracture Toughness

It is not clear at present what effect fracture toughness has on the adhesion of ceramics, but it can influence the wear of ceramics under moderate to heavily loaded sliding contact conditions. Evans and Marshall [84] have shown that under these conditions, the wear rate of polycrystalline ceramics is inversely related to their fracture toughness (K_{IC}). This is a consequence of the interconnection between fracture toughness and the crack propagation process. Such a trend has been observed, as noted earlier, in the experiments of Ishigaki et al. [27]. Both Hannink et al. [29] and Aronov [71] have shown that thermally induced phase transformations during sliding reduce the wear rate of zirconia (Figs. 9a and 9b). The volume changes induced by the phase transformation at a particular temperature have an effect similar to that due to an increase in the fracture toughness.

8. Onset of Surface Damage

Under lightly loaded conditions, the coefficient of friction between sliding surfaces of ceramics can be low ($\mu \sim 0.001$–0.05). Some typical examples in this regard are tungsten on graphite [85,86] and diamond on diamond [46,49, 57,58]. In many of these experiments, when the normal load exceeded a threshold value, the friction coefficients increased by at least an order of magnitude. This increase coincided with the onset of surface damage on one of the sliding interfaces; the induced damage being in the form of cracking, dislocations, and wear particle generation. At present, it is not clear as to what is the phenomenological mechanism linking the onset of surface damage with the simultaneous increase in the friction coefficient. The onset of surface damage would be expected to occur at higher normal loads with solids such as diamond.

9. Percent d-character of the Bond

In the friction of ceramics sliding against transition metals, there is good correlation between the coefficient of friction and the percent d-character of the metallic bond in the metal [3,47]. Figure 10b shows this correlation for the case of ferrite. Similar correlations have also been obtained for a variety of other ceramics [47]. The percent d-character of the bond is a measure of the d-electrons involved in the bond formation in transition metals. The greater the number involved, the more stable and less reactive the metal [87,88]. Hence, adhesion and friction decrease with increasing percent d-character of the bond in the metal.

10. Tribocharging

At sliding interfaces, where at least one of the sliding materials is a poor electrical conductor, electrical charge generation can occur. Forces, both attractive and repulsive, can arise due to this phenomenon called "contact electrification." This can contribute to the adhesion at the sliding interface; for a review of such effects, see References [89–93]. Manipulation of this phenomena could possibly be used to change the adhesion at the sliding interface [90]. In fact, such techniques have been successfully used to control the coalescence (adhesion) and break-up of liquid droplets [94]. Deryagin et al. [89] discuss in great detail the effect of electrical charge generation on the adhesion between sliding surfaces.

IV. IMPLICATIONS FOR MAGNETIC RECORDING SYSTEMS

Currently in magnetic recording systems, Mn–Zn and Ni–Zn ferrite, aluminum oxide–titanium carbide composite, yttria-stabilized zirconia aluminum oxide–titanium carbide, and titanates of barium and calcium are used as slider materials. Diamondlike carbon, SiO_2, and ZrO_2–Y_2O_3 are the commonly used materials for disk overcoats. With the trend toward recording systems with ultra-low flying heights and even continuous contact between the slider and media in the future, the natural question is whether these are the best slider materials. In this section, we shall discuss ceramics with potential for use in lightly loaded, marginally lubricated, sliding-contact applications in light of the experimental results reviewed.

The ceramic slider material and disk overcoats should possess some of the following characteristics for low friction and wear: low adhesion to the mating material; matched crystallaographic planes and directions with the mating material for single crystals; high hardness and Young's modulus; low surface energy for the contacting faces; high thermal conductivity; low chemical reactivity; and matched mechanical properties of the mating member.

Adhesion (which affects friction and wear) cannot be directly quantified in terms of one or more directly measurable properties. Often the selection of materials for tribological applications is based on empirical evidence. Frequently, materials for sliding-contact applications are selected on the basis of their having a high hardness and/or Young's modulus. This in itself does not ensure low friction and wear. For example, zirconia has low friction and wear characteristics but its hardness (~ 1300 kg/mm^2) is at least a factor of two less than many of the nitride and carbide ceramics. In general, oxide ceramics have lower adhesion compared to nitride and carbide ceramics, indicating that factors other than those previously mentioned may be operative. Figure 13 shows the coefficient of friction for various ceramics (considered in this study) when sliding against themselves. Based on the observations made in this chapter and experimental evidence, several ceramics are recommended for lightly loaded, marginally lubricated, sliding-contact applications. Those marked with an asterisk (*) are suitable as a coating material. They are classified in order of priority, with priority 1 offering the most promise. *Priority 1*: single-crystal diamond; single-crystal CBN; sapphire; polycrystalline diamond and CBN*; single-crystal SiC; and polycrystalline SiC*. *Priority 2*: partially stabilized zirconia (PSZ)*; titanium nitride*; Cr_2O_3*; Al_2O_3/TiO_2*; titanium carbide*; sialon*; silicon nitride*; and tungsten carbide*.

In the case of single-crystal materials, it would be preferable to select a close-packed crystallographic system as the slider face. As mentioned earlier, it is critical that the properties of the slider and mating disk material are suitably matched.

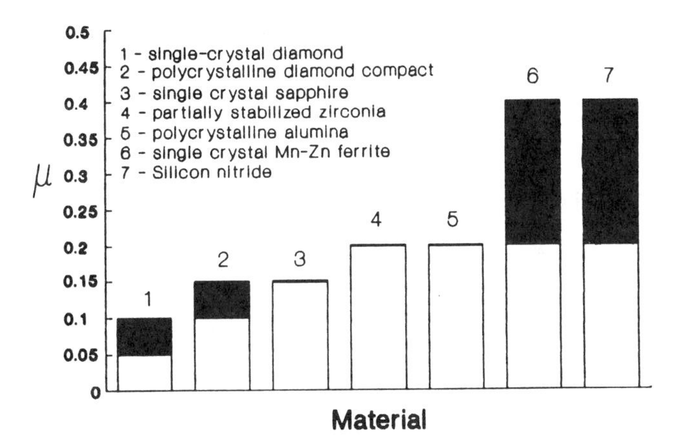

Figure 13 A summary of typical friction coefficients of various materials in contact with themselves in air (note shaded regions denote a range of μ).

V. CONCLUSIONS

The adhesion and friction of single-crystal and polycrystalline ceramics under lightly loaded sliding contact conditions have been reviewed. There are several factors that influence the friction and wear behavior of ceramics, including, among others, the crystallographic system on which sliding takes place, hardness, Young's modulus, thermal conductivity, adsorbed layers, and chemical reactivity of the slider material. Single-crystal diamond and sapphire have low friction coefficients and low wear rates in sliding contact with other ceramics. Based on the tribological behavior of these ceramics and on an understanding of the various factors influencing their adhesion and friction, a composite picture emerges of ceramic material systems with potential for use in lightly loaded, marginally lubricated sliding applications. Several such systems have been identified for future study. It is hoped that a study of the tribology of these selected ceramics would enable the development of a new generation of magnetic recording heads and disks for ultra-low (approaching zero) flying height recording systems.

ACKNOWLEDGMENTS

This work was partially supported by NSF Presidential Young Investigator Awards to Chandrasekar (DDM 9057916) and Farris (MSS 9057082).

REFERENCES

1. C. D. Mee and E. Daniel, *Magnetic Recording I: Technology*, McGraw-Hill: New York, 1987.

2. B. Bhushan, *Tribology and Mechanics of Magnetic Storage Devices*, Springer-Verlag: New York, 1989.
3. D. Buckley, *Surface Effects in Adhesion, Friction and Lubrication*, Elsevier: Amsterdam, 1981.
4. D. Buckley and K. Miyoshi, Friction and Wear of Ceramics, *Wear 100*, 333–353 1984.
5. D. Buckley, Ceramic Microstructure and Adhesion, *J. Vac. Sci. Technol. A 3*, 76 1985.
6. W. Kingery, H. Bowen, and D. Uhlmann, *Introduction to Ceramics*, 3rd ed., John Wiley: New York, 1983.
7. R. Berman, Thermal Conductivity of Diamond, in *Properties of Diamond*, J. E. Field (Ed.), Academic Press: New York, 1979.
8. J. E. Field, *The Properties of Diamond* (Appendix), Academic Press: New York, 1979.
9. R. H. Wentorf, R. C. DeVries, and F. P. Bundy, "Sintered Superhard Materials," *Science 208*, 873–880 1980.
10. R. M. Chrenko and H. M. Strong, (1975), General Electric Corporate Research Tech. Report No. 75CRD 089, Schenectady, New York.
11. J. E. Field, Strength of Diamond, in *The Properties of Diamond*, J. E. Field, (Ed.), Academic Press: New York, 1979.
12. C. A. Brookes, The Hardness of Diamond, in *The Properties of Diamond*, J. Field (Ed), Academic Press: New York, 1979.
13. K. E. Spear, Diamond: Ceramic Coating of the Future, *J. Am. Ceram. Soc. 72*(2), 171–179 1989.
14. P. W. Bridgman, *Studies in Large Plastic Flow and Fracture*, McGraw-Hill: New York, 1952.
15. S. Pepper, Effect of Electronic Structure of the Diamond Surface on the Strength of Diamond–Metal Interface, *J. Vac. Sci. Technol. 20*, 643–647 1982.
16. S. Pepper, Transformation of the Diamond (110) Surface, *J. Vac. Sci. Technol. 20*, 213–217 (1982).
17. P. Lurie and J. M. Wilson, The Diamond Surface: The Structure of the Clean Surface and the Interaction with Gases and Metals, *Surf. Sci. 65*, 453–475 1977.
18. J. C. Angus, P. Koidl, and S. Domitz, Carbon Thin Films, in *Plasma Depositied Thin Films*, J. Mort, and F. Jansen (Eds.), CRC Press, Inc.: Boca Raton, FL, 1986, pp. 87–127.
19. J. E. Sundgren and H. T. Hentzell, A Review of the Present State of the Art in Hard Coatings Grown from the Vapor Phase, *J. Vac. Sci. Technol. 4–5*, 2259–79 1986.
20. M. P. Nadler, T. N. Donovan, and A. K. Greene, Thermal Annealing Study of Carbon Films Formed by Plasma Decomposition of Hydrocarbons, *Thin Solid Films 116*, 241–247 1983.
21. R. C. DeVries, *Cubic Boron Nitride*: *Handbook of Properties*, General Electric Corporate Research Tech. Report, No. 72 CRD 178 1972.
22. P. J. Gielisse, S. S. Mitra, J. N. Plendl, R. D. Griffis, L. C. Mansur, R. Marshall, and E. A. Pascoe, Lattice Infrared Spectra of Boron Nitride and Boron Monophosphade, *Phys. Rev. 155*(3), 1039–1046 1967.

23. P. T. Shaffer, Effect of Crystal Orientation on the Hardness of Silicon-Carbide, *J. Am. Ceram. Soc. 47*, 466–471 1964.
24. E. D. Whitney, "Thermodynamic Properties of Abrasive Materials," in *New Developments in Grinding*, M. C. Shaw (Ed.), Carnegie Press: New York, 1972.
25. E. K. Storms, *The Refractory Carbides*, Academic Press: New York, 1967.
26. B. Bhushan and L. B. Sibley, Silicon Nitride Rolling Bearings for Extreme Operating Conditions, *ASLE Trans. 25*, 417–428 1982.
27. H. Ishigaki, I. Kawaguchi, M. Iwasa, and Y. Toibana, Friction and Wear of Hot Pressed Silicon Nitride and Other Ceramics, *Trans. ASME J. Tribology 108*, 514–520 1986.
28. B. J. Hockey, Indentation Hardness of Alumina, in *The Science of Ceramic Machining and Surface Finishing*, B. J. Hockey, et al. (Eds.), NBS Special Publication, 1979.
29. R. Hannink, J. Murray, and H. Scott, Friction and Wear of Partially Stabilized Zirconia: Basic Science and Practical Applications, *Wear 100*, 355–366 1984.
30. A. G. Evans and A. H. Heuer, Review—Transformation Toughening in Ceramics: Martensitic Transformation in Crack-Tip Stress Fields, *J. Am. Ceram. Soc. 63*, 241–248 1980.
31. A. H. Heuer and L. W. Hobbs, *Science and Technology of Zirconia*, American Ceramic Society: Columbus, OH, 1981.
32. D. Porter and A. Heuer, Mechanisms of Toughening Partially Stabilized Zirconis, *J. Am. Ceram. Soc. 60*, 183–184 1977.
33. R. C. Garvie, Stabilization of the Tetragonal Structure in Zirconia Microcrystals, *J. Phys. Chem. 82*(2), 218 1978.
34. B. Bhushan Overview of Coating Materials, Surface Treatments, and Screening Techniques for Tribological Applications Part 1: Coating Materials and Surface Treatments, ASTM Special Publication No. STP947, 1987, pp. 289–309.
35. B. Bhushan, (1980), High-Temperature Self-Lubricating Coatings and Treatments—A Review, *Metal Finishing 78*, May 1980, pp. 83–88; June, 1980, pp. 71–75.
36. N. E. Cother and P. Hodgson, The Development of Sialon Ceramics and Their Engineering Applications, *J. Br. Ceram. Soc. 21*, 141 1981.
37. P. Spencer, *Characterization of Sialon-Type Materials*, Lawrence Berkeley Laboratory, Rep. LBL 6612 1977.
38. W. Kehr, C. Meldrum, and R. Thornley, The Influence of Grain Size on the Wear of Ni–Zn Ferrite by Flexible Media, *Wear 31*, 109–117 (19xx).
39. K. Miyoshi and D. Buckley, Properties of Ferrites Important to Their Friction and Wear Behavior, in *Tribology and Mechanics of Magnetic Storage Systems*, B. Bhushan, et al. (Eds.), Special Publication, SP-16, ASLE: Park Ridge, IL, 1984.
40. S. Chandrasekar and B. Bhushan, Control of Surface Finishing Residual Stresses in Magnetic Recording Head Materials, *Trans. ASME J. Tribology 110*, 87–92 1988.
41. K. Subramanian and P. Keat, A Parametric Study of the Grindability of Electonic Ceramics, in *Machining of Ceramic Materials and Components*, K. Subramanian, et al. (Eds.), ASME Special Bound Volume, PED-17, New York, 1985.
42. Anon, *Properties of Corning's Glass and Glass Ceramic Families*, Corning Glass Works: Corning, New York, 1979.

43. F. P. Bowden and D. Tabor, *The Friction and Lubrication of Solids: Part 2*, Clarendon Press: Oxford, 1964.
44. M. Seal, The Friction of Single Crystal Diamond, *Proc. R. Soc. London Ser. A 249*, 379–389 1958.
45. M. Seal, The Friction of Diamond, *Ind. Diamond Rev. 25*, 111–116 1965.
46. M. Seal, The Friction of Diamond, *Philos. Mag. A 43*, 587–594 1981.
47. D. Buckley, Surface Films and Metallurgy Related to Friction and Wear, *Prog. Surf. Sci. 12*, 1–154 1982.
48. K. Miyoshi and D. Buckley, Adhesion and Friction of Single-Crystal Diamond in Contact with Transition Metals, *Appl. Surf. Sci. 6*, 161–172 1980.
49. D. Tabor, Adhesion and Friction, in *The Properties of Diamond*, J. E. Field (Eds.), Academic Press: New York, 1979, pp. 325–348.
50. I. Hayward, The Friction of Diamond of Low Loads and for Repeated Traversals, *Proceedings Diamond Conference*, Reading, England 1985.
51. I. P. Hayward and J. Field, Friction Studies of Diamond and Syndite, *Proceedings Diamond Conference*, Bristol, England 1984.
52. I. P. Hayward and J. Field, Friction and Erosion Studies of Diamond and Syndite, *Proceedings Diamond Conference*, Cambridge, England 1983.
53. I. Hayward, Friction Studies of Diamond on Diamond Using Repeated Traversals, *Proceedings Diamond Conference*, Bedford, England 1986.
54. I. Hayward, The Friction and Strength Properties of Diamond, Ph.D. Thesis, Cambridge University, Cambridge, England, 1987.
55. M. Casey and J. Wilks, The Friction of Diamond Sliding on Polished Cube Faces of Diamond, *J. Phys. D. Appl. Phys. 6*, 1772–1781 1973.
56. F. P. Bowden and A. F. Hanwell, The Friction and Wear of Diamond, *Proc. R. Soc. London Ser. A 295*, 233 1966.
57. Y. Enomoto and D. Tabor, The Frictional Anisotropy of Diamond, *Nature 283*, 51 1980.
58. Y. Enomoto and D. Tabor, The Frictional Anisotropy of Diamond, *Proc. R. Soc. London Ser. A 373*, 405–417 1980.
59. Y. Enomoto, Deformation by Scratches and Frictional Properties of MgO Crystals, *Wear 89*, 19–28 1983.
60. B. Samuels and J. Wilks, The Friction of Diamond Sliding on Diamond, *J. Mater. Science 23*, 2846–2864 1988.
61. A. G. Thornton and J. Wilks, in "Diamond Research," A Supplement to the Industrial Diamond Review, 1974, p. 39.
62. R. Mehan and C. Hayden, Friction and Wear of Diamond Materials and Other Ceramics Against Metal, *Wear 74*, 195–212 1981.
63. R. Mehan, Dry Sliding Wear of Diamond Materials, *Wear 78*, 365–383 1982.
64. N. Novikov and P. Kisky, Wear Resistant Coatings of Superhand Materials, *Thin Solid Films 64*, 205–209 1979.
65. M. C. Shaw, *Metal Cutting Principles*, Oxford University Press: Oxford, 1984.
66. K. Miyoshi and D. H. Buckley, The Adhesion, Friction, and Wear of Binary Alloys in Contact with Single-Crystal Silicon Carbide, *Trans. ASME J. Lubr. Tech. 103*, 180–185 1981.

67. K. Miyoshi and D. H. Buckley, "XPS, AES, and Friction Studies of Single-Crystal Silicon Carbide," *Appl. Surf. Sci. 10*, 357–376 1982a.
68. E. M. Trent, *Metal Cutting*, Butterworth: London, 1976.
69. S. Pepper, Effects of Interfacial Species on Shear Strength of Metal–Sapphire Contacts, *J. Appl. Phys. 50*, 8062–8066 1979.
70. H. Scott, Friction and Wear of Zirconia at Low Sliding Speeds, *Proceedings of the International Conference on Wear of Materials*, K. C. Ludema, et al. (Eds.), ASME: New York, 1985.
71. V. Aronov, Friction Induced Strengthening Mechanisms of Magnesia Partially Stabilized Zirconia, *J. Tribology Trans. ASME 109*, 531–537 1987.
72. B. Bhushan, D. Ruscitto, and S. Gray, Hydrodynamic Air-Lubricated Compliant Surface Bearings for an Automotive Gas Turbine Engine, Part II. Materials and Coatings, Report No. CR-135402, NASA Lewis Research Center: Cleveland, OH, 1978.
73. K. Miyoshi and D. Buckley, Friction and Wear of Single-Crystal Mn–Zn Ferrite, *Wear 66*, 157–173 1981.
74. K. Miyoshi and D. H. Buckley, Friction and Surface Chemistry of Some Ferrous-Base Metallic Glasses, NASA Tech. Report 1991, NASA Lewis Research Center: Cleveland, OH, 1982.
75. I. Prigogine and R. Defay, *Thermodynamique Chimique*, Dunod: Paris, 1950.
76. E. Rabinowicz, *Friction and Wear of Materials*, John Wiley: New York, 1966.
77. B. Adamson, *Introduction to the Physical Chemistry of Surfaces*, John Wiley: New York, 1978.
78. J. Hirth and C. Loethe, *The Theory of Dislocations*, McGraw-Hill: New York, 1968.
79. J. Hull and D. Bacon, *Introduction to Dislocations*, 3rd ed., Pergamon: Oxford, 1984.
80. S. Chandrasekar and B. Bhushan, The Role of Environment in the Friction of Diamond for Magnetic Recording Applications, *Wear 153*, 79–89 1992.
81. H. Schlossin, W. Van Reynveld, and W. Harris, *Proc. First Int. Conf. Rock. Mech.*, Lisbon, 1966, p. 119.
82. R. Sappok and H. Boehm, *Carbon 6*, 283–296 1968.
83. S. Chandrasekar and B. Bhushan, Friction and Wear of Ceramics for Magnetic Recording Head Applications—Part 2: Friction Measurements, *Trans. ASME J. Tribology 113*, 313–317 1991.
84. A. G. Evans and D. B. Marshall, Wear of Ceramics, in *Friction and Wear of Materials*, D. Rigney (Ed.), American Society for Metals: Metals Park, OH, 1979.
85. J. Skinner, N. Grane, and D. Tabor, Microfriction of Graphite, *Nature (Physical Science) 232*, 195–196 1971.
86. C. M. Mate, G. M. McClelland, R. Erlandsson, and S. Chiang, Atomic-Scale Friction of a Tungsten Tip on a Graphite Surface, *Phys. Rev. Lett. 59–17*, 1942–1945 1987.
87. L. Pauling, A Resonating Valence Bond Theory of Metals and Intermetallic Compounds, *Proc. R. Soc. London Ser. A 196*, 343–362 1949.
88. L. Pauling, *The Nature of the Chemical Bond*, 3rd ed., Cornell Press: New York, 1960.

89. B. V. Deryagin, N. A. Krotova, and V. P. Smilga, *Adhesion of Solids*, Consultants Bureau: New York, 1978.
90. H. Krupp, Particle Adhesion: Theory and Experiment, *Adv. Colloid Interfacial Sci. 1*, 111–239 1967.
91. F. R. Ruckdeschel and L. P. Hunter, Thermionic Return Currents in Contact Electrification, *J. Appl. Phys. 48–12*, 4898–4902 1977.
92. J. Lowell and A. C. Rose-Innes, Contact Electrification, *Adv. Phys. 29–6*, 949–1023 1980.
93. B. V. Deryagin and M. S. Metsik, Role of Electrical Forces in the Process of Splitting of Mica along Cleavage Planes, Translation from *Fiz. Tverd. 1–1* 1521–28 (October 1959).
94. H. Ochs and R. Czys, Charge Effects in the Coalescence of Water Drops, *Nature 329* pp. 467–471 1987.

Index

Abrasion (*see also* Abrasive wear), 99
  three-body, 100, 108-112
  two-body, 100, 267-269
Abrasives:
  alumina, 105
  boron carbide, 102
  diamond, 108, 266
  hard, 100, 106-107
  silicon carbide, 100, 105, 108
  soft, 100, 107-108
  quartz, 101, 108
Abrasive wear, 99-115
  environmental effects, 103-105
  mechanisms of, 103-108
  models for, 262, 267
Alumina (*see also* Aluminum oxide):
  abrasives, 105
  abrasive wear of, 110-113, 267
  applications of, 10-11
  coated with silver, 145
  coating on cutting tools, 302
  contact temperature, 129-131
  fretting wear of, 94
  lubricated with graphite, 165-175
[Alumina]
  seals, 364
  single crystal (*see* Sapphire)
  sliding against steel, 167-175
  titanium carbide composite, 267-268, 331
  whisker-reinforced, 10, 200-209, 225-257
  zirconia composite, 209-215, 267-268
Aluminum hydroxide film, 19-21, 125
Aluminum oxide (*see also* Alumina):
  adsorption-induced fracture, 53
  chemisorption embrittlement, 57, 125
  friction coefficient of, 16-36
  lubrication in water, 53, 125
  processing of, 4
  properties of, 4, 17
  reaction with hydrocarbons, 54
  surface chemistry of, 54
  tribochemical reaction of, 19-21
  wear coefficient of, 18-36
  wear at elevated temperatures, 24-29, 129

[Aluminum oxide]
wear mechanisms of, 15-49
wear transition in, 33-36, 42-44
diagram of, 18-36

Barium titanate, 397
Bearings, 11, 121, 313-328
heat generation in, 322
hybrid, 320-321, 325
reliability of, 323
silicon carbide, 326
silicon nitride, 314-325
solid lubrication of, 165
zirconia, 315
Boric acid, 139-142
Boric oxide, 138
Boron carbide, 392
Boron nitride:
alumina composite, 184-187
cubic, 296, 386, 390
friction coefficient of, 405
hexagonal, 177,184-187, 390
properties of, 166
silicon nitride composite, 184-187
sliding against steel, 184-187

Calcium titanate, 397
Cam roller followers, 329-356
alumina-titanium carbide, 343, 350-351
silicon nitride, 343-350
valve train tests, 337-353
zirconia, 343, 352
Carbon:
amorphous film, 138
diamondlike film, 139, 383, 389
graphite composite, 358, 363, 364, 367, 377
hydrogenated, 139
Ceramic coatings, 121, 290, 301-306
Ceramic composites:
abrasive wear of, 267-269
alumina-boron nitride, 177
alumina-graphite, 177-184

[Ceramic composites]
alumina-titanium carbide, 331, 383, 397
alumina-zirconia, 209-215, 383
erosive wear of, 269-274
silicon nitride-boron nitride, 184-187
silicon nitride-graphite, 177-184
silicon nitride-titanium carbide, 264
whisker-reinforced, 199-259
zirconia-alumina, 200
zirconia-graphite, 187-194
Chemical vapor deposition, 138, 302
Chromium carbide, applications of, 10
Chromium oxide, 409
Contact stress, 39-42, 81-83, 123, 128
Corrosion resistance, 363
Corrosive wear, 9, 377
Cutting tools:
alumina, 209, 294-296
alumina-titanium carbide, 209, 295
alumina-titanium oxide, 294
cemented carbides, 209, 292, 301
coated, 290, 301-306
cubic boron nitride, 297-299
diamond, 296-299, 306
machining performance of, 266
materials, 10, 289-309
selection of, 290
sialon, 209, 295, 299-301
silicon nitride, 263, 295, 299-301
silicon nitride-titanium carbide composite, 263
titanium carbide coating, 302
wear of, 262, 266, 274-282
whisker-reinforced alumina, 209, 263, 306

Diamond:
film, 389
frictional anisotropy of, 399-401
friction coefficient of, 398-404, 405, 415
polycrystal, 386, 404
single crystal, 385, 398
surface damage in, 401

Erosion (*see* Erosive wear)
Erosive wear, 269-274
  models for, 262, 271
  of silicon nitride composites, 266, 269

Ferrite, 383, 396, 409-411
Fracture toughness, 5, 17, 109, 199-201
Fretting wear, 79-98
  mechanisms of, 90-94
  slip amplitude, effect of, 85-90
  surface damage in, 84-85
Friction coefficient:
  at elevated temperatures, 16, 24, 29, 65, 189, 213, 233
  with lubricants, 52, 54, 138, 164, 217

Glass ceramics, 397
Graphite, intercalated, 165
  alumina composite, 178-184
  silicon nitride composite, 178-184
  sliding against steel, 165-173
  zirconia composite, 187-194

Hardness, 5, 17, 101-103, 107, 109
Hardness ratio, 101-103, 105

Ion beam-assisted deposition, 144-157
  film structure, 147
  process, 144-145
  of silver, 145-157
Ion beam mixing, 142-144
  of molybdenum disulfide, 143
  of niobium, 143
  process, 142
  of titanium and nitrogen, 143
Ion implantation, 164
Ion plating, 138

Lewis acids, 54, 56
Lubrication:
  boundary, 54, 363
  elastohydrodynamic, 318
  with engine oils, 216
  hydrodynamic, 358
  with paraffins, 53, 54, 332
  solid, 138-157, 164
  starvation, 318
  vapor phase, 137-138
  in water, 52

Magnetic recording head:
  calcium titanate, 383
  ferrite, 383, 396
  flying height, 384
  overcoat materials for, 383, 395
Microfracture, 122-123
Molybdenum disulfide, 138-139, 143

Profilometry, 84
PSZ (*see also* Zirconia, partially stabilized):
  applications of, 10-11
  seals, 366

Sapphire:
  fretting wear of, 94
  friction coefficient, 408
  molybdenum disulfide film on, 143
  sliding against sapphire, 408
  sliding wear of, 139
  structure of, 392
Seals, 381
  alumina, 364
  applications, 10-11, 358
  blistering, 377
  boundary lubrication, 363
  carbon-graphite, 358-363, 364, 367, 377
  corrosion rate, 374
  corrosion resistance, 363, 374
  corrosive wear, 377

[Seals]
failure, 375
friction coefficient, 370
hydrodynamic lubrication, 358
materials, 363-368
power consumption, 370
PSZ, 366
PV limit, 358-369, 371
silicon carbide, 364, 367
silicon nitride, 364
temperature margin, 371
thermal shock resistance, 360-363
thermal stress fracture, 362, 377
tribological behavior, 363
tungsten carbide, 364, 367
water pump, 372-374
wear, 370, 375-377
Sialon:
applications of, 10
cutting tools, 299-301
phase diagram of, 395
processing and properties of, 8, 396
Silica (*see* Silicon oxide)
Silicon carbide:
abrasives, 100-103
applications of, 10-11
bearings, 326
chemical vapor deposition, 8, 367
corrosion rate, 374
friction coefficient of, 406
hot-pressed, 8
ion beam mixing with niobium, 143
processing and properties of, 8-9, 367
reaction-bonded, 8, 368, 377
reaction-sintered, 8, 368, 377
seals, 364, 367
single crystal, 391, 406
sintered, 8, 368
surface chemistry of, 55
Silicon hydroxide film, 42-53, 91
Silicon nitride:
abrasive wear of, 110-113, 267-268
applications of, 10-11
bearings, 11
dissolution in water, 52
[Silicon nitride]
erosive wear of, 270
fretting wear of, 84-93
friction coefficient of, 65-66
hot isostatically pressed, 7, 63, 84, 317
hot-pressed, 7, 63, 84, 331
intergranular fracture in, 71
ion beam mixing, 143
lubricated with graphite, 165-175
oxidation of, 7
processing and properties of, 6-7
reacted, 331
reaction-bonded, 7
rolling contact fatigue of, 317
seals, 364
sintered, 7, 331
sliding against steel, 167-175
surface chemistry of, 55
titanium carbide composite, 218
tribochemical reactions, 52-53, 71-76, 91-93
wear mechanisms of, 66-69, 90-93
wear particles, 66-70
wear rate of, 65-66, 88-89
wear transition in, 71-76
whisker-reinforced, 207, 215-218
Silicon oxide:
chemisorption embrittlement of, 57
surface chemistry of, 54
reaction with hydrocarbons, 54
Silicon oxide film:
on alumina, 24-29
on silicon carbide, 8
on silicon nitride, 7, 53, 62, 71-76, 91-93
Sputtering, 138, 139

Temperature:
bulk, 126-130, 251-253
flash, 126-130, 150-153
Titanium carbide, 392
Titanium diboride, 164
Titanium nitride, 392

Transfer film, 168-175, 178-184, 190-193, 227, 360, 363
Tribocharging, 417
Tribochemical reaction:
  of ceramics, 51-60
  films on alumina, 19-21, 44-45
  of silicon nitride, 71-76, 91-93
Tungsten carbide:
  abrasive wear of, 101-113
  application of, 10
  seals, 364, 367
TZP (*see* Zirconia, tetragonal)

Wear:
  abrasive, 9, 99-115
  coefficient, 18, 167
  at elevated temperatures, 10, 24-29, 201-215
  fretting, 79-98
  by intergranular fracture, 29-33
  by local melting, 128-129, 150
  by microfracture, 39, 71, 80, 91, 122
  mild, 18, 29, 201
  by rolling contact fatigue, 317
  severe, 18, 29, 201
  by spalling, 315
  thermomechanical, 125-131, 145-248
  transition, 16, 29-36, 123, 207, 210, 253
  transition diagram, 18-36
[Wear]
  tribochemical, 123, 263
  by whisker pull out, 236, 308

Vacuum evaporation, 147

Zirconia (*see also* Zirconium oxide):
  abrasive wear of, 214-215
  adsorption-induced fracture in, 53
  alumina composite, 209-215
  applications of, 10
  castable, 187-194
  coated with silver, 150-153
  coating, 121
  cubic, 6
  erosive wear of, 214
  graphite composite, 187-194
  intergranular fracture in, 53
  local melting of, 128-129
  lubrication of, 53, 409
  monoclinic, 6
  partially stabilized, 6, 331, 393
  plastic flow in, 131-135
  sliding against steel, 214
  tetragonal, 6
  wear factor of, 133, 151
  wear transition in, 123
Zirconium oxide:
  chemisorption embrittlement of, 57
  processing and properties of, 4-6
  surface chemistry of, 55, 57
  tribochemical reactions, 56